T0226129

THE
DYNAMIC CONCEPTS
OF
PHILOSOPHICAL MATHEMATICS

THE
DYNAMIC CONCEPTS
OF
PHILOSOPHICAL MATHEMATICS

A PHILOSOPHICAL MATHEMATICAL BRITANNICA

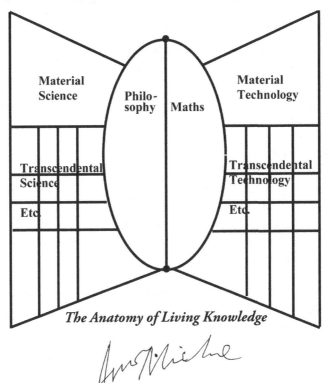

The Anatomy of Living Knowledge

PROF. ANTHONY UGOCHUKWU O. ALICHE

DSC, FSM, FMRG, FPRS, FNIAM, FCIA, FABEN

iUniverse, Inc.
Bloomington

The Dynamic Concepts of Philosophical Mathematics
A Philosophical Mathematical Britannica

iUniverse books may be ordered through booksellers or by contacting:

iUniverse
1663 Liberty Drive
Bloomington, IN 47403
www.iuniverse.com
1-800-Authors (1-800-288-4677)

ISBN: 978-1-4759-6193-5 (sc)
ISBN: 978-1-4759-6192-8 (hc)
ISBN: 978-1-4759-6191-1 (ebk)

Library of Congress Control Number: 2012921691

Printed in the United States of America

iUniverse rev. date: 12/26/2012

CONTENTS

Truth Is Universal Publishers xviii

About Truth Is Universal Publishers xx

Global Praise for Professor Aliche and His Works xxii

Some Comments on One of Aliche's Latest Works xxx

Acknowledgement xlii

Foreword xlvii

Foreword 2 l

Editor's Note liv

Quotes from A. U. Aliche's Philosophical Marble lvii

Preface lix

PART 1

THE DYNAMIC CONCEPTS OF PHILOSOPHICAL MATHEMATICS

Introduction with Abstracts lxxvii

Chapter 1 1

    Philosophical Mathematics and Extracts

Chapter 2 16

    Points and Facts of Dynamic Interests, Which Are Must-Know

    with Philosophic Understanding and Mathematical Appreciation

PART 2
PHILOSOPHICAL MATHEMATICS

Chapter 1 22

    What Is Philosophical Mathematics?

Chapter 2 30

    What Is the Origin of Philosophical Mathematics?

Chapter 3 35

    What Are the Branches of Philosophical Mathematics?

Chapter 4 41

    What Are the Scientific and Internet Dynamics of
Philosophical Mathematics?

Chapter 5 46

    How Can Man Be Explained as an Initiate of Philosophical
Mathematics Using the Formula of P/M = WWW?

Chapter 6 53

    How Can Creation Access the Balanced Wisdom of
PM/WWW Concept?

Chapter 7 61

    How Can Creation Originally Network All Human Endeavours
in the Best Use and Application of Philosophical Mathematics?

Chapter 8 69

    How Can Creation Accept Philosophical Mathematics as a
Perfect Anatomy of Consummate Reality?

Chapter 9 81

    The Philosophical Reality That Lies on the Way Forward

Chapter 10 86

    A Comprehensive Overview of Parts 1 and 2

PART 3

SOME MATHEMATICAL AND PHILOSOPHICAL ARGUMENTS

Introduction and Abstract 92

Chapter 1 97

What Are the Arguments about the Philosophical Relativity
of the Following Mathematical Functions?

Chapter 2 110

The Philosophical and Mathematical Significance of
Algebra and Geometry

Chapter 3 118

Why Philosophy Is Greatly Concerned in Furthering the
Course of Core and Applied Mathematics

Chapter 4 124

Why Is Statistics Scientific in the Best Objectivity of Philosophy?

Chapter 5 132

Why Is Engineering as a Material Science Greatly Interested
in the Use and Application of Statistics and Appreciating the
Philosophy of Science?

Chapter 6 139

How Can Statistics Be Used in Reducing the Errors of
Engineering as a Material Science?

Chapter 7 145

Why Is Statistics Defined as the Databank of Philosophical
Mathematics and the Scientific Basis of Engineering
Mathematics?

Chapter 8 150

A Philosophical Overview on the Mathematical Relativity of
Statistics as the Cheapest Engineering Mathematics: The Concept
of Short Methods

PART 4

Introduction to Some Philosophical Principles with Mathematical Theorems

Chapter 1 163

A Definition of Mathematical Principles with
Their Diversified Theorems

Chapter 2 170

What Are the Functions of These Principles with
Their Theorems?

Chapter 3 179

What Are the Objective Purposes of These Policies?

Chapter 4 183

What Are the Difficulties in Using the Principles and
Policies of Mathematical Philosophy?

Chapter 5 189

Why Are Human Endeavours Purpose-Driven by the
Objective Functions of Principles and Policies of Philosophical
Mathematics, Which Are the Enabling Origin of Research
Methodology?

Chapter 6 194

A Scientific and Technical Look at the Creations of Material
Endeavours, Which Work Against Philosophical Mathematics

Chapter 7 199

Why Is the World Materially Driven against Being
Philosophically Driven, Which Is the Opinion of
Mother Nature?

Chapter 8 206

The Urgent Necessity of Reducing Material Incursion into the
Affairs of the Universe

Chapter 9 211

How Has Man Related and Utilised the Protection and
Perfection of These Principles and Theorems?

Chapter 10 217

An Overview on the Philosophical Mathematics of Part 4

PART 5
PYTHAGORAS' PHILOSOPHICAL MATHEMATICS

Introduction 224

Chapter 1 230

What Are Pythagoras' Contributions to These Eminent and
Scientific Problems?

Chapter 2 236

What Are the Philosophical Relationship of Pythagoras and
Plato, and How Did They Use These Dynamisms to Improve
the World of Mathematics?

Chapter 3 241

Why Is Pythagoras Known as the Prince of Philosophical
Mathematics and the Angelus of Scientific and Mathematical
Theorems?

Chapter 4 246

What Is the Mathematical Metaphysics of Pythagoras?

Chapter 5 250

What Can Be Defined as Pythagoras' Philosophical Legacies to
Mankind, and Why Are These Legacies Translated as the Origin
of Philosophical Mathematics?

Chapter 6 255

What Is the Mathematical Logic of Pythagoras, and Why Was
This Logic the Anatomy of His Originality?

Chapter 7 260

Why Is Pythagoras' Mathematical Logic a Point of
Philosophical Reference?

Chapter 8 264

How Did He Use This Power to Develop a Meta-logic Province?

Chapter 9 268

What Are the Objective Lessons of Pythagoras' Life in
Reference to Philosophical Mathematics?

Chapter 10 273

An Overview in Favour of Part 5

PART 6

PHILOSOPHICAL MATHEMATICS AS RATIONAL AND ABSTRACT SCIENCE

Introduction 279

Chapter 1 284

What Is Abstract Science?

Chapter 2 292

Why Is It Seen and Known as Such by the College of the
Uninformed?

Chapter 3 298

What Is the Logical Reasoning of This Argument?

Chapter 4 304

How Can the Mathematical Reality of the Argument Be
Established?

Chapter 5 310

Plato and Pythagoras Argued That Philosophical Mathematics Is
Not Abstract but a Kingdom of Knowledge, Which Is Meant for
the Well Informed. Why Was This Opinion Formed?

Chapter 6 317
    What Are the Mathematical and Philosophical Proofs to
    This Dialogue?
Chapter 7 323
    Philosophical Review and Explanation of Their Arguments
    about the Concept of Form
Chapter 8 330
    What Was the Purpose of Plato's Metaphysical Community,
    Which Existed as a United Province During the Oriental Era?
Chapter 9 336
    Why Dialogue Was the Beginning and Essence of Balanced Logic
Chapter 10 342
    Why Were They Perfect in the Objective Utilisation of Logical
    Dialogue in Social and Philosophical Argumentations?
Chapter 11 348
    Why Was Their Dialogue in the Areas of Logical Semantics
    Responsible for the Creation of Logical Community, Which
    Extolled Their Visions as a Philosophical, Consummate Village?
Chapter 12 356
    A Comprehensive Overview in Respect of Part 6

PART 7

THE DEMYSTIFICATION OF THE SACREDNESS OF MATHEMATICS BY THE PHILOSOPHICAL WIZARDS

Introduction 361
Chapter 1 370
    What Is Mathematical Demystification?
Chapter 2 376
    What Were the Challenging Visions That Gave Rise to the
    Demystification of Mathematics?

Chapter 3 382

Why Were These Visions Important Mathematical
Achievements to That Era?

Chapter 4 388

What Are the Achievements of the Demystification of
Mathematics as a Sacred Oracle?

Chapter 5 394

What Is the Honour in This Achievement?

Chapter 6 399

How Can We Reconcile the Problems of the Present Era with
Those of the Obscure Era That Mystified Mathematics as a
Sacred Oracle?

Chapter 7 406

Why Is the Demystification of Mathematics a Unique Esteem
That Achieved Plato's Metaphysical Community?

Chapter 8 410

How Can We Believe the Fact That the Demystification of
Mathematics Has Remained the Greatest Challenge of the Past,
Including Being the Best Encouragement of the Present?

Chapter 9 415

Has the Present Embraced the Reward of This Conquest, Which
Was a Philosophical Jubilee?

Chapter 10 420

What Is the Mathematical Mandate of the Demystification,
Which Is Hereby Explained as the Vision of the Oracle?

Chapter 11 426

An Overview to Explain How the Demystification Opened a
New Horizon for Human Elevation

PART 8

Philosophical Mathematics as the Logo and Symbol of the Creative Universe

Introduction..433

Chapter 1 440

What Is Logo and Symbol?

Chapter 2..444

A Look at Their Origin with Philosophical Explanation

Chapter 3..448

What Is the Importance of Logo and Symbol?

Chapter 4..453

Why Is It Known as the Supreme Heartbeat of the Universal

Creativity?

Chapter 5..458

Have These Facts Been Known by the World of Material Science?

Chapter 6..464

Why Is This Logo and Symbol the Living Anatomy of

Philosophical Mathematics?

Chapter 7 470

How Did the Wizards Use the Logo and Symbol as the

Perfect Ornaments of Morals and Ethics?

Chapter 8..475

The Urgent Call for the Modern Era to Appreciate the Use and

Application of the Symbol as an Ingenious Creator

Chapter 9..481

Why Are the Logos and Symbols, the Balanced Keys, That

Can Be Used in Knowing and Appreciating the Philosophical

Rhythms and Challenges of Life?

Chapter 10 490

Why Is This Symbol Known and Defined as the

Ingenious and Illumined Actor to All Actions?

Chapter 11 496

A Balanced Overview of Part 8

PART 9

PHILOSOPHICAL MATHEMATICS AS THE ENGINEERING MODULE, MODEM, AND MECHANISM OF BALANCED AND CONSTRUCTIVE CREATIVITY

Introduction 500

Chapter 1 506

What Is Engineering?

Chapter 2 510

Who Is an Engineer in the Logo and Symbol of Engineering?

Chapter 3 515

What Is the Origin of Engineering?

Chapter 4 521

Has Creation Known and Appreciated Its Roles and Functions in

Directing the Universal Economy?

Chapter 5 528

What Are the Philosophy and Logic of Engineering Mathematics?

Chapter 6 536

Why Does Philosophical Mathematics Control the

Functions of Engineering Technology?

Chapter 7 542

Why Are Engineering Concepts Being Determined with the Use

of Mechanics, Modules, and Methods Known Technologically as

the MMM Concept?

Chapter 8 551

How Does Philosophical Mathematics Determine the Standard
Performance of Engineering and Material Endeavours?

Chapter 9 558

How Can We Explain the Engineering and Philosophical
Mathematics of This Work?

Chapter 10 564

How to Use the Thinking of Philosophical Mathematics in
Determining the Concept of Engineering Creativity

PART 10

INTRODUCTION TO PHILOSOPHICAL MATHEMATICS AS THE BALANCED ANATOMY OF LIVING KNOWLEDGE DYNAMICALLY EXAMINED WITH PHILOSOPHICAL EXPLANATION

Chapter 1 581

What a Living Knowledge Is, Dynamically Examined

Chapter 2 586

The Origin of Living Knowledge, Philosophically Examined

Chapter 3 592

What Are the Parts of Living Knowledge and Their Functions?

Chapter 4 596

What Are the Functions of Living and Transcendental
Knowledge?

Chapter 5 603

When Is Knowledge Dynamically Considered to Be Living?

Chapter 6 607

Why Is Philosophical Mathematics Designated as the
Anatomy of a Living Knowledge?

Chapter 7 614

Why Is the Academic Province Not Using This
Knowledge in the Wisdom of Its Original Nature?

Chapter 8 621

What Are the Consequences of This, and How Can We
Effectively Restructure These Monumental Errors with
Balanced Correction?

Chapter 9 628

How Can We Use Philosophical Mathematics to Appreciate
the Structure and Anatomy of Living Knowledge?

Chapter 10 634

How Can We Define This Knowledge as a Journey in Truth,
Including Being a Pathway to Philosophical Mathematics?

Chapter 11 642

Is It Logical to Explain This as Being the Way to
Immortal Wisdom?

Chapter 12 651

Why Is Living Knowledge Naturally Pure, Perfect, and Dynamic
without Pollution?

Chapter 13 655

How Can We Define This Knowledge as a Perfect and Provisional
Solution to World Crisis?

Chapter 14 662

Why Define Balanced Knowledge as a Consummate Hall of
Immortal Fame?

Chapter 15 666

A Detailed and Dynamic Overview That Blends Part Ten as the
Honest and Practical Tedium of Mother Nature

Chapter 16 670

    A General Overview on the Anatomy, Concept, and Dynamics of
    Philosophical Mathematics, Which Makes It an Ingenious Living
    Knowledge

Afterword 676

    The Urgent Necessity of Using the Infinitude Method in
    Establishing and Solving Mathematical Problems

Appendices 683

Published Books 693

Unpublished Books 696

TRUTH IS UNIVERSAL PUBLISHERS

28 Okpu-Umuobo Rd. PO Box 708, ABA,
Abia State
Nigeria
Tel. 08037065227, 08059227752, 08064718821

www.truthisuniversalng.org;
www.anthonyugochukwuoaliche.com
or by e-mail:
- info@truthisuniversalng.org
- truthisuniversalpublishers@yahoo.com

on Facebook:
- http://www.facebook.com/TruthIsUniversalPublishers
- http://www.facebook.com/profile.php?id= 100001817099148

- http://www.facebook.com/pages/Anthony-Ugochukwu-Aliche/408805502515023
and on Goodreads
- http://www.goodreads.com/user/show/12816453-anthony-ugochukwu-aliche

ABOUT TRUTH IS UNIVERSAL PUBLISHERS

This diverse, dynamic, and highly inspired organization was founded by Anthony U. Aliche, along with several gifted and apostolic scholars. The company was established to carry out the important task of providing humanity with a great line of books; these works are gaining universal recognition.

We are compelled and inspired to establish a mind—and purpose-driven, universal bank whose activities are centred on creating a museum of knowledge for the balanced wisdom of posterity.

Located at Aba in Abia State, Nigeria, with branches in Europe and the United States, Truth Is Universal Publishers is a nonsectarian, non-dogmatic organization that operates principally and tactfully in the medium of the universal mind. It imparts honest and strict principles that foster greater awareness of spiritual tenets in the lives of humans.

This institution is highly involved in propagating wisdom and truth, particularly in the areas of religion, health, psychology, science, engineering, technology, philosophy, and spiritualism—stimulating a colloquium of integrated and balanced education.

We also have begun an online study, research, and development programme, "Truth Is Universal Educational Outreach", with members all over the world. This programme gives people the rare opportunity to explore beyond the boundaries of their present knowledge.

For more information on this organization and its programmes, please write to the following:

Truth Is Universal Publishers
28 Okpu-Umuobo Rd., PO Box 708
Aba, Nigeria

You can reach us by telephone at 082-225217, 08037065227, 08059227752, or 08064718821. You can go to our websites, www.truthisuniversalng.org and www.anthonyugochukwuoaliche.com.
You can contact us by e-mail at any of the following:
- info@truthisuniversalng.org
- truthisuniversalpublishers@yahoo.com
on Facebook:
- http://www.facebook.com/TruthIsUniversalPublishers
- http://www.facebook.com/profile.php?id=100001817099148
- http://www.facebook.com/pages/Anthony-Ugochukwu-Aliche/408805502515023

As well as on Goodreads
http://www.goodreads.com/user/show/12816453-anthony-ugochukwu-aliche

GLOBAL PRAISE FOR
PROFESSOR ALICHE AND HIS WORKS

My beloved son, Anthony Ugo. O. Aliche, follows in the footsteps of my late husband, Dr. Walter Russell, who is widely considered a modern Leonardo da Vinci.

—Dr. Lao Russell, former president,
University of Science and Philosophy, Virginia

You most certainly are an ordained author! Indeed, you are creating a museum of knowledge for the balanced wisdom of posterity. The key word is *balance*. As we both know, we are living in a most unbalanced world of man-woman thinking, and we lack higher consciousness.

—Michael Hudak, president,
University of Science and Philosophy, Virginia

Only an adept writer, and someone of the author's status, could convincingly inject fresh interpretations of existing mathematical and philosophical functions and symbols, as he has done in this book. In doing so, he guides us to the obvious and incontrovertible conclusion that the field of philosophical mathematics is indispensable to the solution of human problems in the past, present, and future. His writings guarantee the continuous relevance, utility, and applicability of the discipline of philosophical mathematics.

Your book *What Makes Great Men* is on the top the bestseller list. I have read it entirely, and it is inspiring. Maintain that spirit.

—Pastor Praise Michael Daniels Abuja, Nigeria

Surely you must be the most published author in Africa! Most important is your continuous effort to awaken and enlighten humanity to the power of loving one another, and the gift of loving service to our fellow man or woman. I have looked at your website and feel a greater appreciation for your love of humanity. We both are committed to the unending effort to raise the consciousness of our fellow world citizens to a new way of thinking and acting as we live our lives.

—Michael Hudak, president,
University of Science and Philosophy, Virginia

Your works aim to restore the dignity of man, and they are destined for global acknowledgement. The fact that it is climbing the bestseller list in Abuja is a sign of the avalanche of global recognition that will follow this selfless and cerebral author of our time. Congratulations, sir, my dear mentor.

—Ibe Edede, Portharcourt Nigeria

I am a fan of your series, and I remain grateful to my friend who bought me a copy of your work *Whose Vessel Are You*. It dawned on me then that I could be an empty vessel, and I really understood what it means to be a true vessel. I decided that I would be a golden vessel, made for the Master's use. Bless you, sir. You are a true vessel.

—Adimchimnobi Ezinne, Lagos, Nigeria

Professor Aliche, you are a genius of the highest order! In all my life as an IT specialist, I have not seen this type of interpretation of information technology and information communication technology. It is a great work.

—Eno Etuk, Port Harcourt, Nigeria

Professor, the northerners in the cabinet of Kastina State government under Governor Shehu Shema are in love with your articles on metaphysics.

—Ovie Ben, editor-in-chief,
The President Afrique Magazine

I know you are an authority in many different fields of human endeavour, and I encourage and congratulate your dynamic wisdom in exploring the boundless world. All of us at the world headquarters are grateful to honour and designate you as our new Nigerian chairman, and to supervise the metaphysical activities in other African countries.

John J. Williamson, president/founder,
College of Metascientists, Archer's Court, England

You are indeed an excellent and brilliant friend whose vast knowledge of nature and its dynamism deserves to be understudied.

—Prof. Ralph Onwuka

Aliche has shown the world that he is a true and perfect breed of the Russellian scientist-philosopher. His achievements, which will be favoured by posterity, must certainly dazzle his contemporaries.

—Laara Lindo, former president,
University of Science and Philosophy, Virginia

Anthony U. Aliche is an illumined mystic from the East. I respect the comprehensiveness of his ingenuity, particularly his rare gift of considering everybody as one.

—Nanik Balani, India

My esteemed friend Anthony U. O. Aliche is best described as a rare gift to humanity, living a life that appears unequalled in our present, chaotic age. To God be the glory for his esteemed and universal intelligence.

—Dr. Singh, India

After studying *What Is Beyond Truth* on my visit to Nigeria, I discovered that great talent can be found in this humble country, the most populous in Africa. I invite noble souls who appreciate the wisdom of truth to buy this book and analyse it in order to appreciate the fact that nothing is beyond truth.

—Prof. Harley Enoch

Your books are greatly inspiring. After studying *The Mystical Powers of Our Lord's Prayer*, I was personally inspired to honour the wisdom of the Celestial Sanctum. Thank God for your person.

—Nick Helen, South Africa

My great friend is a gifted and rare scholar of our race. I respect and love him particularly for his humility, honesty, and intelligence, and for his accomplished, purpose-driven life.

—Sen. Millford Okilo,
former governor of Old Rivers State,
first African president of the University of Science and
Philosophy, Virginia

Surely you are the legend of our race!

—Aigbokhalmode K. A.
senior staff of the Federal Ministry of
Culture and Tourism, Nigeria

You are truly a blessing to our nation. I bought your book *Honey Is Health*, and I am so glad I bought it. I have never thought of honey in this way. God bless you for your knowledge."

—Tony Udedike, Enugu State

You are indeed a mobile encyclopaedia!

—participant at the Garden City Literary Festival,
September 12-17, 2011,
Presidential Hotel, Port Harcourt

Nnana Aliche, you were born great. Many were not aware of this truth; others knew it but refused to accept it. Your life in the service of humanity will be rewarded not only on earth but eternally. Amen.

—Dr. Anthony E. Torty, CEO,
Malic International Services, Ltd.

You are awarded as a role model and the author of the year as one of the twenty distinguished Nigerians who have, through hard work and excellence, put Nigeria on the world map.

—African Child Foundation

You are an epitome of a noble man.

—Erika Ray,
senior publishing consultant,
iUniverse Publishing Company, Indiana

Professor Anthony U. Aliche is an internationally recognized and acclaimed great man, orator, philosopher, scientist, author, farmer, consultant, music and melody composer, sensitizer, management icon and a known noble human who is greatly involved in peace advocacy. He is a metaphysicist/metascientist, he is a rare scientist of our era, and this is why he is endowed with an ingenious knowledge of all areas which benefits human endeavour. This great scholar is a designer, a sculptor, an architect; a technologist including being an engineer of a higher esteem whose creations has always dazzled his contemporaries and the global family. He is indeed the epitome of a God-fearing and wise man. He is the future president the Federal Republic of Nigeria needs. He is a real genius from God in this planet earth.

—Lawrence I. Iwuoha, ACIA,
philosopher, humanist, anti-racist, and
human rights activist

Professor Aliche, I was keen on your authoritative angle on destiny because we had divergent views with no consensus in sight, until it became obvious that there was need for expert rescue. I never knew you had a wonderful book on it already—my instincts were perfect! Am glad I did. You came to mind, and I am glad you came to rescue—and beyond. Thank you.

Please do not hesitate to make the publication of your book(s) known to the house. I wait with childlike anxiety.

—Mr. Vivian Iwuoha, Düsseldorf, Germany

Wow! You sure have some long list here Sir. Versatile, prolific . . . wow! May the future bring you far more plus that distinct fulfilment that seem to run from a lot of people. Best wishes.

Su'eddie Vershima Agema, posted to Aliche Anthony U, principal writer/partner at SAGE Writers Inc.

I thank God for giving you this great wisdom you use in guiding us.

—Jude Obisike, Uyo Akwa Ibom State, Nigeria

I am totally amazed by all the reviews and compliments from people who have witnessed how brilliant you are! Thank you for sharing.

—Marcus Winters
senior marketing consultant,
Book whirl, USA

Aliche dear, you are God's gift . . . No, I mean one of God's greatest gifts to the black race. Even at that deep down I know I have made an understatement. Can you imagine that even being a words smith I still lack enough vocabulary to describe you. I accept my limitations because you are one of the brightest minds of the entire universe at least counting from the last ten centuries. Before anybody picks a fight with me let him or her first google Anthony Ugochukwu Aliche. Even after that I still expect a fight from two set of people; the ignorant mind and the religious bigot.

—Peter Agba Kalu
senior and seasoned writer
—The Daily Sun Newspaper

SOME COMMENTS ON ONE OF ALICHE'S LATEST WORKS:

Have You Discovered Your Assignment with Destiny?

Professor, you are Africa's pride and blessing to the world. You have made us proud through your globalised publishing. Bravo, our hero!

—Dennis Nwankwo, Imo State, Nigeria

Fine paperback uniquely designed, and I believe the content can only be compared to a masterpiece, which of course is second nature to Professor Aliche. A must-read for any with destiny!

—Nwahizu Chijindu Uyo, Akwa Ibom State, Nigeria

Professor Anthony Aliche, you are a great author of the present generation. The Lord who has endowed you with wisdom and intelligence to do all this research will continue to sustain, keep, and strengthen you on this roadmap of greatness. As he changed the name of Jacob to Israel, as he changed the name of Israel, so shall you be called the liberator of the African race. Congratulations!

—Hon. G. C. Iwuoha, FNSIM, Abia State, Nigeria

Life without discovering one's assignment with destiny is subject to tragedy. Congratulations! You are wonderfully made. Sincere love to this iUniverse and Anthony O. Aliche production. God will continue to keep you as you are saving souls for him.

—Uche/Nonye, Abia State, Nigeria

Your self-discipline, hard work, education, and perseverance have metamorphosed into an author of international best-seller status. I know that your talents are transcendental. No wonder you discovered your assignment with destiny! Thank God for your rare, divine intuition and illumination. Congrats, sir.

—Ibe Edede, Port Harcourt, Nigeria

You have made another inroad in the world of academics. Bravo!

—I. G. Akwada, Abia State, Nigeria

Professor A. U. Aliche, shalom! A beloved hip—hip, hurray! Congratulations and kudos for your worthy advancement of knowledge in the province of humanities (Num. 6:24-26). Amen.

—Elder Nkem Agha Okwuagha, Abuja, Nigeria

There are reasons with practical evidence to show that you belong to the topmost echelon of the advanced race.

—Cynthia Herod, England

A great look at your works, which cannot be comprehended by any human being, reveals the dynamism of your genius, which is naturally bestowed in you by no other person than the creator.

—Florence Pam, Switzerland

Professor, you are highly gifted and ordained with a constant and consummate fluent pen, which will continue to flow for the fact that the level of wisdom it passes to humanity is something to acknowledge the creator for the gift of a millicential genius like you.

—Cornelius Uka, Owerri, Imo

Professor, people are celebrating and acknowledging your resource oriented creativity with global ovation, but for me, this is your starting point.

—Chijioke Abuja, Nigeria

A look at your contributions to human development reveals that your presence in our contemporary society is a lesson whose knowledge will continue to dazzle human existence, civilization, evolution, and development. I thank God for you.

—Barr. Nwator, Abia State, Nigeria

All of us are happy to celebrate your ingenuity as the consummate and ordained author of the century.

—Alice Jonah, Minna, Niger State

Bravo! The crown, when placed on a deserving head, fits without adjustment. Your best-selling publication is the deserving head of my dream, wearing this golden fitted crown. More grease to your ever-busy and God-blessed elbows, Professor Aliche.

—Amb. Chief John Onwukwe, MD, IEA, LTD, ABA

"Without philosophy, mathematics makes no meaning."
—Anthony U. Aliche

"Wish thou what you want to desire, and the will, will pave a smooth way for the alignment of that wisdom. Wish thou what you want to will, and the will, with a stronger desire, will emancipate you into the province of immortal wisdom, which warehouses the opinion and functions of wish, will, and wisdom, which the Microsoft and material technology donated to the world of engineering technology as 'www'."
—Anthony U. Aliche

"Science never makes an advance until philosophy authorizes it to do so."
—Thomas Mann

Liberate Yourself with the Power of Wisdom

Oh thou mortal man,
When will you gain immortality?
When will you gain freedom?
When will you come out from your oblivion?
Is it not time for you to appreciate that you are a function
and fulcrum of wisdom?
Liberate yourself with the power of wisdom.

Oh thou mortal man,
Who is destined for so many things!
One of which is to gain immortality from mortality,
To be a true servant with the dynamic powers of wisdom.
Have you not been told?
Have you not known
That time has come when you should liberate yourself with
the power of wisdom?

Oh thou mortal man,
Who has conquered so many things without conquering himself!
Who has created so many things without creating and
recreating himself!
Who is esteemed to be the boisterous king of knowledge with
esteemed dynamism!
Have you ever thought how you came into this universe, and
how you are going to depart from it?
Now is the time for you to liberate yourself with the power
of wisdom.

Oh thou mortal man,

Who is the pinnacle of the quest for the grail;
Who is designed to be part of the celestial pilgrim;
Who is imaged to be the creative thinking of imaginative dynamism;
Who is imaged to continue the multidimensional act of dynamic imagery—
Now is the time for you to liberate yourself with the power of wisdom.

Oh thou mortal man,
Who is sense-possessed with mortal mediocrity;
Who has not known that he is the living aroma of beauty and wisdom;
Who is still in search of his eternal home,
Searching for God in temples, in synagogues, and in the deserts
Without knowing that the search for God's realization starts with his inner self and ends with his inner urge.
Now is the time for you to appreciate the wisdom of immortality and the power that mandates you to understand your assignment with destiny, the true knowledge of yourself, your neighbour, and the consummate universe.

Know ye this: that all these cannot be a rarefied knowledge if you do not strive honestly to liberate yourself with the power of wisdom which, when known, will help you to eternally gain immortality from mortality—which is a cosmic assignment from the divine sanctum of your mother, father, and consummate nature.

—from the book *A. U. Aliche's Exaltation, Celebration, and Inauguration of Wisdom*

Immortal Philosophical Quotes That Came to Man at the Proximity of Professor A. U. O., DSc

- Man is a creative force from the balanced wisdom of electrifying philosophy.
- Philosophy, being the mother of knowledge, is the supreme anatomy of all.
- Every science is electrifyingly compacted into the thinking and balanced wisdom of philosophy. Being an elective and electronic science, it is scientifically philosophic.
- Philosophy, being the supreme way of man, is purposefully and pragmatically driven in becoming the way of all.
- The world can neither know peace nor dynamically progress without the use of the electrifying concept of philosophical dynamism.
- As material science is incoherent at all levels of its existence, hence producing chaotic principles, philosophy is coherent, compact, edified, and knowledgeable in directing the affairs of all.
- The universe, which is a philosophical conclave, is driven by the scientific ideology of philosophy.
- Man, being an extension of philosophy, is still at the fifth kingdom of balanced and electrifying philosophy.
- Philosophy, being the mother of balanced knowledge, is defined as the structure of all, the secret of light, the sacred wisdom of all ages, and the blessed teacher of immortality.

- Philosophy, knowing the truth about all things, will ever remain the only acceptable way of universal civilization, tradition, and religion. It also holds the power in amalgamating all the forces of human dynamism.
- Both truth and beauty are simplified philosophy, and they do not separate their activities from the true knowledge of philosophical dynamism.
- Philosophy, being the scientific net of the universe and the meta-logic hub of the one-world balance, is always contesting what is in the knowledge of man and what will be his future.
- Philosophy is the only science that originated the concept of all websites, all zero powers in the amount of the zero. It also electrifies knowledge with its dynamic balance.
- Man must know that all natural and eternal knowledge originated from the nebula hub of positive philosophy.
- Philosophy is the supreme teacher, which explains what can be known and accepted as the wisdom of a dependable knowledge.
- Philosophy occupies a natural place between theology and science; this is why its extensions are beyond words and thoughts.
- The knowledge of philosophy is the only education that makes the prince of honest knowledge including being the pride of dependable wisdom.

It is important to give these quotes before the introduction of this highly illumined philosophical Britannica, which to the best of my knowledge and understanding can best be defined as an advanced neo-philosophy. These quotes, at all levels of understanding, will always remind the apostolic foundation of philosophical engineering that the goal of philosophy is to produce great and balanced thinkers to run the one-world purpose, appreciating that the universe is one and must be governed with philosophical love.

This work, which is a Britannica of ingenious knowledge, is permanently dedicated to that only One God, the super ingenious universal One, for being in the absolute becoming of the supreme anatomy of philosophical mathematics, which is the illumined dynamism of Mother Nature Monad.

ACKNOWLEDGEMENT AS A PHILO-MATHEMATICAL EXPRESSION TO THE ANATOMY AND WISDOM OF THE INFINITUDE MATHEMATICAL CONCEPT

This is a book to acknowledge and appreciate what can be sanctified as a practical work that must follow the tedium of law with its monumental explanation. Philosophical mathematics can best be designated as an objective, subjective, dynamic, expressive adventure that exists at the proximity of ingenious inspiration. It is important to explain that the scriptural injunction, which states that "in my father's house, there are many mansions", is driven from the anatomy of a super-genius creator who appreciates that creation is a monumental mansion which must be enhanced and driven with empathic and natural growth.

This work acknowledges the ingenious and illumined province of transcendental philosophy with transcendental mathematics, and meta-science with mathematical logic. It is in the enabling postulations of its powers to equally acknowledge the anatomy of **Mother Nature**, which is the creative energy of all things. In the objective rhythm of this segmental acknowledgement, the immortal works, acts, concepts, and monumental epigrams of the Twelve World Teachers deserve to be acknowledged simply because they gave the universe a philosophical, mathematical culture with a scientific and technological expression that humanity could not understand simply because human consciousness and awareness of philo-mathematics was not consciously driven at the proximity of their philo-mathematical ingenuity.

In this respect a lot of ingenious minds, particularly emengards and geniuses that were cultured and nurtured in appreciating the way of Mother Nature, deserve to be acknowledged. It must be honoured for human understanding and empowerment that the concept of philosophical mathematics was the supreme basis by which one can gain the transcendental understanding of science, technology, and magnetic appreciations. It is here that this work is greatly acknowledging the part that developed and discussed the demystification of mathematics as a sacred oracle, including reordering and restructuring its engineering progress and success, which has being greatly affected and effected chaotically by the Microsoft sector.

At this point our roots must be acknowledged, starting with the classic and ingenious philo-mathematical world star designated as Pythagoras of Crotona. Such icon philo-mathematical gurus—like Dr. Walter Russell, Chancellor Alton, Beethoven, Bach, Mozart, Paracelsus, Galileo, Isaac Newton, Sir Isaac Pitman, Euclid, and H. A. Clement, to name but a few—must be given a space in this titanic acknowledgement. I hereby designate these humans as the philosophy and podium of our roots, because they were gifted to appreciate that in spite of man's ingenuity, the Creator stands tall above all ornaments, acts, creations, and concepts of philosophical mathematics.

From the research and academic circle, such ingenious and purpose-driven institutions like the Leningrad Library in the United States, the British Museum of Arts and Knowledge, Oxford University, Harvard University, the University of London, St Peter's Vatican University, the University of South Africa, and the University of Nigeria Nsukka deserve a sensitive acknowledgement in appreciating the synthetic wisdom in the dynamic concepts of philosophical mathematics.

In law, both the plaintiff and the complaint are defined as the originations of what the attorney and the judge are likely to examine with the use and application of logical and luminary technicalities, which must be purpose-driven with legal prowess. In this respect, the immortal and monumental contributions of Rejoice C. Adiele, Queen O. Nwokocha, and Ubani Ikemba, to name a few, must be acknowledged for their immeasurable contributions, which include typesetting, corrections, and ensuring that the work exists naturally at the aesthetic concept of the author.

The need to acknowledge the immortal contributions of the University of Science and Philosophy, the College of Metaphysics, including the great adventure into the world of knowledge from the University of Science and Philosophy and Concept Therapy Institute, must be acknowledged because of the contributions they have collectively made in the life of this author.

In the immortal words of Plato, my wife, particularly my family, is blessed for being the bedrock of my works. In appreciating the dignity of this maxim, I hereby appreciate the immeasurable and unquantifiable ingenuity of my family with my wife, who is hereby donated and designated as the supreme aroma, and who donated all her time in ensuring that the concept and dynamics of philosophical mathematics is gained within a recorded time frame.

A look at what is happening at different research institutions—including the Aeronautic Yard in America; the Airbus University, France,; the Metallurgical Institute, Korea; the Metallurgical Institutes, India and China. The growth and development of the automobile industries in Korea and Japan will certainly make all and sundry to appreciate that philosophy, apart from being the mother of science, is always using engineering mathematics and engineering dynamics in ensuring that humanity does not just advance in material technology but imbibes and cultivates the engineering foundry of Mother Nature, which is the fulcrum of all balanced creativity.

This acknowledgement cannot exist as a comprehensive anatomy of balanced creativity if the Creator—who is honoured and respected as the only holiness in the perfect existence of alpha and omega and in the electrifying objectivity of His omnipotence, omnipresence, and omniscience—is not given his due eternal position and acknowledged forever for being the podium and polyester of the dynamic concepts of philosophical mathematics, which He synchronously upholds with constant and perpetual excellence as the tedium and rhythm of the universal anatomy, which finally designates him as the consummate universal One.

FOREWORD BY MKPA A. MKPA

This book is a product of seasoned scholarship, natural wisdom, empirical research, and inspired originality. It is perhaps one of the most sophisticated intellectual inputs to the world of knowledge. Here, the author delves confidently into his familiar field of philosophical mathematics and navigates the terrain with such an authoritative air that only the specialist or the inspired can easily appreciate it.

The author takes the reader through the authentic origins of philosophical mathematics and relates the concept to the worlds of science, engineering, and human existence. He imputes ingenious interpretations to mathematical and philosophical issues, re-examining their relevance and applicability to contemporary developments. He does so in a most exciting manner that challenges and ignites the thinking mind to accept the unquestionable relationship of the concept to the world of science and technology.

Only an adept of the author's status would convincingly inject fresh interpretations to existing mathematical and philosophical functions and symbolism, as has been done in this book. In doing so, he guides us to the obvious and incontrovertible conclusion of the indispensability of this field in the created world of the past, present, and future. He links the concept of the solution of human problems, and in so doing he guarantees the continuous relevance, utility, and applicability of the discipline of philosophical mathematics.

One interesting element of the book is the structure of the organization. It existed in ten parts, with each part having various numbers of chapters, each of which addressed a particular thrust of the subject matter. Another merit of the book is its comprehensiveness: it delves deeply into the twin subjects of mathematics and philosophy and unifies them to establish the synergy and the complementarity of the two. It then spreads out the discourse in each part and chapter to touch on as many related and relevant concepts as possible—all in an attempt to establish the relevance of the discipline to contemporary living.

Perhaps the most significant merit of this book is its originality. The author does a minimum of quoting from others. He generates much of the information, which is amazingly exotic and therefore can be derived only from the supernatural realm or inspired sources. Not many authors are so gifted!

I must therefore caution that the subject matter is not meant for neophytes or the undeveloped minds. It is for mature individuals who are not intimidated by technical or professional concepts and terminologies in the fields of mathematics, philosophy, and esotericism. I recommend this book to all who can muster the courage to explore the boundless horizon of the author's wisdom, whether or not they are grounded in the discipline of philosophical mathematics.

—Professor Mkpa A. Mkpa
Vice-Chancellor, Abia State University

FOREWORD BY JOSEPH WEEDS

To write a cathartic foreword about this book deserves producing an aggregate and agreeable resume of the whole work. In this respect, this work, which originated from the anatomy of Mother Nature defined and designated as *The Dynamic Concepts of Philosophical Mathematics,* is a wholesome attempt to create a practical structure, including an ingenious mathematical progression that will help future scientists and the global family in appreciating that knowledge is power—it is from God and is the objective and practical seat of wisdom.

Humanity has known a lot about philosophy and mathematics, but this knowledge has existed on the periphery, and it is not structured in presenting philosophy and mathematics as the hall of fame, including being the way of Mother Nature.

It is important to explain that the high level of enmity though which the world is passing is not the best. It will not help the world, which is segmental in arts, in science, and in knowledge, to achieve the desired global family. That is why this book is written: to re-awaken our present chaotic system by ensuring that a galvanized and acceptable raw material is

1

created for the use and application of posterity, which holds the view of brighter minds.

In this respect, the book started with an introduction to philosophical mathematics, which linked to its ontology, expression, and definition of terms before delving into the main body. It will be interesting to know that scholars from all fields will find this book as an ornamented tutor, a research fellow, and a consultant—all depending on the path and approach the person in question wants to follow.

It is not possible to give a comprehensive resume about this book for three reasons.

- The characteristics of the book
- Its style
- Its modest and universal expressions

It must be explained and appreciated that this work is configured as a monumental databank and a mathematical bank that will help all and sundry in appreciating that because life is mathematical and logical, it is equally philosophically driven with the use and application of ingenious wisdom, which is the focus of this book because it hereby represents a scientific and technological galvanization.

This book greatly exposes humanity to a lot of principles, practises, and policies, including abolishing most ancient and primordial mathematical policies and formulas that are not helping the world of science and technology in maintaining a practical and astronomical growth. In this respect, most of these principles, practises, theorems, and methods are strictly and tactfully abolished in order to create a new horizon for the progress of mathematics and philosophy, including opening a new province on how and why humanity should access and appreciate the ornamented technicalities of the infinitude mechanism with its universal methods.

These methods, in their modest and aesthetic postulations, are dynamically explained in the preface and afterword. It is in the enabling structure of this work to appreciate that the universe is naturally and compactly configured in its objective anatomy of philosophical mathematics. That is why it has produced a mathematical table of values with a philosophical databank, which is going to help define a new trend and a new horizon that will certainly acquaint us with the wisdom and tedium of transcendental science. Transcendental engineering and technology will, in my enabling and purpose-driven opinion, help humanity redress the enmities of material and scientific chaos that has eroded humanity, ensuring that science must finally stop locking the Creator out of its creations, which is perpetually responsible for our growth in higher knowledge.

For the record, this book is a consummate resume and is summative of what is in nature's philosophical mathematical laboratory, which is an ingenious technological kit that will help all segmental aspects of human endeavour in appreciating that what is known as science is best defined as nature's mathematical studio with a philosophical logic.

—Prof. Joseph Weeds
University of Science and Philosophy, Virginia

EDITOR'S NOTE

The invitation by an honest and balanced mind presents a compact and tactful work that regulated scientifically, and deregulated practically, the ingenious province of philosophical mathematics as seen and accepted by me as a natural edict in editing a work of this magnitude.

The challenge, which is purpose-driven, can best be defined as a scholastic concept. This is why my acceptance of this honour is best understood from the writer's waves of inviting inventing, including using the systematic conversion mechanism, which is a rare gift.

The dynamic concepts of philosophical mathematics, which started with an introduction and a lot of segmental parts, took a lot of energy and understanding application in order to be able to square myself in the original understanding of what the author has given the mathematical universe as a living basis for the understanding of futuristic mathematics. The mathematics are contained in philosophy as a science of perfection and purity, which is developed as the anatomy of the overall and ingenious calculus. The new horizon that this work has invented will provide scholars with a lot of educational and scholastic statistics. It will certainly knock out the undynamic and unacceptable methods that have degenerated the scholastic province.

Starting from its ontology in part 1—which naturally explains the definition of terms, the concept and purpose of the work, and the nature and its formats—this work is accepted as the logical presentation of philosophical mathematics from the bank of Mother Nature, which exists with the use and application of philosophical mathematics with philosophy, applied mathematics, and logic. The educational input and impact of parts 3-5, include demystifying the oracle of mathematics, is a work that has never been seen or known to humanity and evolution since the dawn of consciousness.

The way he introduced the infinitude mathematical method, including disagreeing with the acceptance of the mathematical stigma of QED, is hereby noted as the mark and format of a living genius.

Taking on the task of knowledge in its originality as not being properly understood by the academia is another great front. The book will certainly open a scientific and technical horizon when all the concepts are purely and practically utilised with the wisdom of the original author.

In this respect, the task of writing a work of this nature is not borne to please the academic sector; it is aimed at enhancing the creative and productive balance of the one-world family through the use and application of philo-mathematical methods, which exist at the proximity of honest and practical knowledge. It must be stated that this book cuts across all bounds of human endeavour with a high level of natural input, which will help the universe in appreciating that creation is the anatomy of honest philosophical mathematics. This is why the idea that spurs the inspired thoughts of the author is hereby dissected with critical and cross-examination as a basis which, in earnest, is deeply rooted in the best and natural understanding of science and technology.

A look at the quantum engineering methods that exist in this work will lead the reader to certainly appreciate that this book is best named a calculative incubator of monumental Britannica.

In the original words of A. U. Aliche, philosophical mathematics took its origin from nature; this is why it naturally extended to all facets of human dynamism, particularly to those rare areas where material science and technology have been finding difficult to merge and submerge.

As an expert in pure and applied mathematics with an electronic bias, I hereby concur with absolute acceptance that the tedious task of writing a work of this magnitude can only be successful if the author adopts the original databank of inspirational techniques both as a guide, as a tutor, and as a Britannica. This is why his gallantry in achieving this feat is best defined as the greatest conquest of modern civilization.

QUOTES FROM A. U. ALICHE'S PHILOSOPHICAL MARBLE

❖ Mathematics exists with the balanced understanding of quantum mechanics.

❖ Mathematics is best defined as an analytical and balanced alchemy, by which the ennoblement of man as the pinnacle of knowledge can be gained.

❖ Mathematics is the alchemy and the objective dynamism of the electrifying force. This is why it empowers the thinking of physics in measuring the length and width of the wave motion and light waves with its reflections and refractions, including appreciating the anatomy of heat, energy formulation, and transference. This is why every wizard is always initiated in the path of mathematical kingdom.

❖ The world will forever remember Pythagoras with his illumined aborigines because he adored mathematics as the living God of the supreme universe.

❖ The illumined words of H. A. Clement, which read, "The rhythm of my heart is arithmetically driven, while the anatomy of my brain is geometrically driven," has proved to be a living mathematical testimony that is in consonance with balanced mathematical dynamism.

❖ "As life is mathematically and logically driven, evolution, and particularly creation, is technically and philosophically ingenious in developing and progressing the noble course of philosophical mathematics."

❖ "If Q is quantious, and if E is practically evolving and evaluating, and if D is futuristic, then it is important to ask what is the proportion, the preponderance, and the position of QED, particularly as donated to the one-world family by the Grecian mathematical wizards, including the English Armadas of which the mathematical emeritus like Euclid, H. A. Clement, and others were very much steady and celebrated at that very era."

❖

❖ If triangular proofs can be utilised in different formats of logic, and if circle theorems can be proved with the use and application of pure reasoning as contained in the *Guinness Book of Pure Mathematics,* it is mockery to believe in the concept of QED, which is used to represent a final and full stop in any mathematical proof—which in honest reasoning cannot further be proved in teaching.

—Argument on the concept of QED

PREFACE

How the Inspired Province of Philosophical Mathematics Is Using the Ingenuous and Multifarious Concepts of Infinitude Postulation to Disprove with Total Abolition the Ancient and Contemporary Issues of QED

Before now, a lot of mathematical proofs have always ended with "Quod Erat Demonstrandum", or QED. This result was the teaching of some mathematical fundamentals that were not nurtured in the best objectivity of the infinitude mathematical concepts. Philosophical mathematics does not logically accept this opinion, which was popular for ancient mathematicians like Pythagoras and Euclid.

For instance, if two triangles are equi-angular, then their corresponding sides are proportional.

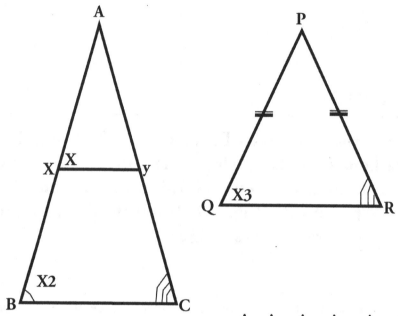

Given $\triangle ABC$ and $\triangle PQR$, in which $\hat{A} = \hat{P}$, $\hat{B} = \hat{Q}$, $\hat{C} = \hat{R}$

To prove that $\underline{AB} = \underline{AC} = \underline{BC}$

$\qquad\qquad$ **PQ** $\quad$ **PR** $\quad$ **QR**

Proof Method: $\underline{AB} = \underline{AC} = \underline{BC}$

$\qquad\qquad$ PQ $\quad$ PR $\quad$ QR

$\quad$ In $\triangle$s AXY, PQR

$\hat{A} = \hat{P}$ $\quad$ (Given)

$\overline{AX} = \overline{PQ}$ (Construction)

$AY = PR$ (Construction)

$\therefore$ $\triangle$s $\underline{AXY}$ = (SAS)

$\qquad$ PQR

$\quad$ X = X2 = X3; but these are corresponding angles on a straight line.

$\quad$ $\therefore$ XY // BC

$\dfrac{AB}{AX} = \dfrac{AC}{AY}$

$$\frac{AB}{PQ} = \frac{AC}{PR} \quad (AX = PQ; AY = PR)$$

Similarly $\frac{AB}{PQ} = \frac{AC}{QR}$

$\therefore \frac{AB}{PQ} = \frac{AC}{PR} = \frac{BC}{QR}$

If a line is drawn parallel to one side of a triangle, it divides the other two sides in the same proportion on ration. How can this theorem be accepted to equally exist at QED, and how can we use the enabling dynamics of philosophical mathematics to disprove these standing concepts?

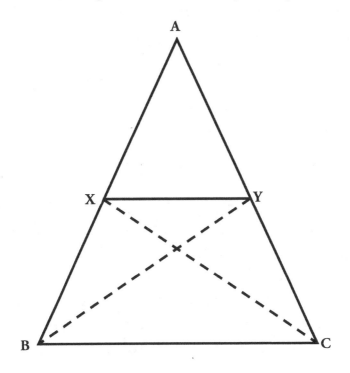

Given $\triangle ABC$ with points X, Y on AB and AC, such that XY//BC.

To prove that $\dfrac{AX}{XB} = \dfrac{AY}{YC}$

Construction: Joint BY and CX

Proof: $\dfrac{AX}{XB} = \dfrac{\text{Area of } \triangle AXY}{\triangle BXY}$,

which is the same altitude from Y to AB

$\dfrac{AY}{YC} = \dfrac{\text{Areas of } \triangle AXY}{\triangle CXY}$

same altitude from X to AC, but

$\dfrac{\triangle AXY}{\triangle BXY} = \dfrac{\triangle AXY}{\triangle CXY} \{ \triangle BXY = \triangle CXY\}$

and the reason is that $\triangle$s on the same // XYX = B, and on the same base:

$$\dfrac{AX}{BX} = \dfrac{AY}{CY} \text{ and that is } \dfrac{AX}{XB} = \dfrac{AY}{YC}$$

The ancient and un-mathematical use and application of QED is one of the serious philosophical mathematical concepts, which this book is going to address with the use and application of pure and applied logic, pure and applied philosophy, and pure and applied and systematic methodological mathematics. This is why it is going to solve this aging question of what QED means.

At the time of such mathematical wizards like Pythagoras, Euclid, H. A. Clement, L. Durell, and S. W. Smith—and including such icons and meta-mathematical emengards like Mozart, Paracelsus, Dr. Walter Russell, and mystical mathematicians from Asian provinces whose records were unknown to evolution and civilization—the use of QED was prevalent within mathematical utilisation. The usage was acceptable because the science, philosophy, ontology, and ingenuous wisdom of the infinitude principle, which science and technology had not been able to identify, was not discovered at that era.

A look at the structure and anatomy of angles, triangles, circles, squares, parallelograms, cuboids, equi-angles, anti-angles, and more will certainly prove that the use and application of QED to either prove or disprove them does not make a philosophical and mathematical reasoning—not even in simultaneous equation, compound interest, simple interest, functional statistics, and all that is contained in algebra and geometry, which can best be defined as the hub of pure and applied mathematics that can be furthered in the advancement of logical reasoning.

QED cannot be attributed as a mathematical issue simply because it can create a fragmented obstruction that can make or mar the subject of mathematics from being a conventional and inferential issue. In this respect, the following questions need to be asked and answered by the followers of QED.

- Is life actually queued in QED?

- If life as believed and understood is mathematical, what is the position of QED in life, which is believed to be purpose-driven with evolutionary concept and dynamic growth?

- If every angle and triangle is embodied with **1-45°**, **45-90°**, **90-180°**, and **180-360°**, is it not possible to reason that QED does not function or feature in the arithmetical progression of pure and applied mathematics?

- If triangular proofs can be utilised in different formats of logic, and if circle theorems can be proved with the use and application of pure reasoning as contained in the *Guinness Book of Pure Mathematics,* is it not mockery to believe in the concept of QED, which is used to represent a final full stop in any mathematical proof that, in honest reasoning, cannot further be proved in teaching?

The challenge to ask this hallmark of mathematical questions is borne out of the fact that the infinitude concept of pure and applied mathematics, which is the hub and anatomy of mathematical progression, is practically set within the era of modern emengards to disprove in totality, and to demystify in context and connection the actuality of acceptance of QED as a non-mathematical issue without basis, with the use of the ingenuous concept of philosophical mathematics. That aspect stands purely and alone in that esteem and aesthetic province, with the use and application of infinitude principles, policies, practises, and a whole lot of methodology that is intertwined in the use and application of mathematical progression.

In this respect, because academic mathematics has developed a province of calculation and calculus, which can take us from zero to ten, ten to a hundred, a hundred to a thousand, et cetera, the same academic province in their industry of mathematics has not been able to give the universe of mathematical conclave, with high levels of advancement in engineering, thoughts of what is considered the anatomy of billions and trillions, the objectivity and concepts of millions and thousands, including the lintious structure of thousands and hundreds. This is why this work is practically demystifying the obnoxious, obscure, and the non-progressive agenda of QED, which creates a mathematical mutiny for the modern technological and mathematical province.

After a look at the features and functions of how and why the Creator is naturally un-mathematical, but perfectly mystical, in His arts and transactions, one can certainly appreciate that the greatest urge and need of modern science and technology is the investigation of what is known and accepted as meta-mathematical philosophy, which is hereby defined and denoted as living knowledge. The objectivity of this section is not to criticize or condemn the contributions of the past eras of mathematics but to awaken the mathematical heartbeat of evolution that is operating unknowingly with the mathematical inquiry of the infinitude method, which was difficult for science and technology to understand because of the high level of the perfection and purity. That purity can best be defined as a universal and cosmic standardized front that made science and technology, with their material profession, neglect and lock the Creator out of its creations.

Before any journey in the field of mathematics, one must understand that the concept of QED fragments the progressive thoughts in believing that mathematics, which is nature's electrical and electronic curvature of defining and redefining the rhythm of everything, does not welcome its existence. For example, if Q is noted as a letter is quantious, and if E is practically evolving and evaluating, and if D is futuristic, it is important to ask what is the proportion, the preponderance, and the position of QED, particularly as donated to the one-world family by the Grecian mathematical wizards.

For the record, this is a mathematical cheat to other factors and units of the universe that were not yet developed in the wisdom and application of mathematics, simply because these enmities used the concept of QED to cause confusion and stagnation in the concepts of mathematics. This is what kept what they termed as third-world countries in the dark about mathematical stagnation.

If QED is acceptable in the real province of pure and applied mathematics, the following questions need to be answered.

- Why are these people developing at an astronomical rate, which is above the reasoning of QED?

- How and why are they experts in science, engineering, and technology?

- Why are they experts in the areas of econometrics, applied and nuclear physics, and universal chemistry with ingenious biomass?

- Why is their civilization at a faster pace when compared to other regions?

- In the area of computer mathematics with electronic and electrical data and statistics, how have they been able to produce the best results, even with their insistence on the concept of QED?

- Is it not right and possible for the overall province of mathematics to investigate the high level of indecency in order to create an acceptable mathematical pathfinder, which will liberate the logical reasoning of mathematics from the whims and caprices of QED?

It is important to explain that what happened during the era of Pythagoras, which worshipped mathematics as an idol and a god, is still prevalent in our modern, global civilization that is sophisticatedly driven with the wisdom and knowledge of extensive and expanded technology. It is in the tedium of this section that can best be defined as a resurrection and an effort to reject, neglect, and abolish the concept of QED, which is not in synchronous agreement with the infinitude mathematical progression; neither does it agree with the enabling dynamics of philosophical mathematics, which is donated to the province of pure reasoning simply because just as life is mathematically and logically driven, evolution—and particularly creation—is technically and philosophically ingenious in developing and progressing the noble course of philosophical mathematics. This is why this work is comprehensive and must be accepted by all engineering sectors as a perfect and purpose-driven legend that will help every sector as a dynamic vanguard.

The understanding of the lintious reasoning of philosophical mathematics is a power that will liberate the academic and commercial sectors from the long existing melancholy that is not acceptable in our modern evolutionary age.

It is here that the author donates the work as a logical and philosophical mathematical dossier that will help all and sundry to appreciate and access the high level of ingenious wisdom contained in the concept of infinitude progressive method, which is hereby defined as a mathematical polyester with a philosophical poser.

The Anatomy of Philosophical Mathematics

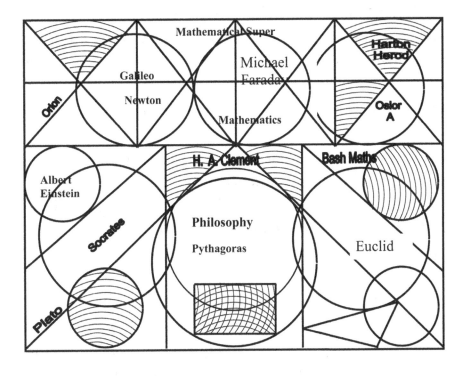

The drawing on the previous page shows the comprehensive anatomy of philosophical mathematics, which creation—and particularly the materially driven academia—have not known as the mutual and dynamic rhythm of the universal mathematical heartbeat. The anatomy of this drawing can be used in accessing a lot of scientific and technological sectors so as to draw an acceptable Internet wisdom from them.

A careful study of the drawing reveals that the universe is philosophically and mathematically driven. This is the scientific basis of universal cosmogony, including the facts and figures that configure the dynamism of one-world balance.

After the first mathematical Philo-Monads, other philosophical mathematical monadians followed. This fact shows that the Creator is an ingenious and sino quantious heartbeat of the supreme universe in His philo-mathematical universe.

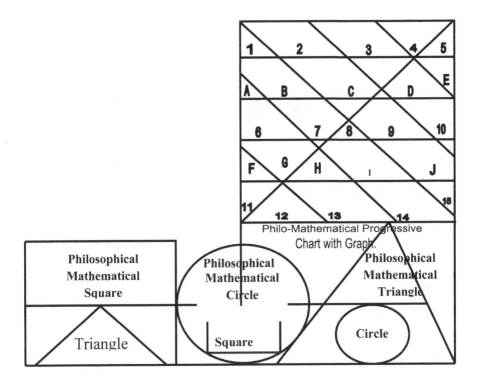

PART 1

THE DYNAMIC CONCEPTS OF PHILOSOPHICAL MATHEMATICS

"Pythagoras was intellectually one of the most important men that ever lived, both when he was wise, and when he was unwise. Mathematics, in the sense of demonstrative deductive argument, begins with him, and in him is intimately connected with a peculiar form of mysticism."

—Bertrand Russell,
On the Mystical Foundations of Mathematics

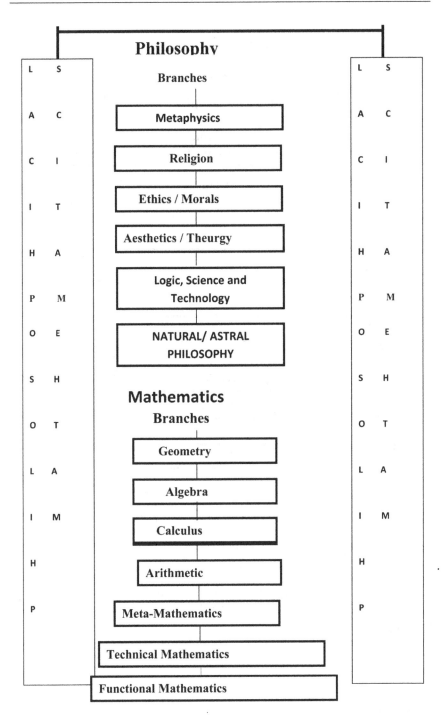

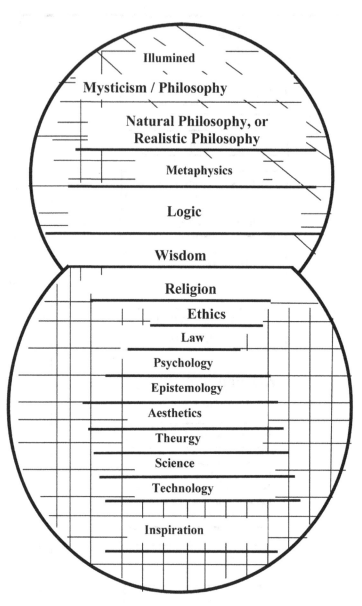

Fig. B

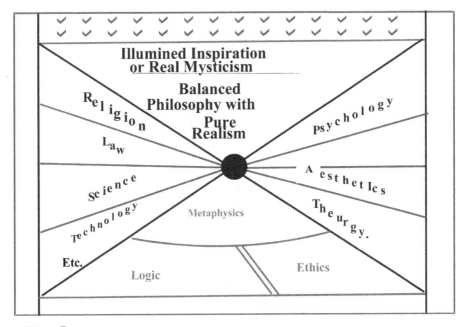

Fig. C

The diagram above shows an empowered second anatomy of philosophy that humanity has not known. This is why philosophy has not been appreciated.

Fig. D

This diagram reveals the most ingenious framework with which philosophy is empowered to perform its multiple and illumined functions.

INTRODUCTION WITH ABSTRACTS

Philosophy—as a science, a balanced art, and an illumined knowledge—is purpose-driven with the use and application of mathematical wisdom. At all levels, it is pragmatic in achieving the illumined desire of perfection and purity, which designates its objective and logical values as the balanced prince of knowledge. Hence, it is defined by an Oriental philosophical apostolate as the mother of science.

It is important to explain for the balanced consumption of the well informed and the philosophical academia that philosophy is quantious and consummate in the best objectivity of calculation, algebra, geometry, arithmetic, trigonometry, data, and statistics. This is why all that is mentioned above compacts their united kingdom as mathematics.

Regrettably, the world of science and conventional psychology has greatly relegated the mathematical functions of philosophy to the background because of its abstract essence—which, when properly understood and utilised, is the central point of the Internet with universal wisdom.

A work to explain the details about philosophical concepts with objective mathematics will certainly appreciate that the Orientals were great scholars and geniuses in the best utilisation of philosophical mathematics, philosophical logic, and legal and luminary philosophy with the abstract use of philosophical mathematics. This is why the objective and electrifying province of further mathematics, which is best known as pure philosophy, accepts that the prince of knowledge and balanced wisdom cannot be relegated to the background. The only illumined philosophical mathematician the world has ever had was Pythagoras. He was initiated without discussing with anybody for five good years by the Great One, and he was philosophically designated for that baptism. He was able to develop theories, theorems, polices, principles, data, and statistics that are particular to philosophy and mathematics. Unfortunately, his aborigines and the primitive province of his time were not in this knowledge. This is why such ordinary scientists like Euclid, Galileo, and Albert Einstein suggested strongly that he was the mathematical element of universal calculus.

In presenting a rare work of this nature, it must be borne in mind that Albert Einstein argued strongly that there is nothing like mathematics without philosophy. I have discussed Einstein's argument as a royal mathematical pyramid that all institutions must contact, consult, and possibly establish as a philosophical laboratory, which will help the world of commerce, the province of science and technology, and the kingdom of balanced arts in knowing that science and technology cannot globalize the universe as a one-world family without the perfect and purest utilisation of the mathematical kit of philosophy.

In writing this book, which was inspired by the great questions, doctorate students asked me at the University of Lagos and the renowned University of Science and Philosophy in Virginia. There is a need to explain that the greatest business of philosophy and mathematics is to give the universe a balanced, logical constitution which, when properly analysed will be purpose-driven in understanding the concept of Internet appreciation with the desires of balanced globalization. It is on this premise that philosophy is mathematically driven, and mathematics at all levels is philosophically united with the wisdom of universal configuration.

Philosophy neither welcomes nor accepts a secondary issue that is not pure and perfect, so mathematics has remained dogged and unchanged in ensuring that the perfect result of $1 + 1 = 2$ and $1 \times 1 = 1$ and $2 - 1 = 1$.

A philosophical review of mathematical anatomy has shown that mathematics and philosophy are perfect but united sciences. Other courses, like the engineering sciences, always wait and look upon them to explore the fundamental issues of engineering mathematics. As presented to the purpose for the utilisation of one-world balance, Richard Ohm, a philosophical mathematician, was specific and purpose-driven in presenting this balanced science as the pinnacle of Mother Nature and the noble price of the well informed. This is why physical sciences are always dynamic in the best utilisation of philosophical mathematics. It is important at this stage of human civilization's chaotic evolution to present this book, which is balanced with the Internet thinking of infinitude principles and illumined philosophical mathematics.

My interaction with these noble students and trustees of philosophical propagation triggered an electrifying thinking whose result and research is purpose-driven in presenting a work that will help the college of philosophy at all levels in the opinion of philosophy. It should aid philosophical professors and mathematical lecturers in knowing and appreciating that the world is dynamically driven with philosophical mathematics.

There is need to explain that all the balanced branches of philosophy—which include religion, metaphysics, ethics and morals, epistemology, aesthetics, and theurgy, and including such other areas like diplomatic philosophy, mathematical philosophy, legal and luminary philosophy, Internet philosophy, and engineering philosophy—have proved without bias that philosophy is truly and nobly the mother of balanced and purest knowledge. This is why its scopes and electrifying functions are purpose-driven in the best objectivity of philosophical dynamism.

Eminent philosophical mathematicians like Anthony Orwell, Pythagoras, Albert Einstein, Euclid, Plato, Socrates, and Aristotle, including the prince of philosophy, Emmanuel Kant, have ever fathomed the philosophical conclave, which is usually known as a mathematical podium.

The need and the urgent philosophical challenge to write an acceptable and balanced philosophical introduction is borne out of the fact that such eminent works written by the author include the following.

- *Philosophy and Applied Logic*

- *The Dynamic Functions of Philosophy*

- *Philosophy: The Maker of Great Thinkers*

- *The Philosophy of the Oriental Wizards*

- *The Fundamentals of Core Logic*

- *The Electrifying Understanding of Philosophy from the Point of Logical Empathy*

This is simultaneous and synchronous in ensuring that philosophy is seen from the objective angle as a pure and practical science that condemns the primitive concept of philosophy as an abstract science. It is here that I feel honoured to state that mathematics is a balanced weaver to philosophy, whereas philosophy is the balanced and illumined ray of light that makes mathematics perfect its alcove. This must be purpose-driven with philosophical reality by using honest logic in presenting its realism with dynamic philosophy, which can be defined as scientific polyester that will help the academia and the institutions of knowledge in appreciating that philosophy and mathematics are quantious and quanta. They utilise all the electrifying dynamics to explain facts and figures from the point of perfection, purity, and balance with an acceptable and dependable united empathy.

CHAPTER 1

PHILOSOPHICAL MATHEMATICS AND EXTRACTS

It is my natural birthright to express myself with the objective and philosophical rays that are fulfilled with infinite dynamism, which in return will bless and make others be philosophically endowed and recharged with mathematical understanding. In presenting this wonderful and balanced philosophical abstract with mathematical extracts, there is a need to appreciate that philosophy and mathematics are interwoven in principles, policies, relativities, and transactions. Both are purpose-driven in ensuring that the concepts of purity and perfection are practised as acceptable standards at all levels, as well as realism of existence and transactions.

Sir Oliver Fox, a mathematical wizard with metaphysical understanding, stated the following.

- The world desires and deserves the perfect and concretized knowledge of philosophical mathematics.

- Every specific agenda with principles and policies, which are not purpose-driven with the use and application of philosophical mathematics, must certainly not perform their functions and duties to an enduring satisfaction.

- Economic principles and policies, which are not focus-driven with the use and application of philosophical mathematics, will serve humanity as orchestrated enmities.

- Science and technology at all levels must be objective with the use of scientific philosophy, including the application of technological mathematics.

The immortal words of this iconic scholar triggered a high level of investigation by the Societies of Philosophers and Metaphysicians, the University of Philosophy, and other renowned institutions of which I am in the forefront with philo-mathematics. These words from this scientific and philosophic genius motivated the likes of Herbert Spencer and Dr. Walter Russell to start thinking of what could be done to reshape the thinking of man, who had remained in the dark since the dawn of consciousness.

The ebonite scholars with philosophical ingenuity were able to draw a scientific and philosophic roadmap that compacted and configured the objective concepts of philosophy and mathematics as the anatomy of creation, including accepting their principles and policies as the alchemy of all balanced creativity.

St Jerome authoritatively stated that there is no separateness between philosophy and mathematics. He explained that because philosophy is purpose-driven in using scientific semantics with objective logic in achieving the purpose and course of purity and perfection, then mathematics is dynamic in utilising figures, formulas, and data to explain the anatomy of philosophy from the point of globalized and scientific empathy.

It is here that the following can be utilised as standards for this extract.

If 1 + 1 = 2, it then means that 2 - 2 = 1. It equally means that A + A = 2A, while $X^2 + X^2 = X^4$.

A look at these mathematical facts, which are embedded with scientific symbolism, will make one appreciate mathematics as a data pot with the use and application of philosophical wisdom. When philosophy is arguing a point with a view to arriving at perfection, it is always making use of mathematical figures in ensuring that all that is contained in the ancient wonders of living antiquity is practically and pragmatically presented in a style and manner that will help the world of knowledge. One must know that the opinion of mathematics to philosophy and other related sciences is a fundamental issue that must be examined and practically obtained as a theory or a theorem. This will help our present civilization in appreciating the monumental relativity of philosophy as the mother of science.

Mathematics, in my best opinion, can be defined as the Internet and electrifying wisdom of Mother Nature, which functions with the use and application of figures, facts, features, functions, data, and statistics with a high level of meta-mathematics and meta-logic. There is need to explain that this extract, which exists as a philosophical abstract, must be seen as a work that will help the colleges of philosophy, the universities of mathematics, and universal research fellows in appreciating that economic problems are a result of the negligence and abuse of philosophical mathematical laws.

For example, if a triangle exists as a diagram ABC, and it is known that A is a capital letter, followed by B and C, then there could be an argument that ABC can equally be extended to triangle DEF.

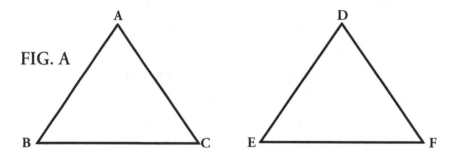

FIG. A

This evidence shows that philosophy and mathematics maintain a synchronous relativity in agenda, illustration, and understanding.

Pythagoras could not hide his feelings when he stated that philosophical mathematics holds the world in appreciating and understanding what is contained in the Internet kit of science and technology. A look at this illustration must certainly reveals that our present generation is living and existing as un-philosophical lumber, because we lack the understanding of mathematical appreciation.

It is unfortunate to explain that Manly Palmer Hall, one of the most luminous beings who has the mandate of philosophy a must-give to civilization gloriously states that the world needs a balanced knowledge of philosophy, an objective wisdom of mathematics, and an appreciation of the enabling and foremost powers of logic, epistemology, metaphysics, morals, and ethics.

This is why a work to examine the intricate relativity of philosophy and mathematics will modestly and scientifically exhibit the facts that make these subjects twin issues and inseparable apostles. The diagrams below show that philosophy is objective to mathematics, whereas mathematics is conceptualised and perfected on its balanced, illumined wisdom.

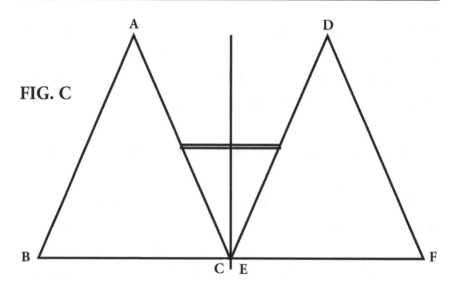

FIG. C

A B = A C   A B C = D E F
D E = D F   A D = C E
B C = E F

That is, $\underline{A\ B} = \underline{D\ E}$ = A B = A C = D E = D F.
$A\ C \quad D\ F$

$\underline{A\ B\ C}$ = A B = A C = D E = D F
D E F

The concept of philosophy to this mathematical expression is that mathematics holds a scientific methodology with its enabling wisdom and other related sciences. This is why the honest approach ends in **A = B = C = D = E = F**, which could be explained as follows.

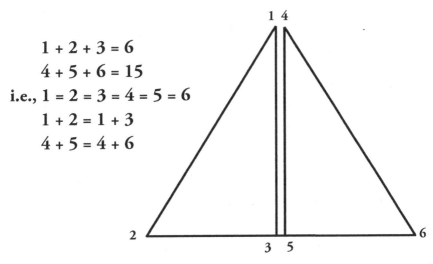

$$1 + 2 + 3 = 6$$
$$4 + 5 + 6 = 15$$
$$\text{i.e., } 1 = 2 = 3 = 4 = 5 = 6$$
$$1 + 2 = 1 + 3$$
$$4 + 5 = 4 + 6$$

A look at this explanation appreciates that the concept of abstract and extract is a philosophical synergy that lives as a mathematical standard. When it is stated that philosophy is an abstract science, it is only a dynamic approval, which is greatly accepted that its abstract nature is centred on the use of mathematical standards with scientific dynamism.

Dr. Walter Russell explained in his highly illumined work, *Secrets of Light*, that there is no separation between philosophy as a science, and mathematics and further mathematics with their objective methodology. This is what informed Stroud to edify engineering mathematics as the scientific and balanced anatomy of that which is contained in scientific objectivity.

There is a need to ask students and professors to always refer to the illumined works *Philosophy: The Maker of Great Thinkers* and *The Fundamentals of Core Logic*, which are other great works that exist on their own.

We need to state that such mathematical experts like Euclid, Pythagoras, H. A. Clement, D. Durell, Einstein, and Edison could be weeping like I do, if they existed in our present, non-mathematical civilization. There is need to explain that when Galileo told the world that the universe was a circle, his province was greatly criticised.

In presenting this book as a work for the well informed, I must not fail to explain that the features of a circle are in philosophical and mathematical agreement with the anatomy of a square. Pythagoras practically and pragmatically demonstrated this living mathematical truth by ennobling a concept in the ever living Pythagoras theorem, which in my opinion is a philosophical principle with a mathematical policy.

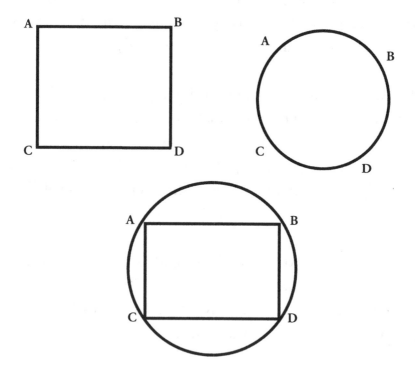

When looking at these drawings, one can appreciate that a circle has a mathematical structure, just as a square. This is why Pythagoras was greatly honest in giving the world what is known and accepted as a circle theorem, including a lot of quadratic equations.

It is here that "to equate" means "to philosophize and to organize". There is need to explain that philosophical equations are the hallmark of scientific quants and quantum. The honest appreciation of these abstracts, which are philosophical extracts, will help the world of science and technology in understanding the Louis operation of Internet dynamism, and how the philosophy of networking concept can best be utilised as a mathematical realism. In this respect, I must conclude this perfect abstract with mathematical extracts in the following idioms, axioms, and balanced methods.

- Scholars at all levels should be friendly with mathematics.

- All institutions should be purpose-driven with the use and application of philosophical dynamism.

- Universities with their curricula should be drawn on the enabling fortitude of philosophical mathematics.

- Methodological policies and principles should be research-driven in the best objectivity of perfection and purity.

- Science and technology should be structured and restructured with the enabling fortitude of philosophical mathematics, which will naturally reopen a new horizon.

- Students at all levels should be taught the philosophy of ABCD of mathematics, which is proven in the objective concept of Pythagoras' theorems, theories, and practises.

- The world of academia should be taught the wisdom of the infinitude principle, which is contained in my book *The First Principles of Science,* and a scientific explanation, which is contained in another book of mine, *Philosophy: The Maker of Great Thinkers.*

At this point, the philosophic universe must be scientifically driven with the use and application of mathematical technology, which is driven by concept objectives. This is why these amalgamated abstracts with extracts are going to serve the whole universe as philosophical and mathematical anatomy, which will solve and remove the falsehood of science and technology and the lapses man has abstracted in science and religion. This work is argued to be on its own, in *Objective Concept of Philosophical Mathematics.*

The Three Great Tetragramatic Elements That Form the Structure of All Creations

1. The triangle △

2. The circle ○

3. The square □

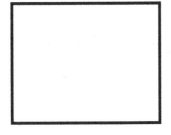

Most electronic elements have the shape of a square. All movable things are shaped in a triangular form, which energizes motion and speed (e.g., ship, aircraft).

<u>PHILOSOPHIE</u> (scientific formula)
MATHEMATICS

i.e. $\dfrac{P}{M} + \dfrac{H}{A} = \dfrac{I}{T} + \dfrac{L}{H} = \dfrac{O}{E} + \dfrac{S}{M} = \dfrac{O}{A} + \dfrac{P}{T} + \dfrac{H}{I} + \dfrac{I}{C} = \dfrac{E}{S}$

$P = M = H = A = I = T = L = H = O = E = S = M = O = A$
$= P = T = H = I = I = C = E = S$

.: PH = MA = IL = TH = OS = EM = OP = AT = HI = IC = ES

This is a philosophic mathematical fraction, or scientific rationale.

$\dfrac{PHI}{MAT} = \dfrac{LOS}{HEM} = \dfrac{OPH}{ATI} = \dfrac{IE}{CS}$

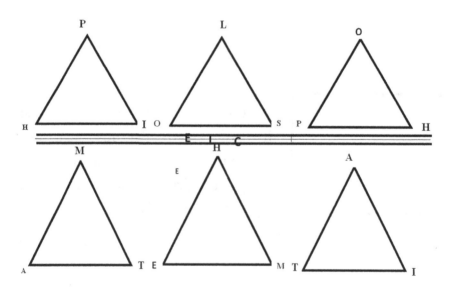

This will be properly examined in the book.

As represented and presented in the above drawings, which are philosophically ingenious in the best mathematical objectivity, the concept and format of

$$\underline{PHI} = \underline{LOS} = \underline{OPH} = \underline{IE}$$
$$MAT \quad HEM \quad ATI \quad CS,$$

explains that philosophy and mathematics are scientifically philosophia. This is why a work to explain the practical and purpose-driven concept of philosophical mathematics of Pythagoras, and the philosophical segmentation of the universe, is the mathematical configuration of data and figures.

As can be seen in the abstract, it is always known and explained that every triangular expression can further be developed to exist as a quadrant. This organized, scientific, mathematical quadrant can further be developed as a cyclic coefficient that, when further proved, can provide the mathematical rationale of the following fractions, including such linear equations as the simultaneous equation and remainder theorem, with its equatic concepts.

A work on the philosophy of mathematics as a science, which exists and behaves on the philosophy of arts, must be calculated and cautious in the presentation and representation of letters, ideas, and concepts. It is unfortunate that what can be defined as the philosophy of mathematics is greatly narrowed down to the little knowledge of empirical understanding. Mathematical philosophy is scientifically a furtherance of the use and application of data with a modest amplitude of the research method. This is why if there is an even multiplication of 000 to infinitude 0, all that is contained in that mathematical zero kit can equally be revisited and reversed to exist as $0 + 0$ or 0×0.

There is a need to explain that mathematical philosophy is calculus oriented. All that was done by Chancellor Simultaneous in explaining his deepest wizardry of the simultaneous equation has immortalized him as a mathematical emengard. This equation has remained the unchallenging mathematical knowledge both in science, engineering, mathematics, economics, statistics, and cost and management appreciation. What the world is utilising at the moment as the zero-point of online relativity can be explained as a mathematical enabling that is both philosophically oriented and logically configured.

In this respect, the mathematical definition of philosophy including the philosophical definition of mathematics, can stand as $P = M$ or $M = P$ and can perfectly be represented in the ratio of P:M and M:P. This explanation is important because it will help students of further mathematics and further engineering calculations, including sciences at all levels, in knowing with objective appreciation that if excluded from the deep thinking of science and technology, philosophical mathematics is an electrifying extermination.

The urgent necessity of this book becomes a must, particularly in our chaotic, un-mathematical world where scientific formats, principles, and policies have neglected the deep thinking of philosophy as both rational and irrational. This is why the challenge in the philosophical presentation of this book will always exist as a mathematical catalyst; further investigations and examinations will help the global world and the technological province in globalizing the ratio of P:M and M:P as the best principle of science and the first principle of mathematics and philosophy.

We can now appreciate why this work is written in the best aesthetic and architectural revelation of mathematics, both as science and philosophy and as the anatomy of mathematical explanation.

CHAPTER 2

Points and Facts of Dynamic Interests, Which Are Must-Know with Philosophic Understanding and Mathematical Appreciation

The dynamic concept of philosophical mathematics, which warehouses a lot of hidden knowledge, will explain the rational and irrational relativity between the following mathematical objects, data, and formulas that are best understood through the direct involvement of philosophy with objective logic.

This explanation is real, important and clear with its empirical estate because the past has created a lot of unscientific and un-mathematical confusion particularly in the Third World Countries in accessing and appreciating their workability and relativity. This has ingloriously affected the invention of cultural science with the utilisation of rational technology. It has equally made info-tech and information data to remain obscure with ambiguous concepts.

My illumined reason for this work is borne out of the fact that the honest harmony which can be seen in the first principle of science and the foremost practises of philosophy is contained in the anatomy of:

PHILOSOPHIE
MATHEMATICS

$$\frac{P}{M} + \frac{H}{A} + \frac{I}{T} + \frac{L}{H} + \frac{O}{E} + \frac{S}{M} + \frac{O}{A} + \frac{P}{T} + \frac{H}{I} + \frac{I}{C} + \frac{E}{S}$$

This can equally be proved and represented as:

$$\frac{P}{M} \times \frac{H}{A} = \frac{PH}{MA} \times \frac{I}{T} \times \frac{L}{H} = \frac{PHIL}{MATH} \times \frac{O}{E} \times \frac{S}{M} = \frac{PHILOS}{MATHEM} \times \frac{O}{A} \times \frac{P}{T} =$$

$$\frac{PHILOS}{MATHEM} \times \frac{O}{A} \quad \frac{P}{T} \quad \frac{PHILOSOP}{MATHEMAT} \times \frac{H}{I} \times \frac{I}{C} \times \frac{E}{S} =$$

PHILOSOPHIE
MATHEMATICS

In this respect, all that is explained here will be represented as follows.

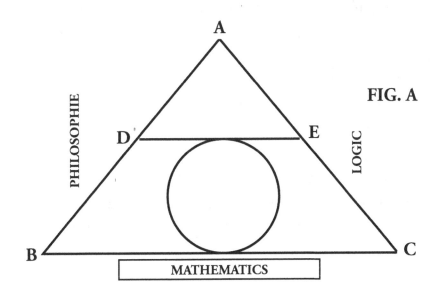

FIG. A

To prove the philosophical and mathematical relativity of this triangle, we need the objectivity of science and logic with an empirical know-how. This is important because the triangle contains another triangle inside with a complete circle, which is inserted in a square that is already holding a visible ADEC—that is,

MATHS or **PHILO**
PHILO   **MATHS**

The proof uses scientific logic.

Prove that ABC = Δ, with AB = AC = BC and with DE dividing the triangle into two unequalled parts to have triangle ADE with square BDEC.

Proof: AD = DB = BD = DA proved. AE = EC = CE = EA, with line BC providing the basis for the logical proof.

It is equally logical that ADB = AEC, with BC as the basis.
. **AB = AC = BC,** and
. . <u>**AD**</u>=<u>**BD**</u> =<u>**AE**</u>=<u>**CE**</u>
  DA   DB   EA   EC.

This shows that ΔADE can be logically reversed to be a square BDEC, which is a perfect square with a circle in it.

One should argue that philosophy is the anatomy of mathematics using the following data with a logical approach, which is both scientific and empirical in conversion.

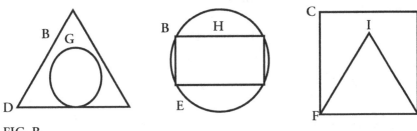

FIG. B

There is every scientific basis that the triangle so proved has been extracted in different formats to show a logical originality. Philosophy and mathematics cannot function without each other but must utilise the honest and dependable wisdom of meta-logic with the balanced use of scientific knowledge. It is easy and empirical to argue with proof that a philosophical work on triangles begins a mathematical knowledge of the concept and occurrence of circles and squares.

The mathematical methods of proving the circle theorem and quadratic equations must be known and traceable to the scientific existence of the triangle theorem, which is hereby designated as the logical way of philosophical reasoning that informed my illumination to state that man is the anatomy of three forces which include the triangle, circle, and quadrangle, or C + T + S = man. The mathematical anatomy has a philosophical structure and a logical physiology.

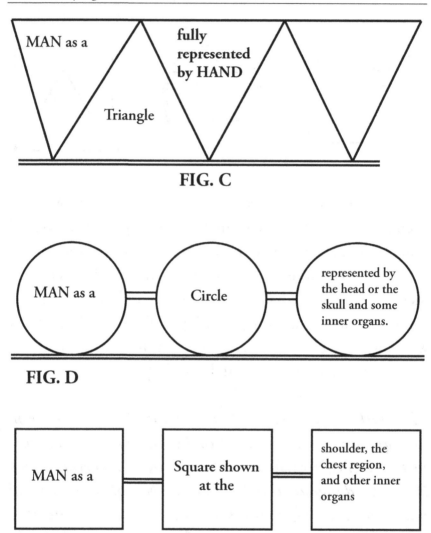

FIG. C

FIG. D

FIG. E

The ingenious world of philosophical mathematics must appreciate that the super-creative ingenuity of the author donated this concept for the enrichment of the universal cosmogony.

PART 2

PHILOSOPHICAL MATHEMATICS

"The errors of Mathematics as a general calculus, which were orchestrated by a lack of in-depth scientific and philosophical knowledge—including the high level of the abuse of the wisdom of the Oriental wizards—have ingloriously affected human dynamism and universal progress."

—A. U. O. Aliche

CHAPTER 1

What Is Philosophical Mathematics?

The concept of philosophical mathematics, which is scientifically represented by P/M × L/X, has remained one of the greatest problems of assessing the inventions that are contained in science, technology, and philosophy, including how and why philosophy is determined and defined as the mother of sciences.

The errors of mathematics as a general calculus, which were orchestrated by an in-depth lack of scientific and philosophical knowledge (including the high level of the abuse of the wisdom of the Oriental wizards), have ingloriously affected human dynamism and universal progress. The concept and the way humans have assessed P/M × L/X have greatly shown that the universe requires scientifically and technologically balanced knowledge.

Mother Nature, in its obvious oscillation, is practical in presenting facts and figures through an organized simultaneous agenda. Philosophical mathematics has remained the anatomy of balanced creativity. This is why meta-logic is of the opinion that the inventions of such illumined mathematicians like Pythagoras, Euclid, Simultaneous, Remainder, H. A. Denat, and others must be recognized and referred to compact a glorious mathematical anatomy which, in the objectivity of alchemy, will remove all the un-abstract problems of philosophical mathematics.

In this respect, philosophical mathematics, or mathematical philosophy, can be defined as:

- The dynamics of engineering mechanics.

- The objective concept of scientific postulations, which are subject to investigation using the propelled and empirical status of philosophy with a view to understanding what can best be defined as mathematical data.

If analysed as a scientific panel, philosophical mathematics stands as the definition of either a Planchette or a Ouija board. In any format and at any time, if fully represented as a conventional science, it can function as an abstract engineer; it can equally warehouse all the concepts of scientific dynamism with a view to investigating the relativity of science and technology.

It must be stated that no existing dictionary can give an acceptable definition of philosophical mathematics. The eminent and unequalled contributions of Pythagoras in his theorems, theories, principles, and policies of calculations, and what Walter Russell defined as the science of light, can assess and provide a philosophical data by which the opinion of philosophical mathematics can be conceptualised.

In this respect, philosophical mathematics is best defined in the following contents and concepts.

- The study of scientific objectivity in relation to using data, statistics, formats, formulas, and arithmetic and philosophical algebra can be used to eliminate many mathematical doubts with their un-dynamic concepts.

- It is defined as an organized and systematic presentation of logical facts, figures, data, formats, and concepts in order to blend with all the scientific and mathematical concepts, which must represent the purpose-driven objectivity of core philosophy.

- It is the scientific bid to harmonize and balance the disunited concepts of mathematics and philosophy into a systematic and acceptable databank that will logically assess all that is contained in the natural anatomy of technology.

- It can equally be defined as the objective input and output of the electronic postulations of science and philosophy, which, when properly compacted and galvanized, will lead science to the invention of permanent principles and policies that must be focus-driven with the best philosophy and philosophia of the universal One. This is why the branches of philosophical mathematics are enormous and encompassing.

- It can equally be seen as an electronic router that can best be used as an information databank.

This is why its branches, functions, and operational mechanisms can be utilised to purposefully and practically postulate, in the best scientific objectivity, how this fragmented concepts can be galvanized into a united scientific format.

In this respect, the definition has engineered and inspired me to develop a philosophical mathematical theorem, which is carefully and logically represented with objective explanation in the following facts and formats.

If philosophy, which is philosophy over mathematics (i.e., P/M) is represented by AUOB, then it is practical and objective to show that

$$\frac{A}{U} = \frac{U}{O} = \frac{O}{B}$$

which means

$$\frac{MATHTHEMATICS}{PHILOSOPHY}$$

can be represented by either a square or a quadratic object. This shows that

$$\frac{A}{U} = \frac{U}{O} = \frac{O}{B}$$

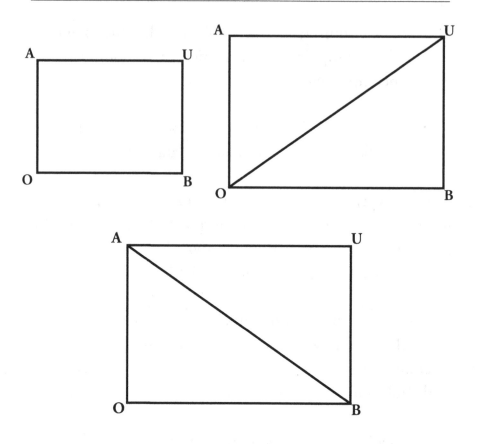

This theorem can equally be proved that

$$\frac{A}{U} = \frac{U}{O} = \frac{O}{A}$$

which means that the concept of philosophical mathematics can be configured as AUOA representing a square and a quadrant. When further divided, the square containing AUOB can give two adjacent triangles, which can be shown as AUA and AOB.

Conversely:

$$\frac{A\ U\ O}{U\ O\ B}$$

is fully seen and known as AUOB.

This philosophical mathematics theorem states that with the use and application of scientific and objective logic, angles can be proved with the use and application of ingenious philosophical methods. This approach can eliminate all the doubts that have been created in the past eras of illogical philosophical mathematics.

With the use of this theorem, the philosophical math reveals that P/M will certainly give AUOB, and A/U = U/O = O/B. This is in both direct application and converse application.

Philosophical mathematics has proved its magnetic and electrifying ability in ensuring that there is relativity that holds philosophy as an in-depth science, which stands in the way of explaining the philosophical concepts of science and technology, including how and for what purpose man can utilise these concepts. It is unfortunate that our civilization has greatly sidelined the objective concept of philosophical mathematics.

The objective explanation of the following ratios, which can be proved under the guidance of logical simultaneous concept, reveals that

$$\frac{1}{2} + \frac{1}{3} + \frac{1}{4}$$

can equally be reversed and reunited in the best objectivity of mathematical logic, which shows that the application, multiplication, and fractional calculation of $1 + 1 + 1 = 3$ must certainly understand the purpose-driven dynamism of mathematical philosophy to arrive at and achieve the same as in the first formats.

The way and manner an elaborate definition has been given with scientific acceleration reveals that mathematical philosophy is a databank—it is the anatomy of science, the structure of technology, the wisdom of honest knowledge and the electrifying magnetism of all creative forces and powers. It is here that an illumined aeronautic philosophical mathematician, John J. Wilson, defined and opinionated that philosophical mathematics is the only science that must be seen and appreciated by all and sundry as the structure of all.

The Philosophic Dynamics of These Theorems

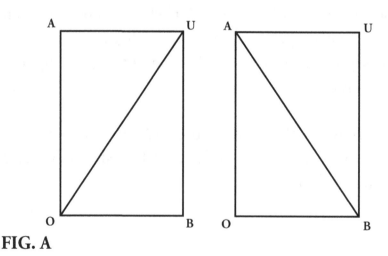

FIG. A

A = U = B = O

AU = OB = AO = BU

∴∆AOU = ∆ BOU

$\underline{AU}$ = $\underline{BO}$

BO   AU

That is, both triangles **AOU** and **UBO** equal triangles **OAU** and **AUB,** which proves that **AUOB** is a square.

It can now be shown that mathematical theorems can only be proved when the employment of objective logic is practically involved. This is so because every theoretical fact is only acceptable through the use and application of scientific and convincing logic. Hence, philosophical mathematics is simple to employ if only man will understand the interwoven relativity of great arts, which are simple and perfectly enduring.

CHAPTER 2

WHAT IS THE ORIGIN OF PHILOSOPHICAL MATHEMATICS?

As balanced and dependable as science is in collectively accepting that philosophy is the mother of knowledge, objective and further mathematics equally agree that philosophy and mathematics are practical and interwoven; they hold a simultaneous agenda in practise, principle, and policy. The way and manner the meta-logical concepts can be utilised with object harnessing is eloquent in appreciating that philosophical mathematics is nurtured and natured. This is why Mother Nature is scientifically and practically accountable to be defined as the alchemy of its transformation with scientific transmutation.

In the blessed words of Chancellor Ozone, the world is mathematically empowered in thinking, while the universe is philosophically principled in practise. Those edified and unequalled words were responsible for further investigation on scientific and philosophical mathematics as a figurative architect and philosopher, and as the mother of knowledge whose in-depth actualization is to achieve purity and perfection through the use and employment of practical and purpose-driven methodology. The questions here are:

- If Mother Nature originated philosophy at the perfect instruction of inspiration, and the Orientals dedicated its establishment as a mandate (which is a must to actualize with scientific vision), how and why has man rejected and neglected all the powerful and purpose-driven ideologies of philosophical mathematics?

- If there is an acceptable and consummate order that fashioned the existence of other forms and creations, why has man abused with contempt the concept of this fulcrum order, which is purpose-driven in arts, policies, principles, and practises?

- Why is creation not in the dynamism of achieving the purpose of philosophy of perfection, which is ordered with the infinitude orderliness without space and time?

- How can the negligence in the amalgamation and understanding of philosophy, as the mother of science and mathematics, be an organized figurative calculator that is both logical in ideology and meta-logic in a culture with scientific tradition?

Within this context, it must be borne in mind that philosophy as a science, with mathematics as a purpose-driven and calculative science, holds the key in appreciating the origination of all things through the use and application of the scientific order, which exists as one, divergent, configured, relative, and transmitted.

A case to establish the origin of philosophical mathematics must represent philosophy and mathematics in the following formulas,

$$MN + P + M = MN \times P \times M = \frac{MN}{P} + \frac{P}{MN} = \frac{M}{MN} = \frac{MN}{N}$$

In this expression, Mother Nature, which started the symbolism and the expression of the origin of philosophical mathematics, is configured as a letter that can be expressed and explained. A look at the anatomy of Mother Nature will always appreciate that it extended as an organized alchemy by which the postulations of philosophy and mathematics are data and compacted as a united kit, which in earnest can further be defined as the kingdom of honest knowledge. Practically stated, it must be appreciated that Pythagoras, who is in my inspired knowledge a mathematical institution, was initiated to achieve the height of philosophical excellence because he gratified and graced the glorious wonders of philosophy as a foremost apostolate of honest knowledge. This later inspired him to be initiated with all that was contained in the letters and provinces of natural mathematics. This is why such eminent scholars like H. A. Clement, L. A. Durell, Euclid, Simultaneous, and Albert, including the contemporary mathematician S. M. Smith, were not on the same level with Pythagoras in blending an acceptable format.

This fact has energized and eulogized him as the god of living mathematics.

Many schools and institutions have greatly argued that Pythagoras was mystically initiated in the best understanding of Neo-mathematical meta-logic; other schools of thought argue that he originated the philosophia of philosophy and mathematics, all of which were sacrament in the proximity of Socrates, who was not mathematically driven but was philosophically configured with all scientific objectivity and an illumined and ornamented, knowledgeable initiate.

The necessity of explaining the origin of philosophical mathematics, which is ingloriously taught in higher institutions as the philosophy of mathematics, is necessitated by the fact that humanity has been greatly incubated with a lot of falsehood and vague knowledge, which in my opinion as an initiate in this dynamic province can be defined as the greatest errors of man.

In this respect, it must be stated that philosophical mathematics is the cord of all things, which include the agenda of Mother Nature, the orderliness of all things, and the compact relativity of science and technology. This is why a work of this nature can be defined as a scientific, objective investigation of what nature as a philosophical mathematician holds for mankind. This aspect includes the urgent necessity of expanding and exploiting what can be acceptable as the powers and forces of Mother Nature, which finally originated with scientific yielding.

The scientific and calculation need of all things simply are represented in this global and objective symbolism as P = M. Also, M = P + M = MN. If conversely revised, it can be explained as MN = P + M.

A work to explain the fulcrum anatomy of what philosophical mathematics represents and originates must certainly appreciate the graceful and balanced oscillation of Mother Nature with the best scientific postulation of an amalgamated concept, which further extolled to what is known today as philosophy, mathematics, engineering, science, and more. The orderly configuration of all things into one remains one of the mysteries of nature, which can only be explained by logic, meta-science, and philosophical mathematics. This is why every origin comes from the main anatomy of the foremost origin, best defined as Mother Nature.

CHAPTER 3

WHAT ARE THE BRANCHES OF PHILOSOPHICAL MATHEMATICS?

Mathematics is perfected in the following objective and scientific branches.

- Arithmetic
- Algebra
- Geometry
- Further mathematics
- Engineering and technological mathematics

There is a need to practically explain that philosophy, on its own, is perfectly diversified with the following extensive extensions.

- Religion
- Culture and tradition
- Metaphysics
- Epistemology
- Ethics and morals

It also includes my practical postulations:

- Internet philosophy
- Diplomatic philosophy
- Oriental philosophy
- Magnetic philosophy

All of these are galvanized and contained in the united, philosophic apostolate foundation of philosophy.

In this regard, mathematical philosophy can perfectly be categorized in the following branches.

- Philosophical mathematics of science
- Philosophical mathematics of technology
- Philosophical mathematics of arts
- Philosophical mathematics of economics (which, when zeroed down, will appreciate the anatomy of micro—and macro-economic theories and principles, including the purpose-driven philosophy of econometrics)

Other branches in this area include

- Philosophical mathematics of culture
- Philosophical mathematics of quantum science
- Mathematical thinking, with its dynamic engineering
- Philosophical mathematics of chemistry, which is dynamically driven in the best explanation of inorganic and organic chemistry

These branches are why mathematical philosophy appreciates perfectly that there is one order, and the existence of that order is dynamic, pragmatically objective, creative, and re-creative with a purpose-driven concept of explaining the principles of transmutation to transmission. The principles are responsible for the true knowledge of alchemy, including the contributions of St. Germain as a renowned philosophical alchemist.

In this purpose-driven and illumined scientific organograms are contained the philosophical mathematics of physics, which explains what physics is all about, as well as the principles and policies of the nature of matter, compounds, and elements, including what philosophical mathematics designates as physics production.

In the areas of chemistry and chemical philosophy, mathematics is detailed to quantify and explain what is best known as oxygen and oxygenated compounds, nitrogen with its elements, and carbon (including all that is contained in carbonated oxides). Such misguided and misdirected elements like HNO^3, HCL, and SO^4, including the versatile areas of chorine and chromium, were first initiated and analysed to be in natural origin by the power and fortress of philosophical mathematics.

A look at this presentation will mandate obvious recognition that philosophy, which has been designated as the mother of science by the Orients, must be recognized as the Paracelsus of original mathematics, and the Emerson of technology. Thus it represents this anatomy as the AUOB of all knowledge.

Paracelsus and Pythagoras could not hide their feelings when they said that philosophy is practically driven in the best knowledge of mathematics, whereas mathematics is practically driven in dealing with issues that are contained in info-tech, geosciences, geology, aquatic science, astrophysics and astral physics, space science, and more.

Mathematical philosophy recognizes the enabling fortitude of the cosmic clock and what empowered Walter Russell to say that genius is inherent in every man, while mediocrity is self-inflicted; the choice is ours.

As a mathematical and philosophical logician, I am practically convinced that the use and application of mathematical philosophy—which serves humanity as an igniter—will help us to appreciate the electrifying dynamism of chemistry, which is explained and treated in biological science with a little contribution of biorhythm and psychological analysis.

It must be stated that the separated but united nature of philosophical mathematics must be seen dynamically by the world of science as the original anatomy of all things, which is scientifically canopied in the existing order of nature. This is why, when it is stated that P/M = philosophical mathematics, the scientific philo-sophia, which is an illumined formula, can be understood by those whose thinking and transactions are practically purpose-driven by the power of philosophical mathematics.

Unfortunately, the globalized universe is scientifically engulfed with the wisdom of mathematical science, which greatly locked the creator out of its creations with the resultant effects of its principles and policies being practically exchanged—simply because mathematical science neglected the fact that one order, which is practical and divine, is the originator of philosophical mathematics and the scientific and enabled structure of all.

As explained earlier, if A/U = U/O =O/B, then it can be appreciated, both significantly as a rule with modest principle and as a policy with diligent and monumental anatomy, that AU + UO + OB will give philosophical mathematics an acceptable format. That format can finally be drawn or investigated in the triangular philosophical matador of locus-lacuna, which can practically be symbolic in the principle theorems. AUOB can be analysed and actualized to resemble A = U = O = B.

It must be kept in mind that the concept of philosophical mathematics and its understanding requires the following.

- The ability to be illumined with inspiration.
- The willingness to empower one reassuringly with objective technicality.
- The use and application of the law of balance, which must be purpose-driven with perfect wisdom as a must-have.
- The concept of living in the now that will make one to think about the mathematical and philosophical

contributions of the Orients, which can be stated to be a must-have, a must-hold and a must-be gotten.

An approach to determine, explain, and investigate what is contained in the anatomy of philosophical mathematics must be asserted as the greatest business of life.

In this respect, the scientific concept of philosophical mathematics is practically and purposefully interested in explaining the concept of laws, orderliness, principles, and policies, including how we can use the system of research method in arriving and concluding that there is scientifically one existing absolute order that, in the opinion of meta-science and meta-logic, is synchronous with the fundamentals of core logic and psychology both as a science, as an art, and as the mother of knowledge; this is why it extends to mathematics. Mathematics, after examining and cross-examining the powers of philosophy, developed other extensive extensions that include formats, formulas, data, statistics, and calculus—including using the log tables with their simultaneous wisdom in explaining the purpose-driven concept of equations.

This chapter is detailed in presenting the branches, their relativity, their orderliness, and their synchronous agreement as a purpose-driven structure of all that is guided and directed by the illumined and consummate order of philosophy, which is known all over the world as the mother of balanced knowledge.

CHAPTER 4

WHAT ARE THE SCIENTIFIC AND INTERNET DYNAMICS OF PHILOSOPHICAL MATHEMATICS?

A look at the balanced expression and explanation of the last chapter must certainly appreciate that philosophical mathematics is not only philosophic to science and technology, but it is also generously agreeable with the Internet dynamism. It is here that the neglected understanding of online real-time, including the Microsoft abuses of website, has strangely demystified the timing of man as an un-philosophical mathematician. Furthermore, it neglects and contravenes all the scientific laws of philosophical mathematics when man does not understand that motion and speed are the same.

A work to explain the scientific dynamics of philosophical mathematics must agree that the universe is configured and compacted in the agenda of philosophical mathematics

Physics is greatly concerned with motion and emotion, light and waves. Philosophical mathematics, in its dynamism, is purpose-driven in explaining that what man names the wave is defined in this content as an electrifying pulsation of additional speed, which in earnest can be understood in the objective philosophy of natural acceleration. This is why the dynamics must be appreciated as the only way forward.

A look at the philosophy of the method and the scientific mode by which people access their mail on the Internet must ascertain all that is contained in electronic VSAT in respect to motion, emotion, and acceleration. When it is stated that "in light dwell all forms", the dynamism is practically and pragmatically clear in ensuring that humans should be able to understand that the evolving man has not known the sciences of motion, emotion, and acceleration, which finally result in speed.

Let's go back to the question, What are the dynamics? The following points will certainly explain that philosophical mathematics has important and impeccable dynamics.

- Able to enable.
- Willing to explain, manufacture, moderate, detect, decongest, and configure all principles and policies into an acceptable order.
- Practically able to explain that in the waves lay the secrets of creation.
- Ability to compact man's thinking and creation as an immortal Rolls-Royce.
- Ability to serve creation as a scientific illuminator with a technological dictator.
- Always willing to explain the scientific concept of emotion, motion, acceleration, and speed. This is why a look at the overleaf construction of automobile

piston and sleeve reveals philosophical mathematics in existence.

- Ability to teach with scientific moderation and practical engineering. This is why the use and application of letters, figures, formats, data, and statistics were at the forefront of philosophical mathematics.

I should explain that the following have philosophical anatomies with mathematical structures.

- Motion
- Emotion
- Acceleration
- Speed

A transcendental investigation of the rays of light, which physics treats in the mathematical dynamism perfectly, reveals the existence of indomitable order, which is scientifically panelled at the proximity of balanced structure. For example, if E + M represents emotion and motion, it is practical to agree that A + S will equally represent Acceleration and Speed, which in earnest has a common purpose with practical dynamism. This is why physics, in this geometric philosophy, is understood to represent the purpose-driven fulcrum of natural energy, including how humans can utilise the transcendent rays in appreciating its philosophy and the dynamism of philosophical mathematics.

Unfortunately, our system is chaotic, un-mathematical, and un-trigonometric. A work to explain what is in the databank of the dynamism of philosophical mathematics must not lock itself out from the perfect and purpose-driven consummate order of the universal One, which is responsible for the manner in which one uses immortalized principles.

Philosophical mathematics is scientifically glorious in achieving a globalized success, which is important for records of past eras, the present museum of knowledge, and the future of human civilization. This is why Manly P. Hall explained that man has not known himself as a philosophical oasis. The illumined words of Palmer Hall can be quantified in its original quantum, as explained in this book, that man is structured and perfected as a data of emotion and motion, and as a bank of acceleration and speed. This is why the dynamics of philosophical mathematics, as an Internet Bill Gates, will certainly help creations to foster a new agenda of human relations. When it is properly blended, it will help all and sundry in knowing that the word "mathematics" is derived from man, whereas philosophy is practically and pragmatically derived in explaining the philosophy of man as a mathematical guru.

It is on this purpose-driven philosophy that the dynamism of philosophical mathematics holds the key in unlocking the wisdom of philosophical understanding, the purpose of motion and emotion. This can only be revealed to the philosophical apostolate who deserves to appreciate deeply that creation, and particularly man, is a wizard of WWW which can be explained by **P/M = WWW,** which means wisdom, wizards, and waves.

Einstein States:

"It has often been said, and certainly not without justification, that the man of science is a poor philosopher. Why then, should it not be the right thing for the physicist to let the philosopher do the philosophizing? Such might indeed be the right thing at a time when the physicist believes he has at his disposal a rigid system of fundamental concepts and fundamental laws which are so well established that waves of doubt cannot reach them; but, it cannot be right at a time when the very foundations of physics itself have become problematic as they are now."

—Albert Einstein, on the relationship between physics and philosophy

"I fully agree with you about the significance and educational value of methodology as well as history and philosophy of science. So many people today—and even professional scientists—seem to me like somebody who has seen thousands of trees but has never seen a forest. A knowledge of the historic and philosophical background gives that kind of independence from prejudices of his generation from which most scientists are suffering. This independence created by philosophical insight is—in my opinion—the mark of distinction between a mere artisan or specialist and a real seeker after truth."

—Albert Einstein to Robert A. Thornton, 7 December 1944

CHAPTER 5

How Can Man Be Explained as an Initiate of Philosophical Mathematics Using the Formula of P/M = WWW?

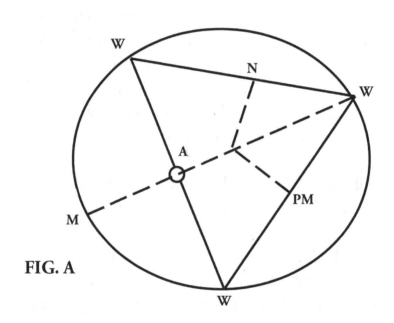

FIG. A

$$MAN = \frac{M}{W} = \frac{A}{W} = \frac{N}{W}$$

i.e., $M = W = A = W = N = W$

$\therefore \dfrac{M}{W} + \dfrac{A}{W} + \dfrac{N}{W} = \dfrac{MAN}{WWW}$

In this respect, the world should be made to understand that WWW is the structure of balanced wisdom, which can be defined as the living anatomy of all. The explained facts will make us to appreciate that P/M = WWW, or simply put, the structure of all known is MAN.

It must be stated that all actions of man, naturally and dynamically, are supposed to be connected to the Internet philosophical mathematics, whose illumined goal is fulcrum in purity, perfection, and globalization. This is what material science galvanized and configured into the world wide web (WWW).

A look at the material postulation of the concept of "world wide web" will certainly explain that the era of material science started the civilization with a high level of mathematical enmity. If material science was truly in the knowledge of "in the beginning was the Word and the Word was with God and nothing was made that was not in that Word", he will certainly come to the true knowledge of appreciating as scientifically accepted by philosophical mathematics that in the waves lie the secrets of creation.

The un-dynamic and unscientific manner by which man has utilised ephemeral material science in establishing a chaotic civilization cannot actually configure the universe because of the understanding of the "*WWW*", which in earnest is not practically globalized and cannot be used to ascertain the perfect and consummate orderliness of all things. This is why this chapter is a necessity in stating a new concept with a monumental agenda that will help to explain and buttress the fact that all the things we observe around us return to their shape, form, and function long enough for us to perceive by our five senses. In addition, we must never forget the mind itself, as clearly and relatively ordered by the consummate order of all orders.

In this respect, perception can also be by any method available to us through the use of physical science—a microscope, a telescope, or a computer. We can use these methods to discover data normally beyond the range of our senses. The necessity to light up this statement from one of the presentations of the Society of Metaphysical Sciences dealing with neo-metaphysics education is important and relevant; it will help us to appreciate most of the lapses of material science. That is why in dealing with this subject, the wonderful and consummate contributions of Pythagoras and Plato, particularly in the areas of quantum mathematics, must be explained and accepted as a work of immortality. In this respect, man must be explained as a balanced initiate of philosophical mathematics because of the following reasons.

- He has the structure of philosophy, including being panelled in the originality of the anatomy of mathematics.

- He naturally lives on the philosophy and concept of the food chain and the food web.

- By nature he is charged and recharged in transforming the following gaseous elements, which include nitrogen, hydrogen, and carbon dioxide.

- He is purpose-driven and is relating and reacting to physical and technical dynamics, which are in accordance with the universal rhythm of "WWW" which is translated as PM = WWW (wisdom, wizard, and waves.)

In this respect, man is an initiate of philosophical mathematics because he is the product of an ordered wave current embedded with wisdom. He is ornamented philosophically as a wizard that he ingloriously or knowingly designated as intellectual, super-intellectual, or consummate intellectual. But the highly illumined appreciate that a wizard is a genius who is purpose-driven with the current of the waves and the order and philosophy of wisdom. This is why man is scientifically designated as a co-creator with the ordered and consummate creativity of the universal one.

It must be stated that a work to explain the synergic wizardry of man, the electrifying concept, and the knowledge of the waves—including how he can access the Internet through the prime factors of philosophy and mathematics—must certainly desire to appreciate that the universe is the province and fulcrum of universal wizards named by the well informed as the perfect consolidation of the arts of the universal One. This is why one needs to explain that philosophy is being named by the Oriental apostolate as the mother of sciences, the order of knowledge, the mission of truth, and the system and concept by which mathematics actualizes its vision.

If X^{10} is simplified as $X \times X \times X \ldots$ in the same order, it is also explaining and buttressing the ingenuity of philosophy as a natural fulcrum in the best and scientific objectivity of mathematics. From the postulation of Richard Allenton, Philosophy which is the mother of knowledge can equally be understood as the original foundry of mathematics.

This idea motivated Sir Louis Lord to assert that—beyond philosophy is nothing. No human being can understand the workings of electrifying anatomy without quantifying his ideas and concepts in the originality and thinking of balanced Philosophy.

The omnibus statement of Sir Louis directed and mandated his era in looking at mathematical figures as representing philosophical data. On the other hand, if X^{10} cannot be mathematically explained, it is a simple statement that it will not be accepted in the domain of philosophy.

In this respect, we need to derive a logical reference from the circle, which is inserted within a triangle, which technically and scientifically explained the philosophical mathematics concept from the empathic originality of WWW. The world is the anatomy of WWW in its natural concept, so we must appreciate that philosophical mathematics can be translated generally to represent the technical and structured anatomy of

WWW, which in earnest can best be defined as an organized philosophical order.

From the inclusive orderliness of the extract with the introduction, it must be explained that man, at any time in his evolution and civilization, is created as a philo-sophia of wisdom, initiated as a mathematical wizard, and illumined to understand the creativity of the waves that in earnest anatomized motion, emotion, speed, calculation, acceleration, and oscillation. The philosophy of this book is borne to present all that is contained in the objective dynamism of the secrets of light, including narrating mathematically that correcting the errors of science is the greatest duty our civilization owes to posterity. This is why a book of this nature must protect and represent their divergent thinking of man not only as a philosophical Sagittarius but also as a mathematical wizard who should choose all the enabling ingenuity in assessing, explaining, importing, and impacting the philosophy of the waves, the dignity of PM = WWW or in converse WWW = PM.

In this respect, the world is configured and globalized under one common law and system, which in this book is presented and represented as an amalgamated philosophical mathematical order, which man considers as being and becoming the ordered initiator who is purpose-driven in the best ideology of electrifying balance, which in earnest is the scientific and pragmatic definition of philosophical mathematics that warehouses the institutionalized concept of online real-time with divergent and permanent natural VSAT, which is above the limit of the Microsoft concept.

When it is stated that in the waves lie the secrets of creation, it is practically explaining that man is a mathematical wizard and a philosophical guru who is yet to understand the purpose-driven concept of the anatomy of WWW synergy.

CHAPTER 6

HOW CAN CREATION ACCESS THE BALANCED WISDOM OF PM/WWW CONCEPT?

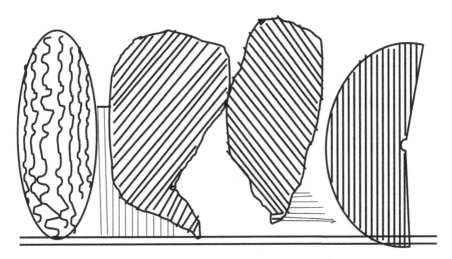

Fig. A

This diagram is a scientific and philosophical mathematical expression to show the wisdom that is contained in the unknown VSAT of P / M = WWW, which will be properly explained in the main chapter.

Scientifically, creation has all the powers and the ingenuity to access the concept of PM = WWW, but unfortunately it is lacking in the following scientific and pragmatic objectivities.

- Undesirable thinking

- Lack of character

- Poverty in the use and application of illumined knowledge

- Lack of understanding of the real technicalities of World Wide Web

- Poverty in the use and application of logic, metascience, metaphysics, and ontology; this is to show that creation is practically material-driven

- Ability to reason in the line of original philosophy, which is the maker of great thinkers

- Lack of investment in the areas of research methods, including the use and application of scientific principles, policies, and practises (known as the P^3 concept)

- Lack of the use and application of logical psychology and the wisdom that is contained in philosophical sociology.

- Inability to practise core mathematics as a fundamental science, and core philosophy as the mother of knowledge.

Creation has not been able to exhume the data contained in statistics, mechanics, quantum theories, and other principles with their scientific methods. This is why the following points suggest practically with engineering and scientific dynamism that philosophical mathematics holds the key in galvanizing and uniting all the monumental diversifications of human transactions in the originality of its ecstasy. The drawings are in synchronous agreement with the illumined concept of how creation can appreciate the wisdom that is contained in **PM** using the **WWW** concept as a database. It must be stated that one of the errors that must be excluded and sidelined is the import of material science, which has a lot of false technicalities with principles and policies with a view to synchronizing or harmonizing with the original anatomy of philosophical mathematics. Philosophical mathematics holds the key in assessing the principles, policies, and practises of the WWW concept, which is a challenging study that is awaiting humans in our present evolution.

The Microsoft concept is internet-driven in accessing the practicality of the electronic network, so this philosophy cannot be amortized either as a permanent science or as a dependable engineering mechanism. This fact is stated because humans started the communication system with the use of telegrams, telegraphs, table phone, manual mailing system, and more, but today material dynamism has greatly researched into the concept of speed and time frame, which finally gave birth to the computer age and the e-mail system. But there is a need to respect and practically regard the old system still in existence.

The high level of ignorance of man and how he is lost with the application of e-mail with its communication lapses can be analysed and appreciated using the dependable concept of philosophical mathematics, which is purpose-driven in the best objectivity of achieving a global perfection with a scientific purity.

For humans to access what is already a free gift from Mother Nature, the following must be achieved.

- The engineering system that humans are already practicing must be charged with a view to understanding practically that man is the engineer of his own soul.

- The existing and practicing scientific methods, which are full of errors and lapses, must be recognized with a view to reassessing the perfect scientific postulations of Mother Nature, which were instrumental to immortalizing humans like Thomas Edison, Albert Einstein, Michael Faraday, Professor Osborne, Sir Isaac Newton, Isaac Pitman, J. J. Williamson, Pythagoras, Socrates, Galileo, H. A. Clement, Herbert Spencer, Mark Twain, Walter Russell, Manly Palmer Hall, and the unknown teacher of India who is designated at all levels of human endeavour as being a Buddha and becoming a Buddhist.

Such other individuals, like the automobile wizards, must not be excluded in this list, which includes Chancellor Mitsubishi, Lord Peugeot, Sir Mercedes Benz, Lord Liverpool,

His Excellency Sir Lord Arthur Guinness, and more. It must be stated that our contemporary civilization has produced the likes of Professor Chike Obi, Dr. Walter Russell, Mozart, Paracelsus, Andrew (the father of modern anatomy), and others. This is why their contributions are purpose-driven and practically dynamic in assessment with natural wisdom and its Internet connectivity, which was at the proximity of philosophical mathematics. This is why a work of this nature must be purpose-driven in the best assessment of technical and ethical concepts.

Apart from this, the present practises of psychology, sociology, government and administration, geology and geophysics, geosciences, and petroleum engineering (and its segmentations like chemical engineering) must be geologized in order to meet the futuristic challenges of the Internet family. That family is all about philosophical globalization, including compacting and configuring the universe as a mathematical province.

When Pythagoras and Plato organized a statistical census of the humans that were already oriented in the use and application of ordinary mathematics, they greatly regretted that their era was enmitied with the use and application of illogical mathematical thinking. This is why a work on philosophical mathematics can best be defined and designated as the supreme anatomy of illumined wisdom that will help the human race—no matter the tumultuous challenges—in creating world balance and world understanding. It is in the ethical and empirical philosophy of this work to exhibit with perfect explanation that human endeavour is purpose-driven in the best use and application of philosophical mathematics, which humanity has not started to assess in the originality of its perfection.

A general look at the development of science and technology in different parts of the world will certainly agree that the use and application of philosophy is practically at a kindergarten level. This situation has created a lot of terrorism, insecurity, and impoverishment, including some advanced countries of the world using their concept of material science to impoverish the third-world countries in the areas of commerce, industry, administration, and international relations. It has equally revealed that the concept of a united order, which is the purpose of balanced creation, is not only neglected, but it is mathematically abused and philosophically relegated to the background.

The concept of world balance, which will achieve the material desire of globalization, is simply defined as a philosophical one-world purpose with a mathematical universal reality. A look at the engineering growth at the north and south poles will certainly explain that the power of the northern star is not relating in the best objectivity of philosophical mathematics. The south, which configures the west and the east, is greatly advancing in science and technology. But regrettably, the invention of nuclear missiles and agreement with the philosophical enabling of the desire of one-world purpose. Respectfully stated, the desire of philosophical mathematics is to achieve a purpose-driven globalization and to enhance the use and application of natural mathematics.

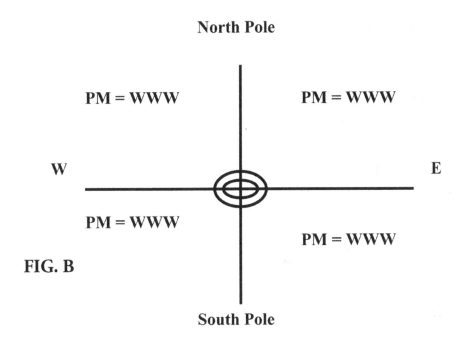

North Pole

PM = WWW   PM = WWW

W   E

PM = WWW

PM = WWW

FIG. B

South Pole

All that is contained in the secrets of light, the secrets of science, and the sacredness of technology are only accessible through the use and application of philosophical mathematics which can be defined as the scientific coefficient of all data, both known and unknown. This is why the human race has a monumental task, the purpose of which is to access the utility and universality of philosophical mathematics both as a school of life, a school of thought, and a canopy to life, including being used as the secret wisdom of all ages responsible for the invention of science and philosophy. That wisdom will invent what posterity will use for the harmony of one-world purpose.

The philosophical mathematical explanation of this can be shown thus.

$$\underline{P} = WWW =$$
$$M$$

$$WWW = \underline{N} + \underline{S} + \underline{E} + \underline{W} = \underline{P}$$
$$ W \quad W \quad W \quad W \quad M$$

i.e., North Pole $= \dfrac{\underline{P}}{M} = WWW$

South Pole $= \dfrac{\underline{P}}{M} = WWW$

East Pole $= \dfrac{\underline{P}}{M} = WWW$

West Pole $= \dfrac{\underline{P}}{M} = WWW$

$$\dfrac{\underline{N}}{S} = \dfrac{\underline{E}}{W} = \dfrac{\underline{P}}{M} = WWW$$

Generally, this method explains that philosophical mathematics can be used to access, compact, and configure all things as an absolute and acceptable order.

CHAPTER 7

HOW CAN CREATION ORIGINALLY NETWORK ALL HUMAN ENDEAVOURS IN THE BEST USE AND APPLICATION OF PHILOSOPHICAL MATHEMATICS?

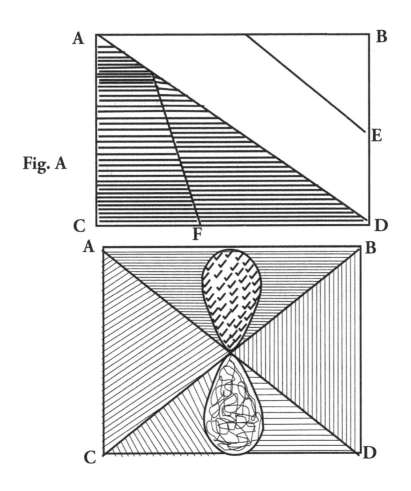

Fig. A

Creation, which is philosophically known as the supreme anatomy of mathematical dynamism, is logically unencumbered in reasoning, knowledge, and understanding, and in assembling and assessing facts. This is why the urgent necessity of attuning through the use and application of philosophical mathematics can further be designated as being and becoming in the living of the computer age, which in earnest is impacted and compacted with the objectivity of electronic and electrical appliances.

A simultaneous understanding of what mathematics holds in relation to appreciating the validity of philosophy will certainly lead one to agree that the universe is a compass and compacts Internet knowledge. This is why logic and philosophical methodology is opinionated that man has not created anything but is known as a co-creator in the original perfection of the philosophical creativity. In the philosophical mode of Pythagoras, who developed a theorem that has remained mathematically immortal, the assessment of an angle in a triangle can be defined as a logical networking of a mathematical concept. This statement by one highly endowed with mathematical philosophy generated a lot of mathematical arguments during his era. To the well informed, Pythagoras, who is indeed a guru of philosophical mathematics, was perfectly intact in his presentations. This is why the world, and particularly the province of science and technology, honoured him with a consummate chair as a mathematical networker.

It is unfortunate to explain that our present era, with its ephemeral scientific dynamism (which is naturally epileptic and myopic), is making boasts of Internet invention and is anatomized in the chemistry and alchemy of Mother Nature, who is hereby defined as the anatomy of philosophical mathematics. When Stroud and Smith were inspired to impact what is known today as ontology of engineering mathematics, the uninformed among them looked at these honourable gentlemen as humans who possessed the knowledge that was too difficult to understand. But unfortunately, it dawned on creation, particularly the province of engineering knowledge, that Stroud and Smith were compactly eloquent in presenting the concept of further mathematics from engineering mechanics, which is quantum-oriented in the best objectivity of Internet knowledge, which is becoming a home drive. This situation is why globalization cannot and will not be achieved through the use and application of material science.

If creation will accept that philosophy is the mother of knowledge, and that mathematics can be defined as the fluorescent of philosophy with logic as its enabling anatomy, then it is important to know and to greatly believe that creation can only assess all that is unknown using the powers of the secrets of light and the wisdom that is contained in the omnipotent vanguard of philosophical mathematics.

I am practically and pragmatically assessing consummate knowledge, opening up the universal Internet website—which in reality must be accepted by the province of posterity as existing in the formula of PM = WWW, or PM = P/W = M/W = P/W = P/M, which can equally be W+W+W = 3W or $W \times W \times W = W^3$.

In this simultaneous and objective sign, I assess philosophical mathematics using the original Internet website. My inspired drawings, which can be defined as philosophical polyesters with a mathematical wizardry, will certainly help creation in appreciating that the world is naturally globalized with Internet thinking, which is visibly seen in the powers of that which originates from human words and voices. It is here that Internet technology must be philosophically driven and at the same time understood.

There is need to explain that Bill Gates, who is known all over the world as a dropout at Harvard University, has been able to give the world computer and Internet cosmogony, simply because his intelligence was deep both in the knowledge of nature and in the power of mathematics. These lapses can be known and defined as the greatest undoing of the likes of Charles Darwin in his theory of evolution, and Adam Smith in his theory of supply and demand.

It must be understood that the likes of Darwin, Smith, and others were purpose-driven in the best understanding of philosophical mathematics. This is why, for example, automobile wizardry serves the whole universe as a medium of private and commercialized transportation.

In this respect, creation can network all that is presently known to human endeavour with the use and application of the following knowledge.

- One understands with balanced appreciation that philosophy is the mother of consummate knowledge.

- One appreciates greatly that mathematics cannot be understood without the use and application of balanced logic.

- Creation must appreciate this totality as the hallmark of scientific ontology.

- Science at all levels must be approached without the lapses of the mental incubus, in order to galvanize the universe as a common village.

- There is need to critically examine the wisdom of past eras and the magnanimous contributions of the Oriental wizards in order to compare and contrast with a view to drawing a standard benchmark.

- Most investments in material science should be drastically reduced and refocused to natural science.

- Globalization at all levels is not possible in a system that has a high level of poverty, terrorism, and human massacre. All philosophical codes are a fulcrum in the objectivity of "Man, know thyself".

- Our present universities should be able to have humans who should be nearer the rank and file of Pythagoras in respect to mathematics, and Euclid and Einstein in respect to statistics and data. The same civilization should greatly invest on humans who will help to lead the interested ones to the anatomy of a natural Internet village. Unfortunately, the system is chaotic and unfriendly. A work to present what can be done as the desired goal of philosophical mathematics must exist as electrifying polyester, which can be utilised in driving the force of other areas of human endeavour. This is why this book is practical in nature and structure, and is principled in understanding and practise.

Philosophy has monumental extensions with consummate understanding, and so mathematics, which maintains a simultaneous agreement with philosophy, is designated in the best utilisation of logical reasoning. Creation should assess all human endeavours with the use and application of philosophical mathematics, which can best be defined as a consummate ontology of human endeavour. It must be explained that philosophical mathematics, as a science, an art, and an occupation, has the following universal agenda in bettering human standards. This is where the practicality of using it to assess all human endeavours becomes an obvious assignment with a target-driven objectivity.

- Existing as a perfect study. It is going to be utilised in restructuring the lapses of science and the enmities of technology, including the orchestrated danger of using atomic weapons.

- It will serve posterity as a standard economical vanguard, including being utilised as a commercial guide.

- The province of science will certainly assign a lot of responsibilities to this endeavour because it warehouses a lot of knowledge, which will provide the answer to human relations.

- It will be used to solve the problems of the banking industries, which have rendered the zenith of crisis because reforms at all ages are greatly imposing more problems for the industries.

- Philosophical mathematics will be utilised in assessing and restructuring all financial principles, which have greatly hindered the growth of universal economy.

- Philosophical mathematics will greatly serve as a balanced incubator of knowledge to psychology, sociology, and ontology. Both government and administration at all levels will benefit from the undying standards, which are principled in their unchanging enabling.

The urgent necessity of explaining these points is borne out of the fact that human dynamism is greatly infected with obscure, un-guaranteed knowledge. This work can be seen and defined as a teacher of truth that, when properly understood, helps to reinstitute the world not only to fulfil its agenda as a global village but also as a home for all and sundry where peace, love, and understanding (including the practise of the brotherhood of man's principle) will be extolled and exalted because creation is rooted physically from the anatomy of consummate mathematical dynamism in explaining its mathematical formulas in the best utilisation of the multiplication principle, including what is known by man as a philosophical transition or scientific transmutation.

CHAPTER 8

How Can Creation Accept Philosophical Mathematics as a Perfect Anatomy of Consummate Reality?

It is important, and at the same time philosophically relative, to state with objective explanation that apart from being the mother of all sciences, philosophy can best be defined as the dynamic structure of all mathematics. This is why there is an acceptable realism in my invented concept of WWW. A work to explain the urgent need of creation as a mathematical apostolate must be purpose-driven in the use and application of data and constructive thinking, with statistical explanation. Creation can only be accepted as a consummate realism if all forces and factors of its extensive reality are practically down with scientific understanding of philosophical mathematics.

During the ancient era, such ennobled souls like Socrates, Aristotle, Pythagoras, Plato were purpose-driven in the best use and application of mathematical philosophy. All that Pythagoras invented was creative in the best acceptability of philosophical mathematics. The questions therefore are:

- If science is configuring the world as a global castle, why is philosophy mystifying the world as a cosmic province?

- If meta-science is greatly inventing a lot of apostolate facts using the logic of objective reasoning, what actually is the position of material science in relation to the great works of meta-logic, philosophy, and mathematics?

Both Smith and Stroud, in engineering mathematics, were eloquent in accepting that in engineering logic, **X** can be utilised as a power point in approaching the mathematical acceptability of **A** and this can best be explained as $X = A$ or $A = X$ or conversely $X \times A = XA$. This is why creation must be purpose-driven in the best understanding of philosophical mathematics with its extensive reality, named meta-logic.

A work to explain the reality of creative consummation must be purpose-driven in presenting creation as the structure of all the anatomy of wisdom and the philosophy of true knowledge. This aspect must be explained in the best objectivity of what is beyond truth mathematically. In the objectivity of a common thinker, philosophical mathematics is technically and scientifically consummated in the reality of all forms, formats, features, and figures, which can be represented in the opinion of a quadrant or a circle with the inscription of a square.

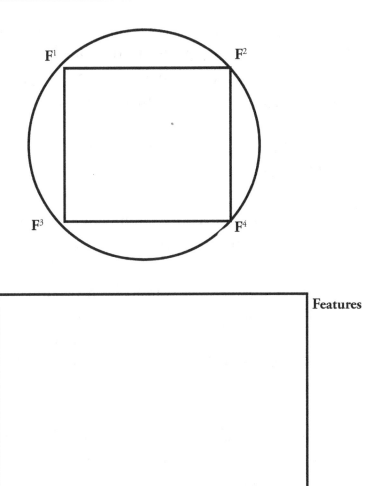

A look at these drawings, which are practical and technical, will certainly show that F + F + F + F = 4F. Conversely, $F^2 \times F^2$ will appreciate the philosophy of F^4, or technically F × F × F × F = F^4. This is why the figurative concepts of philosophical mathematics is technically centred on presenting creation as the structure of all, the anatomy of balance, the philosophy of truth, and the mathematics of great thinking, including being and becoming the scientific enabling of posterity.

In this respect, creation will appreciate philosophical mathematics as a consummate creation of a balanced reality that will be accomplished through the use and application of the following philosophical mathematics dynamics.

- Our present era needs to revisit the inventions of the Oriental wizards.

- The practise of man's inhumanity to man, such as slave trade, must be abolished.

- The church as a cardinal institution must be philosophically driven in the use and application of religious mathematics, which when properly understood will expose man to the reality of his divinity.

- The social circle and its studies need to be reengineered purposefully and practically in the use and application of philosophy.

- The academic province at all levels needs to be recognized with the use and application of philosophical mathematics, which must be the fulcrum in the best objectivity of research methods and data processing for statistics. The academia must be practical both in obtaining issues and explaining concepts, because the province of mathematics is essential for the best use and application of scientific realities.

- The democratic institutions need to be recognized with a view to enhancing the use and application of democratic ethics, which is centred on ensuring that the practise of the brotherhood of man's principle is established as a standard for government and administration.

- The scientific province, which has created a lot of enmities in the things of creation, requires a universal and philosophical reorganization. This is so because material science has created a lot of viruses through the practise of acts that are not in synchronous agreement with philosophical mathematics.

- If science in earnest is globalizing the universe as a common village, its first approach to relate the issues could have been to establish a philosophical mathematical structure that must be galvanized with the emotional postulation, with the world's positive thoughts and universal positive thinking.

Unfortunately, science has greatly degenerated not only in formats but also in ensuring that this F^4 concept is evolved with imbalance. This is why the reality of the world as a globalized province is not only shaking but also un-philosophically and un-mathematically acceptable.

I should mention that this chapter is centred on the best opinion of presenting facts and figures that, when properly understood, will help the world to understand the purpose of globalization and the enabling forces and features directing the powers of philosophical mathematics. This is why philosophy technically accepts that the world being against being rooted in the concept of a circle is consummated in the best reality of a triangle, which was defined as a rock by the one blessed and initiated in the philosophical opinion of creative reality. Hence this chapter can be defined as the Hanson of philosophical presentation because it utilises the features, formats, figures, and realities of mathematics, which at all ages and in all forms are essential in ensuring that creation exists as a philosophical alchemy. When properly explained and investigated, that alchemy will reveal and review the secrets of philosophy as purpose-driven in the balanced anatomy of a triangle. This idea prompted the Oriental wizards to strongly and strictly edify the philosophical opinion that the pyramid is the central point of all knowledge, which in earnest explained and extended the reality of creation as a consummate scientific and mathematical province that is willing to appreciate philosophical mathematics as the centre point of its hub.

The Concept of F^4 Explained—B

A look at the powers of philosophical mathematics, including its use and application, will appreciate the following points as philosophical laws, which are practically dependent on the use and application of mathematical Sagittarius.

- Every institution has an enabling anatomy by which its structure can best be appreciated in the best objectivity of natural mathematics.

- As the mind drives the brain, philosophy rules the world.

- As the function of the liver is responsible for perfect and balanced urination, mathematics and philosophy have always served creation as data points and bankable estates, by which the opinion of science is best explained after being identified as the anatomy of philosophy.

- The universe cannot be one if creation is not oriented in the use and application of philosophical mathematics.

- Globalization is not achievable if creation does not institute a philosophical mathematics college, which will serve humanity as a strategic and dynamic institution for empowerment, capacity building, and intellectual transformation.

This is why this work exists as another version of universal frequency modulation.

It is unfortunate to explain that our philosophic thinkers have not explained the concept of further mathematics with the use and application of philosophical and mathematical communication. If it is accepted that the world is a globe, as Galileo discovered, how and why is it that the following, which are going to be outlined here, are structured in triangular anatomy?

- The ship

- The aircraft

- Assorted automobiles

- Engine boats

- Communication antennas (which are structured in the proximity of a triangle)

It is a common opinion to accept that the world is not a circle per se, but an anatomy of an infinite triangle that must be approached with the use and application of A, B, and C and that warehouses a 90° or an 180° angle depending on the point, the environment, the culture, and the philosophical tradition that started its voyage.

The world is configured with the infinite anatomy of a triangle, and creation, to the best knowledge of the powers of philosophical mathematics and the wisdom of its usage, must accept that configuration. This is why a philosophical look at the aquatic estate, which is dynamic with mathematical segmentation, reveals that the province is triangularly anatomized with all the aquatic ethics and aesthetics.

The F^4 concept must be seen and appreciated as a converse principle with a philosophical theorem. This is why material science could not finish its work in explaining the purpose-driven philosophy of the WWW concept, using what I mystified in this chapter, as an enabling power of the use and application of philosophical mathematics. In acceptable reality, the WWW concept is triangular in the best objectivity of a food web or food chain. A look at human relations with its transactions reveals that man is really in search of honest reality, whereas science is in need of balanced philosophical mathematics, which it will use to assess all things with acceptable reality. This is why creation is greatly in need of philosophical mathematics as a consummate reality, which will help the urgent need of man to achieve the concept of acceptable globalization.

The concept of this section is essential in the best objectivity of the universe existing and being directed by the powers of a WWW world star.

The Concept of F⁴ Explained—C

A look at the way and manner the world is being driven with the forces of human dermatology, which includes emigration and immigration, will certainly appreciate the reality of mathematics as a population census-diver. It uses philosophy in numbering, data, and statistics; this fact is why the world has not appreciated what is contained in philosophical mathematics as a way by which the naturally and consummately globalized universe can be accepted as a purpose-driven province, or PDP. It is important to explain that the WWW concept can equally be renamed PDP when man is scientifically educated in the use and application of philosophical mathematics.

The illumined words of Albert Einstein and his creations, including those of Euclid, can be explained logically using the power of philosophical mathematics. Those concepts have remained one of the greatest assignments of the college of scholars, mathematicians, and philosophers.

The embedding word of Pythagoras is greatly synchronous with the wisdom of purpose-driven Province.

I should explain that this book is not written for the uninformed; it has no business with apostasy. Its technical foundation is purpose-driven with the use and application of philosophical mathematics and apostolate foundation of wisdom. It is important for the records and for creation to understand that a work at this level is practically original in the best opinion of the anatomy of science and the anatomy of philosophy, including applying the psychokinetic concept of mathematics. Unfortunately, the way and manner we have utilised mathematics and philosophy from the kindergarten stage of human evolution to the university level is greatly regrettable, unappreciable, and unexplained. The material science at all levels has greatly demystified the illumined wisdom of philosophy, both in its task as the mother of knowledge and its foundation as the anatomy of science.

I am highly involved with a scientific assignment to present this book, which will help the universe to appreciate that the creative powers of life is the fulcrum in the anatomy of philosophical mathematics. All human transactions at all levels are practically dependent on the use and application of philosophy and mathematics. Mother Nature, with its tangent wisdom, can be assessed if the human desire to live in a one-world family with a one-world purpose can be achieved with the use and application of practical globalization, which is indeed the anatomy of the WWW concept, of F^4 globalization, or what I have already explained as a purpose-driven province.

It is in my deep thinking that is borne out of my philosophical thought to give this aesthetical and epistemological explanation of the F^4 modulation concept, even though I have purposefully avoided some rational issues of mathematics (which I am going to introduce in my next work, *The Anatomy of Philosophy and the Anatomy of Knowledge*).

In this concept, philosophical mathematics is dependent on a purpose-driven province—or what an ordinary thinker can define as Philosophy, the mother of honest knowledge, including mathematics existing as the electrifying philosophy of creative explanation, which uses figures, formats, and features with philosophical forms.

CHAPTER 9

THE PHILOSOPHICAL REALITY THAT LIES ON THE WAY FORWARD

It is important to explain that science desires empathically to explain theories and theorems using the data contained in log tables. Material science, which has greatly stagnated the thinking of man, has avoided the use and application of philosophical logarithm in presenting facts and figures as they naturally exist and originate. Mathematics is generally known as the philosophical will of wisdom; Philosophy in turn works to correct all the errors of mathematics, which have stagnated human endeavour, particularly in the areas of science and technological icons with econometrics.

The features of engineering mechanics, with its mathematical know-how, have been greatly neglected by the objectivity of scientific facts. The philosophical reality must dynamically remain and be well accepted and assessed by human thinking. All echoes of humans are configured into voicemail in the presentation of the voice box. This is what empowered Galileo, in his material thinking, to present the universe as a mathematical and scientific globe, which is scientifically designated as a circle. It is here that the scientific achievement of Albert Einstein, including other renowned physicists and astronomers, were practically configured in assessing what can scientifically be defined as the sacred roots, which lie in the anatomy of philosophical mathematics.

The invitation to write this work can best be defined as an Internet agenda, which is maintaining a universal and cosmic balance. This is why this book can best and further be described as the anatomy of engineering mathematics, which will help to prove all that is incubated in science and the anatomy of technology.

In this respect, the mathematics that will actually provide the way forward, both in philosophical reality and in calculative realism, lies in the use and application of calculative logic. A work of this nature can be defined as the meta-logical presentation of philosophical Hansen.

In my abstract understanding as a philosophical sailor, the letter **A** can be converted to represent the letter **I**. It can further give engineering the letter and orbit of letter **1**, which can equally be translated into $A + I = I + A,$ or mathematically conversed to represent $(I + A) / (A + I) = 1$.

There is a need to figuratively appreciate that the way forward lies in the best application of honest facts and balanced figures. It is unfortunate that our present, chaotic civilization is not in tune with the use and application of philosophical mathematics; this is why national and international budgets, appropriation bills, and economic analyses (including what is defined as technological know-how) are practically insufficient and not solving the problems of commerce and industries, which must be approached in order to provide the way forward. In the objective and illumined words of Santome, the anatomy of man, if well examined, is practically constituted into two organs, which are naturally blended with philosophical mathematics, including the biology of the spirit. Santome, who was an abstract scholar of his era, further stated that the world has remained inglorious because of the abuse and inordinate ambitions of humans, which are not in synchronous agreement with the dynamism of mathematics, instrumentation techniques, numerology, and geometry. He said that the world has created a vacuum for the appreciation of data that, when properly configured into philosophical reality, will help the future of our civilization particularly science to appreciate one-world purpose, one-world balance, and globalization. That globalization, including existing as an Internet community, is unrealistic if people ignore the reality of the powers of philosophical mathematics.

It is here that the task of presenting this mathematical asset with the umbrella and canopy of philosophical engineering becomes a truism, particularly on how global problems can be solved. The way and manner Einstein spoke authoritatively during his era inspired and prompted Edwin Markham to scientifically state that in vain we build the city without building the man. If man is mind, as we are told, there is a need to appreciate his blending factors as a spiritual entity. Being and becoming in the reality of that entity, the mathematical goal is essential in the achievement of purity and perfection, which is the definition of philosophy.

Knowledge can logically be defined as apostasy if it is not keyed from the pattern of Internet wisdom, which is greatly spirited in the presentation of a united truth. This reason is why philosophical mathematics can be defined as spiritual salt that understands the mathematics of Internet solvent.

Soropulus, a Jewish physicist, analysed with perfect understanding that the world must be galvanized with the use and application of philosophical mathematics. Kenneth Anselm, and his niece who worked with him in his laboratory, defined the concept of Soropulus as being and becoming in the databank of Internet originality. It is here that I present my opinions.

- That every reality with logic can be accepted as a truism.

- That realism with meta-logic can be accepted as futuristic assets.

- Mathematics, if pure and un-tampered, can be accepted as the natural bank of all configured statistics.

- Philosophy exists in this originality as a perfect and pure science warehouse, different and divergent knowledge both in its core nature. It is a prudent force, is purpose-driven, and has empathic knowledge with its metaphysics. This is why the reality in the way forward can be defined as the science of mind, the science of the soul, and the postulations that are contained in the sacred agenda of Mother Nature.

This chapter can be defined as a living and pragmatic abstract, which is technically and logically blended with the use and application of philosophical mathematics. It is here that the anatomy of science, including the technology of philosophy is purpose-driven in the best knowledge of appreciating facts and figures through the inputs and imports of algebraic fractions, linear equations, and simultaneous equations. This work is patterned to tabulate the Internet thinking of man in order to cooperate and comprehend the philosophy of Mother Nature, which can best be defined as the hub-nub of philosophical mathematics.

CHAPTER 10

A COMPREHENSIVE OVERVIEW OF PARTS 1 AND 2

Philosophy with mathematics, calculus, and technology are naturally configured as a quantum reality. This quantification has remained the genesis of all creativity that material science and ephemeral knowledge have not been able to prove as being and becoming the basis of all anatomy. A scientific and balanced look at the scope of philosophy with its mathematical dynamism suggests that humans must appreciate the principles in perfection and policies in purity.

Unfortunately, humans started approaching life with an ephemeral understanding of the realities of nature. It is here that scientific concepts and principles, no matter how new they may be, instantly become obsolete with obscure ambiguities. These lapses stand to be addressed by the future of science and technology through the use and application of mathematical philosophy, applying the use of Internet wisdom and appreciating that the mind is the curvature of online real-time, which naturally is beyond the scope of Internet science. It is here that the tiers of natural law, material science, and living philosophy are concerned regarding how to correct the errors of material mathematics, which have affected the growth, evolution, regeneration, and dynamism of civilization. A wonderful look at what Dalton presented as his theorem makes man to know that humans stand the chance of committing grievous mistakes out of the existence of the real philosophical mind.

Abstractly, if $X + 1 = X + A = X + B = X + C$, the philosophical concept in its reverse mathematical dynamisms requires that

$$\frac{C}{X} = \frac{X}{1} = \frac{B}{X} = \frac{X}{A}$$

which in consonance can be accepted as a simultaneous dynamical mathematics, with its anatomy originating from the perfect and purest dynamism of philosophy. A scientific look at that linear expression will make us to appreciate that philosophical mathematics can be accepted as an abstract science.

Mathematics has maintained an engineering position with its technicalities, arguments, illustrations, and facts and figures. It is here that logarithm use, which can be defined as a stroud in philosophy, becomes the reality that lies in the hub of scientific dynamism. The position of material science with its errors will certainly appreciate the powers of philosophy and the wisdom of mathematics including neglecting how these errors could be reversed.

As material science greatly progresses at an algebraic level, philosophical mathematics expands its horizon in building and rebuilding the anatomy of technology, the principles of chemistry and alchemy. This is why a comprehensive analysis of the Orion who originated from the Orientals will certainly appreciate that both philosophy and mathematics originated from the roots of compact Internet wizardry.

A look at the purpose and dynamic-driven formats and formulas of arithmetic, algebra, and geometry include the following: linear equation, simultaneous equation, philosophical equation, and the occult Analysis with its statistics, all of which maintain a tangent dynamism with the above, will certainly appreciate that philosophical mathematics is practically configured in representing facts with the use and application of the realities that are scientifically driven with equations. It is unfortunate to explain that the world is toying with the reality of truth. This is why the machine age could not endure, and the Renaissance could not achieve its desire. Presently, the lapses of material science with the hollows of its technology will certainly remake and mimic human engineering as a wager board.

Reasonably, this is un-philosophic and unacceptable by mathematics. Empathy, which is a consummate law, is perfectly explained in the purity of philosophy as the mother of knowledge. Science is a thinker of thinkers, and technology is creative in its objective creativity. Therefore P/M, which represents philosophical mathematics, can be zeroed to accept all that is contained in the anatomy of science. The physiology of technology, which humans deal and accept without investigating technological contents in nature's technological laboratory, can be defined and designated as the greatest errors in human evolution.

In my humble opinion as a disciple of inspiration, the following needs a great reengineering with a philosophical regeneration.

- Cross-examination of the facts at all levels of mathematical dynamism.

- Investigation of the purpose and dynamic-driven concept of philosophy, and how other sciences perfected agendas that opinionated philosophy as the mother of sciences.

- Metaphysical investigation of how man appreciates the visible universe, including his opinion about the invisible universe.

- Investigation of the truism of the logic that is contained in occult philosophy, including how the principles and policies of meta-logic can be utilised by science in moving the universe forward.

- Cross-examination of all scientific principles and policies in order to ascertain their paralleled and unparalleled relativity with the laws of economics, applied geophysics, numerology, and astronomy.

The greatest necessity of science lies in the facts presented to it by Albert Einstein, Isaac Newton, Galileo, Pythagoras, Emmanuel Kant, Max Miller, and Thomas Edison, as well as Leonardo De Vinci.

A look at the benefits and contributions of philosophical mathematics will certainly approve that it is the basis of engineering mechanics and engineering mathematics, including all that can be assessed in the anatomy of alchemy. There is need to explain that humans are not careful and calculative in the objectivity of how Mother Nature uses Internet mathematics, Internet technology, and Internet philosophy to enhance the dynamism of the cosmos.

It is here that the comprehensive overview of parts 1 and 2 can be defined as being the wish, will, and honest opinion of philosophical mathematics, which can neither be altered nor defaced by scientific principles and policies. This is why a look at the anatomy and alchemy of technology reveals that it is powerful and purpose-driven with the honest and pragmatic wisdom of philosophical mathematics.

PART 3

SOME MATHEMATICAL AND PHILOSOPHICAL ARGUMENTS

"I also ask you my friends not to condemn me entirely to the mill of mathematical calculations, and allow me time for philosophical speculations, my only pleasures."

—Johannes Kepler

INTRODUCTION AND ABSTRACT

It has become an urgent necessity that every human endeavour, particularly those that deal with knowledge, must be opinionated in the objective balance of a dynamic abstract. As an abstract science, philosophy is mathematically driven with the use and application of technical concepts and engineering technicalities. It is here that mathematical formulas are utilised in accessing and appreciating the formats mathematics needs. The scientific engineering of philosophy can neither be appreciated nor understood by those who are not Internet-driven by the objectivity of natural wisdom.

When Archimedes, a renowned alchemist, presented his principle in relation to chemical formulas, his era was confused and declared him infected with spiritual madness. Arguably, every era has always had humans who have been appointed at the level of Archimedes, who is renowned for his postulations in chemistry, alchemy, segmentology, and chemical force (which can be defined as recycling the thoughts of philosophy in the opinion of chemical dynamism). This abstract has decided to present him as a philosophical and chemical star.

It is regrettable that our present civilization, including our educational system, is not universally driven with the thinking of science and mathematics. A look at the mathematics of Stroud and Smith can be defined as advancement in the objectivity of philosophical mathematics. For example, if a mathematical table is lock-jammed with sine and cosine formulas, it must be appreciated that the table of value can only be understood when all the letters are aligned in the best understanding of simultaneous arrangement. It is a philosophical realism that sine and cosine formulas, including all that are contained in the knowledge of algebra and geometry, are most of the time configured in the best opinion of the following objects: triangle, circle, quadrant, and square.

There is need to explain that the concepts of sine and cosine formulas at all engineering mechanics are purpose-driven in assessing and appreciating the advancement of philosophy as the science of purity and perfection. A look at the concept of Archimedes' principles, including Dalton's atomic theory, must appreciate that Dalton, Archimedes, Isaac Newton, Smith, Holderness, and Lambert—and including our modern physicists like Nelkon, Albert, and Philip—were gingered to utilise mathematical letters in explaining the powers of philosophy, practically and scientifically, from the honest meta-logic of its principles including assessing the realism of its empathy.

The onerous responsibilities of writing this abstract with a mathematical introduction, which is sequential with the objective of philosophy, must be seen, known, understood, and appreciated as a work at the proximity of inspired and Internet wisdom. A look at the abstract contribution of Walter Russell in relation to the secrets of light, including his scientific cosmic cosmogony, is cosined in presenting facts and figures not for the ordinary man. It is here that this genius opinionated that genius is inherent in everyman, and mediocrity is self-inflicted.

Mathematics has remained the geophysical anatomy of human reasoning. This is why philosophy involves mathematics in everything, particularly in issues that have to explain some technicalities in engineering, science, technology, and quantum dynamism and abstract knowledge.

This work is going to look at the configuration of philosophy and mathematics as enabling meta-logic endeavours. It is important to explain that this book can best be understood when utilising the sacred teachings of all ages—the secrets of science, the secrets of light, and the sacredness that lies in the life of creation.

A look at the ontology of Plato's metaphysics will at best authenticate the work existing at the proximity of philosophical mathematics. For the human world to appreciate this, he needs an illumination, a balanced wisdom, and a purpose-driven life. If in a triangle that has ABC as symbol letters, it must be stated that all angles must exist and remain parallel to each other, both in converse and transverse. Then the scientific formulas of cosine and sine, as tabulated by the Oriental apostolate, would be accepted with the use and application of log tables. It is here that the concept of philosophical mathematics started to gain ground, particularly with the acceptance of further mathematics, engineering techniques, and meta-logic concepts.

A look at the achievements of the globe, which is symbolized in the configured quadrant of north, south, east, and west, reminds us that the cosmogony of the universe is practically quadrant, dynamically triangled, and philosophically consummated.

The intake of mathematics as an electrifying philosophy of all human endeavour was best ornamented in the opinion of letters, sines, cosines, formats, formulas, and equations. The anatomy of philosophical mathematics is purpose-driven in assessing purity and perfection. This abstract and introduction is technical with the mechanics of the electrifying understanding of science, technology, and abstraction. It is here that the introduction stands alone as harmonizing the difference that humans have created in the province of philosophy and the dynamic sector of core mathematics.

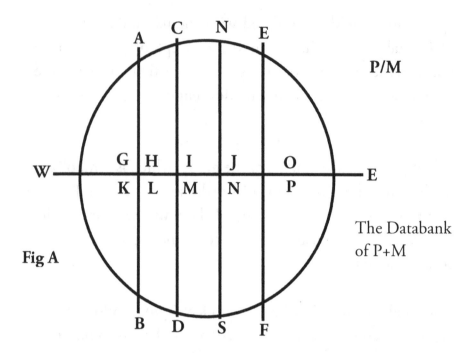

P/M

The Databank
of P+M

Fig A

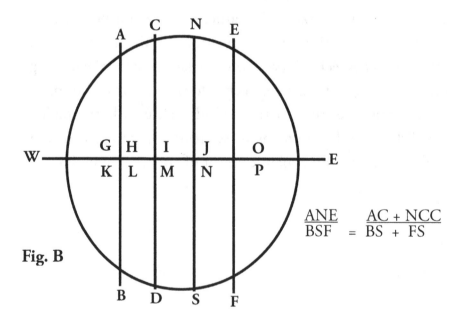

$$\frac{ANE}{BSF} = \frac{AC + NCC}{BS + FS}$$

Fig. B

CHAPTER 1

WHAT ARE THE ARGUMENTS ABOUT THE PHILOSOPHICAL RELATIVITY OF THE FOLLOWING MATHEMATICAL FUNCTIONS?

Sine Formula

It must be stated that philosophy is factual and aesthetic in the usage of empirical knowledge, scientific principles, and policies. Its objectives are essential in the assessments of mathematical facts and figures. The academia understands the primary functions of sine formula, but to translate these functions to a secondary usage has greatly instituted a philosophical stigma both in the creations of the academia and in what the scientific professionals have contributed.

The purpose of the sine formula is to connote and relate all facts and figures to an understandable level. The sine formula at any point in time has remained the heart of mathematics. This is why all simultaneous equations, linear equations, geometric technicalities, and algebraic equations adhere strictly to the philosophical formats of the sine formula.

H. A. Clement and Howells explained the abstract philosophy of the sine formula, which is a mathematical ingredient. They reveal that we must design and philosophically fall back to our roots. The abuse and misuse of the sine formula during the era of Pythagoras, who was designated as a wizard, made him consider his roots and aborigines as humans that could best be defined as unscientific.

For the record, the purpose by which the sine formula can be used is to buttress mathematical letters in order to synchronize with philosophical facts. The use and application of the formula, when configured with the facts of philosophy, can help engineering, science, and technology at all levels to develop facts and figures, which can be purpose-driven with practical endurance.

Material science is not falling back on the alchemy of natural chemistry, so it must be explained that the opinion and wisdom in the sine formula, particularly as a mathematical format, cannot be determined as an institutionalized and philosophical factor. This lack of knowledge, which is an error, has affected man's deep knowledge of arithmetic, further mathematics, algebra, geometry, and the rhythm of remainder theorem and linear equations. The result has philosophically affected the position and deep knowledge of simultaneous equation.

At this point, the technicalities that are involved in the use and application of tangents, which are symbolic in $\tan^{\theta}$, is erroneously applied in the areas of science and technology. In assessing the simultaneous relativity of philosophical mathematics with other sciences, one needs to appreciate that the world can best be defined as the anatomy of the mathematical sine formula, including being and becoming the alchemy of all its cosine formats and figures. This concept is why engineering mathematics is practically enabled in the objective of the use and application of philosophical mathematics.

It must be stated that the prominence of the sine formula in relating to and solving some mathematical problems was practically instrumental to the knowledge of statistics, the wisdom of logic, and the philosophy of meta-logic. Its mathematical core, was designated by the apostolate mathematical wizards as the supreme foundry of facts and figures. It is important to explain this fact, particularly in an era where man is at ease with knowledge, wisdom, science, and technology.

A hibernated understanding of the Crawnwell Lectures was formatted to present the rhythms of core mathematics purely and perfectly regarding the proximity of the sine formula. It is here that meta-logic designed and designated the sine formula as the structure of all. Unfortunately, man has developed an unacceptable tabular for its usage, which has greatly affected the deep thinking of engineering at all levels.

It is my opinion that we should co-ordinate and relate all these missing gaps so that humans will appreciate the fact that there is a balanced unity in the divergent unity of all, which is the supreme ecstasy, including being the ornamented fact that established the electrifying unison of philosophical mathematics. This is why the structure is natural, real, and perfect with purity, and why it must forever serve the province of mathematics as a leader, taking mathematical knowledge to the pinnacle of illumined wisdom.

Conclusively, the invention of the sine formula by the Orients is to teach creation with balanced mathematics so that the unity of all is a determinant factor. What is in the anatomy of Internet sine formula can be bequeathed to creation as existing at the ornamental position of a philosophical boulevard.

This is the truth, which is sine formatted in the best objectivity of the structure of mathematics and statistics. The philosophy of every data respects the opinion of the sine formula, which when properly used helps us to know the legends in the secrets of core mathematics with a view to investigating what is in the context and concepts of applied logic.

Sine Formula = $\dfrac{\text{Opposite}}{\text{Hypotenuse}}$

Cosine Formula = $\dfrac{\text{Adjacent}}{\text{Hypotenuse}}$

Tangent Formula = $\dfrac{\text{Opposite}}{\text{Adjacent}}$

A look at these formulas reveals the scientific and technological relativity of mathematics to philosophy, philosophy to applied logic, applied logic to mechanics, and mechanics to quantum understanding. Further investigation takes us to the anatomy of wisdom, which can be defined as the supreme mother and father of all the mathematical formulas.

The Argument about the Mathematical Format of the Cosine Formula:

The following is a look at the way the cosine formula (which originated from the deep thinking of philosophical mathematics) operates, protects, and defines a statistical agenda, which is essential for the best objectivity of facts and data.

If the cosine formula is propelled to represent the adjacent divided by the hypotenuse of an angle, square, triangle, or circle, to name but a few, it must be stated that philosophical mathematics from the point of the sine formula is empathic and dynamic in coordinating the meaning of sine and cosine, using tangent as a base mark. The understanding here is logical, strategic, and empirical with scientific evidence. Empirical knowledge is mathematically oriented, and so scientific knowledge has always been purpose-driven with the best ideas and facts. It is here that science is purpose-driven with tests and evidences. The cosine formula, which is a philosophical fort, is mathematically protected in the best use of its data. Every table of value must be analysed and at the same time proved with the use and application of the log table.

Unfortunately, this process is materially handled with errors and lapses. The cosine formula has helped in the propagation of mathematical agenda; this is why every geometric equation is cosine-formatted as an Oriental and electrifying course, which cannot perform its functions if the facts and figures that are contained in cosine formulas are neglected.

The cosine formula, which inspired Pythagoras to define and designate mathematics as a core science, always cooperates in perfect coordination with the use and application of empirical philosophy. Empirical philosophy means a knowledge that is provable, logical, conventional, and arguable. If one is asked to produce the cosine formula of a triangle, he must appreciate that every triangle exists perfectly in the agenda of 90 degrees. This is why the fraction of 90 degrees must zero itself to 45 degrees. It is here that the cosine formula of mathematical data has remained as knowledge for the well informed.

In the illumined knowledge of H. A. Clement, designated as a wizard of mathematics and philosophy, the cosine formula is related to engineering mathematics, scientific mathematics, and technical data. Clement stated strongly during his era that the universe is a crescent of mathematics anatomy. If this is true, the mathematical logic will help us to appreciate the following facts.

- That the cosine formula is a mathematical databank.

- It has a scientific relativity in all areas of science.

- It has a greater impact with engineering mechanics.

- Technology is pragmatic in the use and application of the cosine formula.

- Engineering logic cannot be understood or practised if the data that are contained in cosine formulas are neglected.

- The challenges of core mathematics, with its purpose-driven nature, cannot be averted if the opinion of cosine formula is neglected.

- The adjacent and hypotenuse nature of the cosine formula is practically structured to explain the logical and philosophical heritage of all sciences, using technology with its technicalities as a benchmark.

- It is a formula that works to harmonize with balance the electrifying engineering of mathematical facts. This is why A + A is perfect in producing 2A, and A × A is in a synchronous cosine agreement with A^2.

Cosine formula as a mathematical reality is a purpose-driven mechanics. This is why the concept of philosophical mathematics, which remains incomprehensive of this mathematical think-tank, is not philosophically explained in order to help engineering and scientific students assess what is contained in alchemy and astral knowledge. This work exists on its own in order to help research fellows and overall academic community in knowing that the science of philosophical mathematics, if neglected, is capable of eroding knowledge beyond the knowledge of the fifth kingdom.

The Argument on the Mathematical Symbolism of the Tangent Formula

It is important to explain that the tangent formula, which great mathematical wizards define and accept as *tan0* (representing opposite divided by adjacent), is purpose-driven in the true explanation of the sine and cosine formulas.

While at the College of Science and Philosophy, I investigated strongly that the tangent formula is a strong and undivided mathematical postulation. This is why it can be used as a hub-nub in empowering the sine and cosine formulas.

Regrettably, the ephemeral world of mathematics has neglected its powers, which include:

- Directing the wisdom of mathematics to agree with the philosophy of statistics.

- Utilising the figures that are available in core mathematics as a blending forces in ensuring that mathematical formats are simplified.

- Co-coordinating the activities of the sine and cosine formulas in order to exist as one mutual entity.

- Ensuring that the use and application of engineering mathematics is practically united in achieving a configured goal.

- Ensuring that the tangent data that are contained in the log table are not turned as a mathematical logjam by the uninformed.

- Ensuring that every letter of mathematical symbolism must maintain a synchronous realism with other letters.

This is why the tangent formula is closely related to statistics and is synchronously related to the use and application of data. It must be stated that the tangent formula is the core logic of mathematics—the fundamentals of its propagation, the philosophy of its work, and the dynamics of its thinking. This is why mathematics has maintained a standing chair with science, technology, philosophy, physics, geophysics, geosciences, and astronomy.

Renowned astrologer Akin King stated with all mathematical conviction that the astral plane is made up of the thinking of mathematical tangent. Unfortunately, his era, the fourteenth century, could not investigate his finding, particularly the impending position of its postulation by believing that all scientific facts are provable in the lab.

A look at the balanced formation of the tangent reveals that this aspect of mathematics is the anatomy of engineering philosophy, particularly those that have to deal with mechanics, civil and construction technology, methodology, and polymer and textile engineering. It must be appreciated that every aspect of human creativity is purpose-driven in assessing the concept of opposite divided by adjacent. This fact buttresses the tangent formula as another action and version of two-way thinking; that is why it is in formidable agreement with the sine and cosine formulas.

A look at the networking concept of Internet technology, including its electromechanical mechanism, reveals that the science of the tangent formula is actually at work. The concept of online real-time is purpose-driven with the use and application of opposite-to-adjacent symbolism with the tangent formation.

It is important to state that philosophical mathematics is blended with the use of the following.

- Sine formula
- Cosine formula
- Tangent formula
- Cosec formula
- Sec formula

The philosophia of the above is the origin of what is defined in this context as the logic of science and the meta-logic of technology, including appreciating all that is contained within the Internet table of value. This work exists as another postulation from a universal Internet kit.

A look at the performance of the tangent formula will certainly serve a card that philosophy is the mother of knowledge. Professor Ebenezer Bid of the University of Canada argued strongly that the tangent formula could be used as purpose-driven data in assessing and appreciating all the core mathematics problems that have greatly defiled human solutions in his technological endeavour. My argument, which is well understood and explained, is in synchronous agreement with that of King. It is here that I am placed at the proximity of explaining the tangent of philosophical mathematics, and I make the following suggestions.

- The world of philosophical mathematics is in search of balance.

- It is in need of advanced knowledge.

- It requires a reengineering, which will help philosophy to regenerate from the obscure understanding of human concepts.

- It must be used to create and establish an Internet lab, a technical lab, and a resource lab, including a scientific databank.

- It must be used to appreciate the electrolysis of the universe, particularly from its alchemy as a crystal anatomy.

- It must be used in correcting the lapses of scientific principles and policies that quickly become obscured simply because they are not originated from the bank of inspiration.

- It must be used in creating a united industry for the use of a united economy, which will greatly uplift humans in the best objectivity and appreciation that against making industries, man should determine to make man.

- It must be used in explaining the anode and cathode concepts of electrolysis, which are synchronous with the opposite and adjacent philosophy.

With these pieces of advice, it is hoped that the world will certainly be renowned and reengineered in the best use of mathematical formula, including perfecting its agenda with the opinion of philosophy—which is known as the mother of knowledge and whose dynamic agenda is achieving purity and perfection, something that is in synchronous agreement with the desire of mathematics, statistics, mechanics, and quantum wisdom. This work will certainly uplift humans from the oblivion of philosophical mathematical lapses.

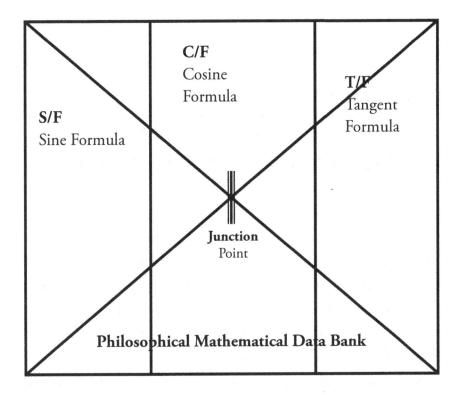

FIG. A

This drawing is known as a philosophical mathematical databank. It can be used to access the agreement of all mathematical formulae and their relativity with Internet technology.

CHAPTER 2

THE PHILOSOPHICAL AND MATHEMATICAL SIGNIFICANCE OF ALGEBRA AND GEOMETRY

A group of dictionaries and encyclopaedias, including some advanced Britannica, define algebra as a type of mathematics in which letters and symbols are used to represent quantities. They equally term it as the measurements and relationships of lines, angles, and other aspects of a particular object or shape. The definition of Algebra maintains a simultaneous relativity with the purpose-driven dynamism of philosophy. It is here that this branch of mathematics postulated the deep thinking of quantum theorem, including the philosophy of quants (which is a deep and dependable science). That explains the workability of mathematics with a view to knowing its anatomy and bases and how the utility of the bases can be used in assessing data, facts, and figures.

The mathematics of algebra will represent its relativity in the best opinion of sine, cosine, and tangent formulas so as to be able to draw what is known as the hypotenuse philosophical position. For example, algebra as a branch of mathematics is purpose-driven with the use of letters ranging from A to Z. This is why simultaneous equations, remainder theorem, linear equations, and cosine and sine formulas are purpose-driven with the practicality and thinking of algebra.

The mathematics of algebra states that $A + A + A = 3A$ and that $A \times A \times A = A^3$. This can be represented as $1 + 1 + 1 = 3$ or $2 \times 2 \times 2 = 8$. It is here that algebra, mathematically defined as the statistics of letter representation, is greatly involving the deep thinking of letters, assessing and assembling mathematical opinion.

It can equally be stated that

$$\frac{1}{A} + \frac{2}{B} + \frac{3}{C}$$

will give

$$\frac{BC + 2AC + 3AB}{ABC}$$

Conversely,

$$\frac{1}{A} + \frac{2}{B} + \frac{3}{C}$$

can logically be represented as

$$\frac{6}{ABC}$$

which can equally be explained as $A = 1$, $B = 2$, $C = 3$.

It is here that the concept of algebra as a mathematical branch that deals with letters, symbols, and quantities was at the proximity of Professor Quants, who developed a simultaneous mathematical philosophy that can be defined as the beginning and origin of engineering mechanics and engineering as a science, including science as engineering postulations.

It is important to explain the functions of this mathematical branch, with its relativity to philosophical thinking. Professor Louis of Cambridge, who is renowned for his understanding of algebra, opinionated strongly that mathematics has adopted a confused state because scholars, professors, and research institutions do not understand that letter A can equally be used to represent 1, and B can equally be used to represent 2. Algebra is greatly interwoven with geometry.

The opinion of H. A. Clement, who is designated as the father of modern geometry and is the epitome of modern algebra, technically differed from what material science understands to mean the science of mathematics and algebra. He stated that Quant's theorem, with its inspired principles, can be understood if mathematics is not defined as a course that understands the use and application of data and statistics.

I am in agreement with H. A. Clement, Durell, Louis, and Pythagoras, as well as all enabled professors emeritus of algebra. It is important to explain that algebra, as a mathematical podium, is philosophical and logical. This is why it is purpose-driven with the use and application of the lowest common multiple (LCM), the highest common factor (HCF), and the sine, cosine, tangent, and sec formulas.

It must be appreciated that 0, if purpose-driven at the proximity of infinitude algebra, must certainly return to 0 if the institutions of logic and philosophy are mathematically driven to propel and propose its meaning.

A look at the philosophy of engineering mathematics appreciates its enabling and its Internet position with the use and application of algebra. The anatomy of a mathematical equation cannot be defined or understood if algebra is neglected. For example, the wonderful contributions of simultaneous equation, linear equation, functional fractions, and others are purpose-driven with the use and application of algebra. This is why a work on the mathematics of algebra must present the lowest common multiple of all letters in the best opinion of its thinking. This explanation is certified and configured as being in the creation of Algebra, which is a figurative and futuristic subject and which uses letters, quantities, and qualities. It is here that the architects, surveyors, and engineers that deal with methodology and constructions are best initiated in the objective use of letters and figures.

Algebra as a mathematical branch is always investigative. It is here that it maintains a synchronous agreement with the wisdom of philosophy, including the use of core logic as a way and means of explaining the use of letters from the point of electrifying balance. This explanation with a multiple definition stands alone as a mathematical fort.

Geometry

Geometry is a branch of mathematics that deals with the measurement of lines, angles, squares, and solids, and it employs the deep thinking of algebra in formatting and formulating its data. The concepts of sine, cosine, tangent, and sec are postulations that mathematics uses to buttress the philosophy of geometry.

It must be stated that geometry is a branch of mathematics that greatly deals with circle theorem, squares, angles, degrees, and variation. In construction and particularly in civil engineering, geometric projection has remained the databank that surveyors, architects, craftsmen, and designers use to present what can be accepted as an environmental survey.

When Galileo told the world that the universe is a circle, he was practically geometric and scientifically algebraic. The postulations of Albert Einstein—including Worley, who is defined as the anatomy of modern physics—are essential to the structure of geometry. It is unfortunate that our modern scientists are not naturally geometric, algebraic, and figurative, engineering with mechanics in presenting facts and figures.

A look at the networking concepts of sine, cosine, and tangent formulas philosophically reveal the synchronous agreement of algebra. It is important to explain that the purpose-driven mathematical understanding of geometry is philosophically equated and mathematically explained with the reality of logic. Every human endeavour is configured in the best opinion, which exists between geometry and algebra, technical calculus, core mathematics, further mathematics, and pure Statistics. A look at the philosophy of research is technically geometric in the involvement of algebra.

The philosophy and ontology of this works is to explain the following mathematical points.

- As algebra is well represented with the use of letter, geometry is well galvanized with the use and application of measurements, lines, angles, squares, solids, and surfaces.

- It is the prism and the pinnacle of philosophical wisdom.

- It is embedded with the pure knowledge of core geophysics; this is why geoscientists at all levels employ geometry as a mathematical agenda.

- Geometry as a mathematical branch is always used to determine and explain the physics of solids. A look at the opinion of core physics reveals that its alchemy is quantious in explaining physics, both as a physical science and a natural science.

- Geometry as a mathematical branch is a fraction of construction and an ornament in creative expression.

- It is philosophic in structure, mathematical in data, and logical in explanation. That is why it is greatly interested in the use and application of solids.

The urgent necessity of explaining these points becomes a requirement because engineering philosophy and engineering science have suffered a lot of humiliations—simply because humans are not geometrically driven or oriented with the best use of algebra.

The sine formula represents opposite divided by hypotenuse, so it must be instituted and explained that both algebra and geometry are greatly allied in the structure of this formula. This is why a work in the best opinion of core philosophy and core mathematics will only produce an acceptable Britannica and what is defined here as the anatomy of technological fort.

It is important that the will and wish of students who have neglected philosophy and mathematics should be restructured and reengineered in accepting that the courses are for their ennoblement, understanding, and growth. A work on geometry and algebra must not and cannot neglect the use of letters, measurements, solids, and squares. It is here that a triangle can be represented with letters and that a square can be explained with the use of lines. This is why philosophical mathematics is purpose-driven in creating ideas and concepts that will help the world of science and technology in appreciating the wisdom of Galileo, which can best be defined as having originated the philosophy of Internet thinking (which is strongly imbibed by material science).

This work will help humanity to access issues both practically and objectively, which in earnest have remained the philosophy of mathematics, philosophy of logic, and philosophy of creative thinking with its methodology. It is here that geometry is greatly interwoven and scientifically related to algebra, both as a mathematical kit and as a philosophical asset.

CHAPTER 3

WHY PHILOSOPHY IS GREATLY CONCERNED IN FURTHERING THE COURSE OF CORE AND APPLIED MATHEMATICS

The Orientals, who are universally known as the creators of philosophical wisdom, designated this noble prince of science as the mother of all knowledge. It is symbolic and dynamic in furthering particularly the workings of physics, applied chemistry, and core mathematics with its applied ontology; this is why it holds a simultaneous agenda with the most deep-thinking of mathematics. By networking the understanding of electrical-electronic physics, philosophy appreciates that mathematics is purpose-driven with the use and application of core logic.

Figuratively stated, this chapter can be neither avoided nor neglected because of the following philosophical reasons.

- Mathematics as a science is figurative and constructive.

- Philosophy as the prince of knowledge is equally logical, constructive, objective, and dynamic.

- Mathematics is a science that understands the use and application of figures; philosophy is seriously interested in ensuring that mathematical laws, rules, methods, principles, and policies are scientifically explained in the best understanding of philosophical dynamism.

- Applied mathematics is greatly interested in converting all data as pluralized factors, and so philosophy uses these factors in accepting an acceptable and dependable databank.

- Mathematics has diversified branches, ranging from abstract mathematics, engineering mathematics, Internet mathematics, pure and applied mathematics, algebra, and geometry. Philosophy is scientifically divergent with extensions ranging from logic, metaphysics, moral and ethics, thurgy, aesthetics, and epistemology. Philosophy equally has other branches that are added in my book entitled *Philosophy: The Maker of Great Thinkers*. These new branches include technical philosophy, engineering philosophy, Internet philosophy, strategic philosophy, logistical philosophy, administrative philosophy, psychological philosophy, and ontological philosophy.

Philosophy is purpose-driven with the use and application of nature. It is here that the fundamentals of core mathematics, pure and applied mathematics, principally and purposely utilises the essence of empathic philosophy.

For example, if the letter P stands for philosophy and the letter M stands for mathematics, it can be appreciated that philosophy furthers the dynamic concept of pure and applied mathematics when it is used as a symbolic representation that reads thus: P + A + M = Pure and Applied Mathematics. This can be represented as P = A = M. That is why both mathematics and philosophy are greatly interwoven, interconnected, and inter-anatomized. The way and manner that applied physics, in its ontology, utilises the data of pure and applied mathematics in representing and explaining its philosophical dynamism has remained one of the greatest tasks of science and technology.

Philosophy is energized and pragmatic in achieving the concept of purity and perfection, and applied mathematics is scientifically determined to utilise the electrifying system of perfection and purity in observing, explaining, and understanding the mechanism of data.

Unfortunately, the world of science, which is greatly polluted with material thinking, is not purpose-driven in the use and application of applied mathematics. This fact is why scientific principles, inventions, and creations, to name a few, easily fail with time. It has invariably created a melancholic situation in the annals of universal cosmogony.

The world of philosophy is greatly concerned about protecting the institution of pure and applied mathematics. A look at Plato's metaphysics and pure and applied mathematics reveals that humans have greatly defeated and destroyed the purpose-driven philosophy of mathematics. This is the cause of what this book is going to treat as engineering virus and technical cancer.

With the illumined words of Plato, core mathematics, with its applied ontology, is a human endeavour that can only be understood and appreciated when humans decide to make philosophy a journey of life.

It is here that the argument between Plato and Pythagoras can be defined as the opinion of all, the voltage of philosophy, and the voltage of Internet understanding, including being and becoming the anatomy of pure and applied mathematics.

It is important to explain that applied mathematics is scientifically relative to all human endeavours, whereas applied philosophy networks this endeavour in order to achieve a balanced and dependable relation. As Manly P. Hall said, philosophy, either as a science or an art, must be objective in its symbolic determination. This statement necessitated the electrifying wisdom of Dr. Walter Russell to examine the networking concept of philosophy in view of what is contained in applied philosophy, applied engineering, and mechanics. This noble prince established that the combination of applied mathematics holds the wisdom in the best objectivity of appreciating the RBI, which gave the world its foremost and dependable objectivity.

Unfortunately, the way and manner in which science has degenerated the thinking of humans has affected the growth of pure and applied mathematics, philosophy, and technology.

The greatest woe of our era is that man has not appreciated himself as the symbol of metaphysical mathematics. By metaphysical mathematics, I mean an institution or an image that is perfect, pure, and greatly energized to perform all that is natural and neutral.

It is here that the world is greatly in search of emengards and geniuses who will propel and be concerned with philosophical labs and philosophical kits.

In this respect, all that is explained in the lowest common multiple of sine, tangent, cosine, sec, and cosecs are practically original in the wisdom of applied mathematics, but their understanding has remained one of the greatest problems of man simply because he eroded his thinking from the principles of philosophy—the way of the illumined, the journey to life, and the symbol of truth. This is why it is important to explain this fact using what I may consider, in this context, as what is beyond truth, including appreciating that philosophy is the way of all creations. It performs its functions using what is defined as an electrifying network, including facts, figures, data, and formats with scientific formulas.

For example, if mathematics is perfect in the use and application of letters with philosophical ideas, then applied mathematics is technical and scientific in the best use of log table, which contains the sine and cosine formulas. The significance in this context is that philosophy can be defined as an extensive and figurative applied mathematics, applied science, and core geophysics.

A look at the postulation of universal anatomy must certainly blend the ideas of astronomy with the symbolism of astrology. This is why Professor Paul Piro of Cambridge University stated that the world institutions of higher learning cease if research does not involve the deep thinking of applied philosophy with modest use of core logic. Piro, in his efforts to examine the structure and anatomy of applied mathematics, harmonized his thinking and scientific postulation by stating that $2 \times 2 = 4$ and $4 - 2 = 2$. He furthered his explanation by stating that human thinking must be purpose-driven with the use and application of scientific philosophy. Professor Piro is practically in tune with the metaphysical mathematics of Plato, which can be understood as being and becoming the hallmark of mathematics, applied ontology, and metaphysics.

It is the intension of this book to reveal that human endeavour should neglect the networking concepts of mathematics, which stands with posterity as a work without basis, agenda, and ontology. This is why the concept of philosophical mathematics greatly energizes the material concept of science and technology with a view to configuring a scientific lapse in order to blend with the wisdom of Mother Nature, which holds the key in educating us to know, appreciate, and understand that applied mathematics can be used in expanding human knowledge. That human knowledge is always there in order to bring us to the true knowledge of what the science of true perfection is all about. This is why this chapter is defined as "The Anatomy of Philosophical Mathematical Truth".

CHAPTER 4

WHY IS STATISTICS SCIENTIFIC IN THE BEST OBJECTIVITY OF PHILOSOPHY?

It is important to explain that statistics is practical and honest in the use and application of scientific logic and in assessing and understanding the objectivity of philosophy. In this context, statistics can be defined in the following patterns, panels, formats, and formulas.

- As an extensive analysis of pure mathematics.

- As a collection of information that is purpose-driven in the best opinion of the use and application of numbers.

- It is the science of collecting, collating, and analyzing data, particularly those that have to do with research and methodology, and this is why it can be defined as an extensive and pragmatic structure of logic.

- It can be defined as an objective science, which deals with records, numbers, and data. This is why it is structured in figurative analysis.

- Statistics can equally be defined as an objective and amalgamated structure of mathematics, with a view to understanding the enabling of philosophy as the prince of knowledge including being the anatomy of sciences.

- It can equally be defined as the geometric analysis of facts and figures in order to determine their scientific base marks.

It is here that philosophy determined and explained the science of statistics as the rotor of mathematics, the podium of physics, the enabler of all world stars including being the panel by which Internet technology can be understood. In this respect, statistics scientifically understands the best objectivity of philosophy. This is why every institution, community, and data with its research methodology remains compacted and configured in the best thinking of philosophical statistics. It works practically in protecting the province of statistics and enabling that it is contained in core and applied mathematics. This is why statistics scientifically remains harmonious and objective as the practical philosophy, as the hallmark of balanced and dependable knowledge.

The simultaneous relationship that exists between mathematics as a science and statistics as a collector and reservoir of mathematical data must be seen and accepted by the world of science and technology as having its rotor from the deep thinking of science and philosophy.

Unfortunately, philosophy is purpose-driven with perfection and purity, and statistics is dynamic in the use and application of figures, particularly those that are greatly concerned with evolution, revolution, history, anthropology, general scientific details (including examining the purpose of technology), the province of commerce, and industry. A look at the constructive mechanism of applied economics understands with philosophical revolution the presence of statistics, including what can be considered as its electronic mission.

It is here that Lohom defined statistics as the philosophy of figures and the science of letters, including being and becoming the enabling world star that can be utilised to develop concepts and strategies.

It is important to reveal with practical explanation that statistics is dynamically objective with the use and application of philosophical prowess. It is here that aesthetic philosophy, including the province of meta-logic, is databased with the bankable understanding of statistics.

Philosophical mathematics is a living science that appreciates:

- The first principle of science.

- The first principle of technology.

- The first principle of philosophy.

- The first principle of aesthetic mechanics, data, resource methods, and strategic management, which are greatly concerned with the formula of SEC, including all that are formatted in the deep thinking of algebra, geometry, and linear and simultaneous equations.

The inglorious understanding of statistics as a scientific gate is traceable to the lapses of material science. Statistics, both as a science and an art, is a dynamic human endeavour that is figurative, calculative, and concerned with data collection and collation; this is why it does not sever its relationship with philosophy and applied mathematics.

A modern statistician named James Young states that statistics is a human endeavour that is scientific and practically oriented in the deep thinking of data, numbers, and tables of values, which can be protected and preserved. In his thinking, Mr Young is of the opinion that the province of statistics, if properly allied with philosophy and mathematics, can be utilised in determining community, national, and international population censuses. He believes strongly that statistics can be utilised in analyzing the fortunes of businesses, including the failures of industries.

It is here that a statistical psychologist, Pauline Part, must be reengineered to provide a database for commerce, industry, technology, science, and economics. Professor Part upholds strongly that philosophy, with its purpose-driven portion with statistics, will help human engineering achieve a technological prowess.

Based on these facts, which are real in the realism of philosophical statistics with their technological impact, I present the following scientific advice for the use and application of the one-world family.

- Philosophical mathematics must utilise the enabling power of psychological statistics.

- The database of statistics at any point in time must be consulted to perform its task as a strategic and dynamic manager.

- Industry and institution must understand the networking concepts of statistics as an information collector, as a science, and as an analyst. This is why its functions must be taken seriously.

- Resource institutions, including universities, must invest in the philosophy of statistics in order to access the perfection in its deep thinking as an information technology.

- Technology must be purpose-driven in establishing statistical laboratory, which can be defined as a philosophical museum.

- Science must utilise all the data and figures that form the purpose-driven nature of statistics as a fulcrum of structure of analyzing scientific principles.

- Statistical principles must be energized in accessing what is contained in all the universal websites. Both commercial and individual websites must be structured with the opinion of philosophical statistics.

- Population figures must be original in the best thinking of statistics. Population instruments must be factual with statistical figures.

- Nations and multinational institutions must understand the scientific institution of statistics in relation to formulating principles, which will help them to appreciate that the world remains reengineered with the fabric and dependable features of statistics.

- Medical engineering at all levels must be structured and at the same time be purpose-driven with the policies and principles of statistics.

- Industrial philosophy and its psychology must be philosophic and statistical in the use and application of scientific technicalities.

- Individuals must be prepared at all times to compute the statistics of their computations with a view to creating a statistical balance sheet, which will help them with analysing the future of tomorrow.

I hold the view that website creation with negligence to statistical philosophy must certainly produce pores that will result in technological viruses. These facts are structured with the best objectivity of balanced thinking, understanding, and dependable knowledge. A look at philosophy and applied mathematics will reveal the scientific understanding of statistics from the existing institution of these practical endeavours.

It is a pity that humans are not driven with statistical ingenuity. The high negligence of philosophical psychology has remained one of the greatest woes of human dynamism, and the consequences of this are enormous. It is clearly shown and written in the faces of science and technology that this is why the technological institution only succeeds in giving students the ray of statistics without its philosophy and practicality.

In my deep thinking as a scientist, the urgent necessity of reversing this trend remains one of the greatest challenges of philosophy, including being the greatest need of statistics. Obviously, the opinion of mathematics and core and applied geometry cannot differ from the deep thinking of statistics and mathematics.

Professor Ebenezer Bill Gate of Harvard University has constantly maintained that the concept of globalization without involving the scientific thinking of statistics will certainly jeopardize human progress. He advocates strongly that the use and application of applied mathematics with philosophy must be purpose-driven in the best objectivity of involving statistics, which holds the key to ascertaining the purpose of information, its collection, and the principle of science with its analysis. This gentleman is working to save the world from another chaos and calamity of un-statistical materialization.

It is on this note that I boldly say that because mathematics is symbolic, practical, and informative in the use and application of figures, the numbering objectivity of statistics as an information databank must not be neglected. It is here that the concept and opinion of philosophical mathematics is practically quantified in the best thinking of how, why, and what, including the way forward by which the universe can be statistically managed with the use and application of philosophical economics, mathematical understanding, and psychology with econometric prowess. This chapter can be defined and considered as a bankable databank.

CHAPTER 5

WHY IS ENGINEERING AS A MATERIAL SCIENCE GREATLY INTERESTED IN THE USE AND APPLICATION OF STATISTICS AND APPRECIATING THE PHILOSOPHY OF SCIENCE?

It must be explained that engineering as a material science is focus-driven in the use and application of statistics. This is why all that is conceptualised in the philosophy of science cannot separate its diversified agenda from the information collective of statistics.

As defined by different encyclopaedias, statistic is interested in analyzing, documenting, and proofreading. This is why engineering is purpose-driven in the best fact of statistics.

In the illumined work of Thomas Edison that is designated to be the foundation of civil engineering, the science and art of statistics is philosophically driven in the best objectivity of facts and figures. One of the cardinal errors of material science has remained not being purpose-driven with the use and application of philosophical mathematics. This unholy error can be considered as a monumental massacre of ideas and creations.

The philosophy of science is statistically dependent on the use and application of ideas and creations. For example, a civil engineer is always seriously driven with the use and application of statistics, data, roadmaps, research methodology, and qualitative and quantitative analyses. This is why material science at all levels greatly affects the objective of statistics naturally and scientifically, from the point of empathy.

Simultaneously, polymeric science, with its engineering dynamism, is practically driven with the use and application of statistics, algebra, geometry, and further mathematics. What is considered as engineering mathematics with its mechanics cannot be accepted as a dependable and organized science if the illumined and conceptualised knowledge of the philosophy of science is not configured and directed as being and becoming the originality of its agenda. It is here that the wonderful contributions of Albert Einstein, Michael Faraday, Chancellor Mitsubishi, and others can be regenerated and respected as being in the best objectivity of engineering dynamism.

Donald Emengard and Euclid were practically concerned in the philosophy of science and the philosophy of material dynamism. This is why a look at the monumental contributions of statistics, in furthering the concept of engineering as a material science, will practically appreciate the following.

- That engineering as a material science is tactfully interwoven with statistics.

- That the functions of statistics are to provide an acceptable databank, which is very important and must use engineering as a practical material science.

- That the application of philosophy as a science cannot be galvanized to achieve its goals if statistics does not collect its data, which is used for information in the use and application of material science.

- That the science of philosophy and mathematics is practically galvanized in the best opinion of engineering science.

- That it is simultaneous to state that EMS, which stands for engineering as a material science, must equate with PS = SS, which can be cosined to represent engineering mathematics.

Philosophy as a science must certainly be purpose-driven in the best objectivity of statistics, which can be defined as a data collector and information storage. This is why statistics at all levels is determined and dynamic in the best objectivity of reserving the formats of engineering mechanics.

It must be stated that statistics and its province is greatly interested in compacting all the achievements of medical engineering, medical laboratory, and medical databanks. In the illumined words of Professor Andrew, who is designated as the father of modern anatomy, the science of engineering must be philosophical in the best objectivity of statistics.

It is unfortunate that the Internet era, with its un-globalized ideas, are not statically driven; neither can the creations be determined in the best objectivity of statistics. For example, globalization is all about familiarization of ideas in the use and application of statistics—but unfortunately, this familiarization is already having an eroded and un-engineering stigma.

This is why I am attempting to unfold an information asphalt that will help the province of engineering as a material science; philosophy as a core scientific agenda; meta-logic with its objective reasoning; and psychology, logic, and core mathematics with their fundamental principles and policies, including the onus and statistical philosophy as the mother of sciences.

Designated as the way of perfection and purity, the greatest concern of philosophy is to create a balance between the achievements of engineering as a material science—including statistics as the cheapest mathematical model, which is extensive and pragmatic in modulations and data processing, and is research focused with its vision, tactful management, and objective reasoning with information data. Its synchronous objective as a science must be instituted in the best use and application of $PS = PE = PM = PP,P$ which can be purpose-driven in accepting what the philosophy of statistics is all about given that the format is designated as the anatomy of engineering as a material science.

It is important to appreciate that the desire of science to give the world a material cosmogony can be argued as being imperative and non-imperative. This is why this chapter can be defined as an engineering polyester, simply because all that is contained in engineering mathematics and statistics. Quantum explanation with its electrifying dynamism is technically propelled to protect a globalized VSAT and a universal voice-box with its voicemail. The world of science is greatly reengineering its resources with methodology in order to create a standard statistical formula, which can be represented as SSF.

Professor Piro, who is known as a determined astronaut, saw all that is contained in statistics in respect to its use and application defined by astronomical science as a world of its own. That world can only be appreciated with the use and application of SSF with PS, PE, and PM, including galvanizing the philosophy of PP, which is earnestly quantious in the best application of engineering mathematics. Engineering mathematics as a material science is tactful in creations, production mechanics, mathematical hibernation, and philosophical intake.

There is an urgent need to recommend the following, which will help the world of material science in lifting the problems it has accumulated over the years of not caring to know the secrets of science, the sacredness of philosophy, and the benefits of wisdom, including being able to provide a dependable statistical databank that must be purpose-driven in the best dynamism of creating a balance.

- All aspects of material science such as engineering mechanics must be rooted in truth.

- The concern of environmental engineering, which is greatly focus-driven in creating a societal and sociological order, must be accepted, propelled, and galvanized as a way by which environmental hazards, with its model and technical hazards, can be solved.

- All aspects of material science with its engineering mechanics must be driven with a modest standard.

- The use and application of philosophical mathematics with its universal scientific incubator must not be neglected as a philosophic incubus.

- Website developers must be purpose-driven in appreciating that material techniques must have a standard format with the use and application of methodological voice box.

- The philosophy of online real-time must be determined by the wisdom of the Internet hibernation using what is defined as the opinion of all.

- The concept of electronic logic must be greatly utilised in accessing and appreciating the wisdom of philosophy as a science.

- The wish of logic as a science of honest and dependable reasoning must be purpose-driven in creating concepts and ideas, which must work in synchronous agreement with the objectives of electronic logic.

- Because material science is greatly dependent on the purpose-driven creations of engineering, every aspect of it must be dependent and pragmatic in achieving a globalized goal, which must aim in reducing poverty, enmity, and more.

It is simultaneous equation in its luminous understanding that can examine X^4 to mean $X \times X \times X \times X$. The honest converse of this can be appreciated as a philosophical tangent which, when properly understood, will help us to know that engineering at all levels is practically eloquent with the creation of philosophical mathematics.

It is in the interest of this chapter to register and record that the universe is a simultaneous alchemy, practically working in the use and application of engineering as a material science, as an actor, and as the anatomy of statistics, which works to blend the originality of its concept with philosophy as a science. It is here that the inclusion in its appreciation can best be defined as being and becoming the way of honest and practical thinking and practical creativity.

Every aspect of human endeavour is practically driven with engineering creativity, without which the honest thinking of philosophical mathematics as the alchemy and philosophy of all stands to be defeated. This is why this chapter can be defined as what lies ahead of engineering as a material science.

CHAPTER 6

How Can Statistics Be Used in Reducing the Errors of Engineering as a Material Science?

A look at the explanation of statistics with its compact definition must be appreciated as a work that is in synchronous agreement with detects and directives of engineering mathematics. Statistics are part and parcel of engineering mathematics, engineering mechanics, and engineering dynamics because it has a lot of inputs that have not been known by the province of engineering science. Globally, science has already degenerated the thinking of man through the use and application of false data, which technically investigated against the use of statistics. This is why the functions of statistics as a databank, including being an information archive, must be respected, honoured, and utilised in salvaging further errors of engineering.

It is necessary to explain that material science has produced so many products to the benefit of man, but it has not well understood the networking concepts of statistics as a stockbroker, a stock taker, and the stock bank, which understands the relativity of material phenomena. This fact is why the dynamic powers of statistics must be utilised in reducing the lapses of engineering as a material science, as a human endeavour, and as a dynamic provider of resources through what can be defined in this context as a technical and empathic conversion process.

A look at the creations of engineering as a material science will certainly let one understand the lapses. As Professor Osborne stated, engineering has succeeded in transforming natural objects to material ones, but it has not succeeded in reconverting these natural objects to their natural, compact originality. This is part of the onus of chemistry and physics. A look at the contributions of statistics as a mathematical bank must understand the networking of philosophy with mathematics. A look at the functions of the cost mechanism of statistics must certainly appreciate that it works with business mathematics, business dynamics, business focus, and vision. It is here that every aspect of human endeavour is being statistically driven with the use and application of management at all levels.

A look at the concept and constitution of project management with its technicalities will certainly lead one to discover the enormous contributions of statistics, particularly in the following areas.

- Statistics provides effective management of materials.

- It reduces cost, including ensuring that cost mechanism is applied to the letter.

- It is used to counteract the effects of inflation, which is one of the greatest problems of engineering simply because projects are determined by the philosophy of cost mechanism.

- Statistics is used in providing data and information, including being used to develop engineering vision with material focus.

- It is a mathematical think-tank, which is used in advancing the cost of data processing and technical development. Most of the time statistics is best defined as the stadium of engineering laboratory.

- It helps engineering as a material science to develop new products through the use and application of structural and mechanical dynamics.

- Statistics is used in ensuring that balance exists between the provinces of engineering and other aesthetic sciences.

- The concept of feasibility studies, which has to deal with project development management and how these projects could be consolidated, does not lack or neglect the mathematical contributions of statistics in ensuring that data at all levels (including the inputs of computer hardware) are practically galvanized.

- Because material science is focus-driven with stocking items and products, it can only be protected through the use and application of statistics, which can best be defined as a managerial gate, including being a scientific VSAT that has remained unavoidable in the conception of projects.

It is important to explain that the lapses of engineering as a material science greatly disturb the province of philosophy. By looking at the principles and policies of engineering, one will discover that this policy has contributed to the non-performance of statistics as a mathematical data.

A look at what happened during the mathematical era of Pythagoras and Plato in a community will certainly let one observe the lapses and ingloriousness of engineering as a material science in order to determine why Aristotle, in his luminous wisdom, eloquently stated that engineering at all levels and faculties will certainly be a force that will demystify the thinking of man, simply because its creations are not philosophic.

The mathematical computation of figures and data, including its utilisation, reveals the powers and forces of statistics, including how these creative ideas are used in favouring the wide province of engineering. For example, such engineering faculties like mechanical, civil, meta-logic, architectural, polymer and textile, structural, atomic, and nuclear, and including the newly invented engineering that is at the proximity of meta-science designated as wave length engineering, is always on the input and output of statistics. This is so because the universe is technically driven by the wisdom of philosophy, mathematics, logic, and engineering concepts with their wide technicalities—but it is on record that statistics cannot give the world a globalized cosmogony if it is neglected.

It is important to explain that in writing this book for research fellows, research institutions, and professional bodies including postgraduate colleagues, I was tactful in the opinion of engineering as a material science, human endeavour, and transformational asset with its mathematical and conventional objectives. This is why all efforts are reengineered towards ensuring that the lapses of engineering as a material science can equally be contained, controlled, minimized, and corrected with strict adherence to the rules and regulations of statistics, which in the opinion of mathematics can best be defined as the anatomy of project concepts and consolidations.

A work to determine the immeasurable contributions of statistics must certainly appreciate that the creation of engineering as a material science, which does not respect the rules and regulations of statistics, must certainly dehumanize the acts of all creations. This work is technically and practically written at the proximity of engineering and mathematical website with the use and application of statistical balance, which must be guided by the wide wisdom of mathematics through the use of geometry, algebra, ratio, simultaneous equation, linear equations, and what Euclid gave the world as a balanced and dynamic anatomy of statistics, which is a glorious and important factory that helps engineering to perform its duties using data and information as bankable assets.

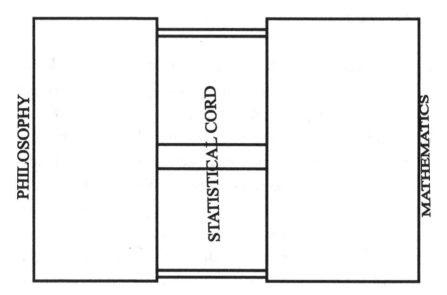

Fig A:

This drawing shows that statistics has always been there to provide a balanced service between philosophy and mathematics, which engineering always uses as a material science to perform and perfect its works.

CHAPTER 7

WHY IS STATISTICS DEFINED AS THE DATABANK OF PHILOSOPHICAL MATHEMATICS AND THE SCIENTIFIC BASIS OF ENGINEERING MATHEMATICS?

It is important to explain scientifically (and with relative objectivity) the mechanics, including the orderliness, that have designated statistics as the databank of philosophical mathematics. This has become an urgent necessity because all that is contained in the anatomy of statistics is essential in saving information with a view to providing it when needed. This service, which can only be performed by statistics, has greatly obtained all the data that engineering as a material science, including mathematics, utilises in formatting principles, polices, and guides. Statistics has always remained a databank of philosophy, including being and becoming a midwife to mathematics. Unfortunately, the high level of material understanding of philosophy and mathematics has greatly debased and demystified the linkage of mathematics to philosophy and philosophy to mathematics.

It is important to explain that the functions of a base that is in synchronous agreement with mathematical philosophy is to provide a panel and a structure that can serve to objectively enabling and ensuring that ideas and concepts are technically galvanized as a structure. It is in this aspect of human endeavour that statistics has always being defined as the databank of mathematics, including being the aging basis of engineering mathematics. A work to explain the objectives of a databank must not fail to utilise the following as the most important supporting reasons.

- Statistics is greatly interested in the usage of figures, particularly those that understand the synchronous language of mathematics.

- It is a great databank that warehouses information with a view to storing and retrieval, and most of the time it gives both a philosophical and mathematical analysis of such information.

- Statistics is greatly skilful in the usage of records. It uses the philosophy of the records to provide a scientific enabling for the utilisation of mathematics.

- Engineering is compact and configured in the usage of material science, and so statistics is always being involved in the best protection of these materials by serving engineering as a guide. This is why most engineering formats, principles, and orders are simultaneous in the best base of statistical arrangements.

As philosophy works to achieve perfection and purity, it utilises all the standards of statistics in ensuring that excessive philosophical recesses can be reduced in order to ensure that everything that has to do with standards, records, formats, codes, morals, aesthetics, and theurgy, is statistically driven. In the best words of Italian, Leonardo da Vinci, every data cannot be appreciated without involving statistics. This is why engineering mathematics considers statistics using the anatomy of the aesthetic architecture.

In the best opinion of data banking, which is philosophically quadrant, statistics is regarded as a server; this is why our present material age, which is computerized with a view to achieving globalisation, is arithmetically configured in using statistics as a moderator, engineering consultant, and scientific recorder with its bankable objectivity.

A look at the synchronous arrangement of further mathematics appreciates the networking arrangement of statistics as an overall databank. A work to explain the databank that is contained in AB = C and D = AD can simultaneously utilise statistics to explain the data as easily as in the objectivity of A = B = C =D, using A as a databank while B, C, and D can simultaneously represent the networking concept of philosophy, mathematics, and engineering.

It is important to explain that statistics utilises the deep thinking of logic in explaining the engineering contents of its postulation. It is unfortunate to observe with scientific explanation that material science, and particularly engineering, has relegated the functions of statistics to the background. This is why the extent and the esteem of the work is to reawaken the mathematical philosophy and dynamics of statistics. If this error is not corrected, it will greatly spell doom for the world of science, thereby making it difficult for the millennium goal to be achieved and for the world to exist as a globalized family.

For the record, if the world can be neither accessed nor appreciated by the ephemeral enmitied mathematics, engineering, and material mathematics, then it will certainly consult and constitute a rational anarchy. The best way by which humans can live is to make philosophy a journey with a view to moderating the thinking of mathematics as an engineering technicality.

This chapter is written to present with dynamic objection that any effort of science and technology, which relegates the bankable functions of statistics to the background, must certainly create an imbalance that can be technically contagious and scientifically chaotic. To appreciate the magnanimity of this chapter, it must be appreciated that science is always thinking for the welfare of the universe, whereas technology is purpose-driven in creating new ideas. This is why the cheapest aspect of engineering mathematics, designated as statistics, has always provided these institutions with simplified concepts with which all the galvanized concepts can be retrieved, restored, compacted, and configured into a single-unit estate designated as a databank.

A look at the functions of mechanical, civil and structural engineering, and aesthetic and metallurgical engineering—with metallic concepts and architectural engineering, which exists on its own as being and becoming the mathematical thinking of scientific alchemy, including polymer textile, chemical and petroleum engineering, geophysics, and geology—must certainly observe that these human endeavours are purpose-driven with the use and application of philosophical mathematics, including standing on the facts and figures of mathematical statistics.

In a wider opinion, it is not gainsaying to define statistics as the simultaneous logic of engineering mathematics, principles, and policies. It is in my best objectivity to appreciate that $1 + 1 = 2$, $2 + 2 = 4$, $4 + 4 = 8$, and $8 + 8 = 16$, all existing on the ascending order of statistical philosophy. In the same ratio, $16 - 8 = 8$, $8 - 4 = 4$, $4 - 2 = 2$, and $2 - 1 = 1$. The objectivity of this format reveals that statistics is a databank which, when utilised with scientific appreciation, will certainly network all the creations of engineering mathematics to perform its functions with the use and application of philosophical mathematics. This is why this chapter can best be defined as understanding statistics in the best objectivity of philosophical mathematics.

CHAPTER 8

A PHILOSOPHICAL OVERVIEW ON THE MATHEMATICAL RELATIVITY OF STATISTICS AS THE CHEAPEST ENGINEERING MATHEMATICS: THE CONCEPT OF SHORT METHODS

Statistics exists on its quantum as the cheapest aspect of engineering mathematics, which is not understood or utilised by the uninformed experts. It must be stated that statistics utilises the concept of ratio, data, and short methods. These are its simultaneous agenda to achieving a moderate concept, which will ease all the difficulties in engineering as a material science. The concept of my short method is specifically ignited in the use and application of statistical logic with its mathematical and philosophical enabling, which is information on VSAT anatomy and utilises mathematical statistics on the aspect of mathematical science.

If one looks at the illumined Pythagoras and his simultaneous and remainder equations with their theoretical formats, one must certainly agree that the incubation originated from statistical method. By this I mean the process or procedure on which there could be an asset and harmonized order. This strongly inspired me to review the elongated mathematical engineering and mathematical philosophy, including statistics as the cheapest mathematics, in order to provide and produce what could be described as the cheapest statistical mathematical method.

It must be stated that all that is contained in simple interest and compound interest, including the elongated interest ratios (which the banks and financial institutions have adopted in duping their clients), do not synchronize with the original point of statistics as a databank for commerce, and being and becoming the origin of commercial industry. When it is stated that commercial industry cannot globalize the universe as a one-unit family, it is simplified that commerce and industry have rejected statistics as an industrial and commercial moderator.

It is on this premise that accounting principles and policies have greatly remained stagnated, while industrial and commercial growth have myopically been driven to create a lot of economic and structural problems that have greatly demystified the growth of the social order. By demystifying the social order, I mean the inglorious neglect of the principles of sociology, psychology, culture and tradition, government and administration, strategy, and diplomacy. It is important to appreciate that statistics remained a co-director and co-coordinator in assessing and handling facts, figures, data processing, and engineering technicalities. This is why other aspects of dynamic endeavours recognize statistics as the consummate and dynamic warehouse of data.

The purpose for the concept of the cheapest method is to achieve the following.

- To reduce stress.

- To reduce the difficulties associated with engineering mechanism and its technical difficulties.

- To strengthen and appreciate all information as being the way by which the concept of the millennium goal can be achieved with scientific achievements.

- To provide a standard ratio that can be utilised as a national and international format.

- To increase and encourage the use of figures both objectively and logically.

- To appreciate engineering mathematics as the best calculative concept that appreciates how philosophy is related to mathematics, including how mathematics exists as the illumined thinking of statistics. It also has a view to using statistics to harness the function of data, information studies, and records.

The science and concept of the short method is simply engineering models and mechanics. Engineering as a material science is encumbered with a lot of difficult mathematical principles and policies, and that fact has greatly frightened students, experts, and professors. There is need to straighten the ugly trend in order to make the overall province of engineering know with appreciation that statistics can best be appreciated when seen from the angle of a moderator.

In this premise, the cheapest method will help us to appreciate that $2 \times 2 \times 2 = 8$ and that $8 \times 2 = 16$. In the same vein, $16 \times 2 = 32$. Thus, 2 and 8 can be utilised to reduce 32. It can be mathematically argued that the ratio of 32 can be a simplified method of 2×16 or $(8 \times 3) + 8 = 32$.

A look at this format has greatly shown that unnecessary arguments, mathematical elongation, and engineering difficulties have been reduced using statistical method with the concept of additional principle with additional methodology. It can equally be argued that $2 - 1 = 1$ and $1 + 1 = 2$. Statistics, in its cheapest method, is always concerned in ensuring that mathematical uncertainties are removed and that engineering difficulties can equally be redressed with a view to using the perfection of philosophy as being and becoming the incandescent technology. It is on this note that the concept of science and technology cannot be utilised in achieving the desire of statistical globalization if unnecessary difficulties (which are aged and incubated by the uninformed because they have constituted ingloriously in retrogressing statistics as the cheapest in engineering mathematics) are not addressed.

It is important that all scientific enabling, with their philosophical modes, are not in a balanced and agreeable data with the retrieved engineering format of statistics. My aim and purpose in developing the statistical method is to bring to bear a simplified method that can be utilised in appreciating the functions of philosophical and engineering mathematics, as well as creative science. This is why the opinion of statistics has remained and configured all information as a databank: it will help in appreciating that the world can only be globalized if all units of human endeavour appreciate that the world is logical and rational in the best objectivity of statistical mathematics.

As Albert Einstein stated, science cannot exist on its own if there is no statistically derived basis that will relate the functions of each other in assessing and ensuring the universe is not reduced to the unknown thinking of material science. Einstein's argument is understood to be a world-class system commonly recognized as a wakeup call by which the lapses of engineering technology with mathematical know-how can be appreciated.

Unfortunately, the incubus that orchestrated his era to be myopic in the knowledge of science and technology, particularly engineering, is still revolving with mathematical rotation to our present era. It is on this note that statistics as the cheapest engineering mathematics—a moderate scientific method and a philosophical and technical apostolate, including being the simultaneous anatomy of mathematics—is said to exist on the use and application of methods for information warehousing with its technology. This is why I strongly feel that the science of statistics must be well explained as being and becoming the way of human logical ratios.

The opinion of statistics as the cheapest engineering mathematics is to provide technical and compact mathematical insurance by which all facts and figures can be galvanized as existing in a united but extensive mathematical estate.

This is why my overview can best be defined as being the foremost engineering think-tank, including providing a philosophical insurance that will help all endeavours in accessing technically and practically the proximity of statistical dynamism, which in earnest is cheap and purpose-driven. The concept and idea of the short method is to reduce waste in order to encourage the enabling concept of occupational growth, an invention with engineering mechanics. Thus, this work can best be defined as how and why engineering can be used in assessing and solving the problems of material science using philosophical statistics as an organized and compact databank, which can be appreciated as a monumental moderator.

PART 4

INTRODUCTION TO SOME PHILOSOPHICAL PRINCIPLES WITH MATHEMATICAL THEOREMS

"Pure mathematics is, in its way, the poetry of logical ideas."

—Albert Einstein

Quotes

❖ *"For a noble man with a noble idea, Philosophy must be his journey. For a living soul, who is on the ascent of philosophical empathy, such a soul must desire to achieve the concept of purity and perfection."*

❖ A. U. O Aliche

❖ *"Thoughts create a new heaven, a new firmament, and a new source of energy from which new arts flow."*
—Paracelsus

❖ *Satisfaction lies in the efforts not in the attainment. Full effort is full victory.*
—Mahatma Gandhi

❖ *"If you hate forgiveness, stop saying our Lord's Prayer."*
—Big Ben, a philosophical icon who educated the universe in the 17th and 18th century

❖ *"The mind is not a vessel to be filled but a fire to be kindled."*
—Plutarch

❖ *"Don't make us make imaginary evils when we know we have so many real ones to encounter."*
—Oliver Goldsmith

❖ *"Being broke is a temporary situation; being poor is a mental state."*
—Mike Todd

❖ *"A man can fail many times but he isn't a failure until he begins to blame someone else."*
—John Burroughs

❖ *"Politics and the pulpit are terms that have little agreement."*
—Edmond Burke.

❖ *"To be kind to everyone you meet is fighting a hard battle."*
—Plato

❖ *"If philosophy is the bread of the soul, I will ever desire for the eating of that bread; if it is the only wine that can stimulate an objective desire in me, I will always make that wine part of my living bread. But if it is the way to heaven, I will rather praise the wisdom of the gods simply because all illumined institutions exist perfectly and purely at the proximity of philosophical wisdom with its illumined dynamism."*
—A. U. O Aliche

It is important to present what could be considered as would-be ageless but dynamic philosophical principles with a pragmatic and scientific mathematical theorem, which is known by the world of knowledge to be responsible for the consummate exaltations of such gurus like Socrates, Plato, Pythagoras, Einstein, Euclid, Michael Faraday, Emmanuel Kant, Aristotle, Max Miller, Herbert Spencer, Desmond Oliver, and more. Such great theories and theorems were practically perfect in the use and application of scientific mathematics with its illumined philosophical protection and perfection. A look at the dynamism of philosophical mathematics will certainly appreciate that the soul of the dead is still controlling and monitoring the spirit of the living. This is why the science of the continuity principle is technically realistic in educating the universe that the world is a globalized family. It must be appreciated that the Orientals were great super-geniuses in sowing what the modern man is utilising as a ladder of success. The dynamic ecstasy of the universe is purpose-driven in the best utilisation of historical, mathematical, and philosophical ontology, including what can be defined in this philosophical mathematics as the ontology of Mother Nature.

This introduction will explain and exhibit how and why mathematics is quadrant, whereas philosophy is quadratic. Their parallel and oblong natures can be defined as consummate curvatures that are realistic in producing facts and figures which, when scientifically certified and approved, can be accepted to exist as principles, policies, theorems, and methods. This is why philosophical mathematics can be defined in this introduction as the simultaneous agenda of the electrifying power of the universe.

If a theorem is propagated and presented that $1 + 4 = 5$, the same theorem can conversely be changed to mean that $5 - 1 = 4$.

In the objectivity of simultaneous equation with its electrifying remainder theorem, it is practically understood that mathematics is figurative, dynamic, objective, pragmatic, and constant with activities. This is why honest mathematical method has always remained a scientific and technical mandamus. By mandamus in this context, I mean a mass of ideas, a factory with fathoms, an architect with lintious and luminous structures. This definition is why a work to look at the principles and theorems of mathematical philosophy cannot and must not elude the illumined propagations of Pythagoras, Albert Einstein, Euclid, and others. This context is going to propagate the ideology regarding how and why Pythagoras is quantious in his original QED, which has greatly baffled the world of knowledge, particularly the province of mathematics and calculus. A lot of people are not yet in the knowledge of appreciating that QED simply means "Quantify your ideas with Examination and examine your ideas for Development". It can equally be translated to mean the consummate confirmation of quads theorem with its scientific theorems.

Among all mathematical and philosophical geniuses the world has ever had, Pythagoras is the most blessed, and I define him as an immortal goras. By immortal goras, I mean one who is blessed with rational thinking. My illumined definition of goras is one who is in the original soul. Mathematics used the principles of philosophy to authenticate his mathematical perfections, which gained him the knowledge of his principles, policies, theorems, ideas, with lintious concepts. This is why an introduction of this nature can be designated as a philosophical mathematical abstract.

CHAPTER 1

A DEFINITION OF MATHEMATICAL PRINCIPLES WITH THEIR DIVERSIFIED THEOREMS

Mathematics—which is naturally oriented in the use and application of facts, figures, principles, and policies, including being the databank of research methodology and statistics as the cheapest engineering mathematics—functions dynamically with the use and application of philosophy with its consummate and purpose-driven principles. It is here that mathematics can be defined as a philosophical trailblazer, as an engineering anatomy including determining how the electrifying principles of science, with their balanced methodology, are transmitted in the reality of appreciating that $A + A = 2A$ and $B + B + B + B + B = 5B$. With this ratio, one can now understand the mathematical principle and appreciate it as living in the best objectivity of logical mathematics.

Principle as we know has lots of definitions, but mathematically it can be defined as a natural and believable action plan that can be utilised in assessing the quantum of theories using what is noted as inference, to explain how and why it is used in the best opinion of scientific and technological understanding. I argue strongly that the dynamic functions of philosophical mathematics must embrace the tedium of mathematical realities. This is why principles and theorems, if utilised at the objectivity of material science, cannot stand the test of time; neither can they be used in permanently underscoring the philosophy of mathematics, whose action plan is to achieve purpose-driven purity and perfection.

Mathematical principles are scientifically oriented in the best opinion of facts and figures, including harnessing the purpose of purity and perfection. In the illumined words of the goras, who is defined as a mathematical principle, the world of science and technology, including industries and commerce, cannot appreciate the nurtured and natural mathematical concepts if the electrifying understanding of mathematics is not gained using the perfect infinitude principle. The infinite principle is mathematically dated in accepting that theorems can exist as policies, while policies can equally be accepted as mathematical living laws (MLL).

It is unfortunate to explain that humanity does not understand the tedium and rhythm of mathematical principles. It is here that the compartmentality of theorems is scientifically galvanized in appreciating what is defined in the context as the futuristic nature of mathematical principles using the law of parallelism, notation, correlation, and rhythmic ecstasy with balanced interchange. This is why mathematics and philosophy are tactfully amalgamated as existing in the scientific objectivity of twin and twined orbits. In earnest, mathematical theorems use philosophical principles; philosophical principles equally use mathematical principles. Philosophy and mathematics can synchronously be illustrated in the best understanding of Mother Nature as PH, where PH = MA. It is here that the mathematical relativity of commercial theorem becomes a triangular truism when the formula of T + T + T is mathematically utilised.

In earnest, mathematical theorems are usually log-jammed, and they can only be separated, protected, and proven using philosophical input and output systems. It is in the scope of this book to explain that the purpose for which mathematical principles are being utilised to access the use of theorems was tactfully Euclid—the Oriental mathematical apostolate that was inspired and exposed to know that philosophy cannot exist without mathematics, and that Mathematics cannot perform its functions without philosophy. It is here that the philosophy of mathematics is best explained in that orbital meta-logic compartmentalization. By meta-logic compartmentalization, I mean an organized compact disc or system that is tactfully united with mathematical and philosophical diversification. It must be explained that material science at all levels has greatly misunderstood the ethical logic of mathematics, including the Oriental logic of philosophy.

A look at Stroud's opinion on engineering mathematics reveals that he utilised the province of meta-logic, meta-philosophy, meta-mathematics, and meta-statistics. This is why that book, with its capital letters, is perfectly structured in furthering the course of "Advanced Mathematics". In my opinion as a lintious and fluorescent genius of mathematical philosophy, I do not hold any misgiving on the functions and contributions of material science, but I am greatly against the limited nature of its functions, including how these non-structured and unscientific concepts have deceived man, who is designated as the anatomy of philosophical mathematics simply because his actions are as dynamic as the universe. Man's desire at any point in time is to advance the course of knowledge. By conviction and convention, and by the use and application of what is permanent, man in essence metamorphosed himself in mathematics. This is why **MA** stands for "man" and for "mathematics", which in earnest opens the corridor of new knowledge on how and why MA and MA are always looking for PH, which is "philosophy" permanently designated as the science of all knowledge and the fathers of all great makers with permanent ideas.

Pythagoras, who is a founding pillar of mathematics, told the Grecian world (including the world of consummate knowledge)that his concept of mathematics was blended with the principles and policies that originated from the kingdom of Mother Nature in guiding man who, was the opium of mathematical and philosophical dynamism. A look at the engineering of this chapter will certainly appreciate that mathematical principles with man as a case study are the first principles of science, including being the anatomy of the infinitude method. This chapter is going to be technically inspired in narrating with objective conviction that mathematical and philosophical principles are equally the tedium of natural facts and figures. That aspect of the book is not a theory per se, but it is a objective and technical theorem of understanding.

At this point, it is important to establish the functions of mathematical theorems using Pythagoras' lintious triangular theorem as a case study. This light is under a monumental task to appreciate that 180 degrees inserted into a triangle is both the tedium of that triangle, the radius, the converse, the moment, and the feature of its nature. This is why theorems can be defined as mathematical letters, figures, and mathematical explanations of what is known as philosophical reality.

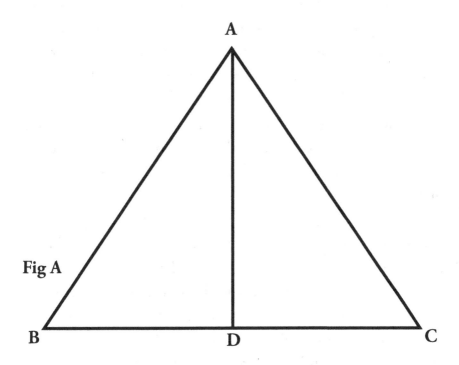

A look at the structure of this diagram will let one appreciate practically the mathematical and philosophical configuration of what the theorem is. For example, if ABD = ADC, then it is gained in the objectivity of mathematical theorems that AB = AC = AD, AB / AD = AD / AC, and BD = DC = CD = DB. The principled theorem of this triangle explains practically that ADB can be symbolic to represent "man", that ADC is tactfully symbolic in representing "mathematics", and that the bottom line approach the two parts of the triangle to be designated as the structure of the anatomy that lies in BDC.

The principles of mathematics are permanently galvanized in the law of constancy, whereas the theorem is purpose-driven in the law of convention and conviction. This explanation can be defined and accepted as the principles that underscore the philosophy of mathematics and the principles of philosophy. That is why a work on this nature must always be accepted as what can be designated as mathematical and philosophical principles. This method can equally be used in explaining the philosophy of a circle theorem, quadrant, and square, including what is accepted in the world of illumined science as a purpose and practical punctuality of philosophical mathematics.

There could be more to this, but it must be understood that the purpose for which **MA + MA = PHIL** which is philosophy, is borne specifically and technically on the podium of what is considered regarding how and why these principles and policies of philosophical mathematics can be used in attacking and reducing the influx of wrong scientific ideas, which have not helped man to appreciate that in him and only in him is he designated to know why philosophy is the orchestra of balanced knowledge. Mathematics is the Oscar of luminous principles with the use and application of its theorems. It is clearly understood that the two are twined with the use and application of mortal laws, which are galvanized in the objectivity of its technical relativity. The dynamic functions of philosophical mathematics will function as a scientific and technological guide, particularly for the use and application of honest knowledge, which is indeed the simultaneous and synchronous way of MA = MA = PH with more and much to be harnessed.

CHAPTER 2

WHAT ARE THE FUNCTIONS OF THESE PRINCIPLES WITH THEIR THEOREMS?

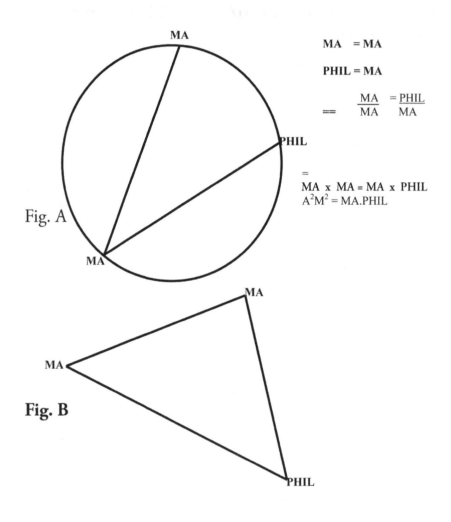

MA = MA

PHIL = MA

$$== \quad \frac{MA}{MA} = \frac{PHIL}{MA}$$

=

MA x MA = MA x PHIL
$A^2M^2 = MA.PHIL$

Fig. A

Fig. B

These diagrams explain that the purpose of mathematics in relating with man is to achieve purity and perfection, which is the law of constant relativity, or scientific naturalization of great concepts, which exists in the proximity of philosophical determination. At all times, mathematical principles are purpose-driven with blended facts, figures, and understanding. The illumined theorems are practically empathic with the will of philosophy. This system or concept holds the key in appreciating the anatomy of mathematics, the structure of philosophy, and the mechanism of the emulations of the two. One cannot determine practical and functional proposition of mathematical functions if the functions of mathematics are not explained and examined using its theoretical and theorem factors.

It is here that mathematics remains the mandate of man, the mystery of life, and the abstract ontology of science, including providing both rational and irrational factors, figures, data, and statistics. It includes what is designated as a scientific and Internet enabling, by which the mathematical functions are galvanized in the logical province and scenario of online real time.

It is important to explain that the part and parcel of mathematical functions are dated in the use and application of motion, emotion, waves, light travelling and transference, thermal expansion of objects, radiation, convention, and compression. There is a need to explain that mathematics is a factor of functional mechanism that furthers scientific and technological enabling by which creations, particularly the world of commerce, are structured in the objectivity of its data with statistics.

When the likes of Isaac Newton, Euclid, and others originated their inventions using mathematics as a keyword, they were only blending the mathematical postulation of Pythagoras in order to realize the functional mechanics, including the functional dynamism. It is here that Stroud and his aborigines designated engineering mathematics as the anatomy of the frequency of mechanics, motion, and emotion. I decided to present a scholastic agenda of the reality of mathematics using the enabling functional realities of philosophy. It is here that the principle with its illumined theorems can be galvanized both as a table of value, a databank, a philosophical eye camp, and an inspired statistical rendezvous that indeed can be defined as the anatomy of olive tree.

For instance, if light travels at the speed of 500 km/hr, it is believed that mathematically, $500 = 1$ m, and this formula can be utilised in knowing the ratio of the journey of light per second, including what is defined as the momentous and obsequious abstract, which was not explained in physics' particulars because it affects prism binoculars, the mirage on the road, and the reality of the reaction of light on hard surfaces. Philosophically this theorem, if galvanized to exist as a principle, can best be understood as the chancery and chancellor of honest principles.

A look at the Internet thinking and functions of mathematics reveals that its knowledge, utilisation, and appreciation cannot be well understood if the principles and theorems are not nurtured and philosophically completed. In this respect, the following points exist as the functions of mathematical principles using theorem and theoretical principles as measures of value.

- Mathematical principles are proof-rhythm, proof-oriented, and proof-valued with theoretical co-existence.

- They are the eye of all letters, particularly those that fall under algebra, geometry, data, statistics, information technology, introductory technology, and what is designated in this chapter as the XY concept of mathematical principles, which can be explained as $X \times 1 = X$, $Y \times 1 = Y$, and the ratio of this converse mathematical understanding, $X \times Y = XY$. To denote this formula as a factor of mathematical fathoms, $X = 1$ and $Y = 1$; conversely, $X \times 1 = X$ and $Y \times 1 = Y$. This is why $X + Y = Y + X$.

- The mathematical principles, with their theorems, are the centre point of dependable and balanced reasoning.

- They exist as key factors in accessing and appreciating all data.

- Their existence, along with mutual philosophical cooperation. serves the province of statistics and records by carrying out its functions with determined actuality.

- They hold the philosophy of further mathematics, engineering mathematics, technological mathematics, and being used to examine what can be defined as the anatomy of Internet mathematics.

- The functions are permanent and realistic; this is why they are always standing the test of time despite the modelling, the converse, and the ambiguity, including being used to underscore the philosophy of quantum and functional mechanism.

- The functions exist practically as balanced creation, designated to be a co-creator when understood from the point of man, mass, mathematics, and medium.

In this respect, the concept of the 4M sequences, as explained above, has remained the opium and includes a positive stigma that has greatly elucidated and illumined the obvious knowledge of modern thinkers.

The functions of the principles have always been designated as letters A to Z of the key factors. The way and manner mathematics rejoins with philosophy has always worried the uninformed; this is why a look at the existence of mathematics, man, mass, and medium signifies that the natural union of M/P is purpose-driven in obtaining and examining the electrifying structure of M + M + M + M, including M/P = M/M + M/M. Conversely, the tedium including the radius of this understanding can be logically understood when a drawn square is utilised to capital the magnanimity of these letters. This is why this book cannot be written below the present standard—there could be gained a lot of meanings on the functions of mathematical principles using their theorem and policy polyesters, and this is why philosophy is mathematically driven with quantious understanding.

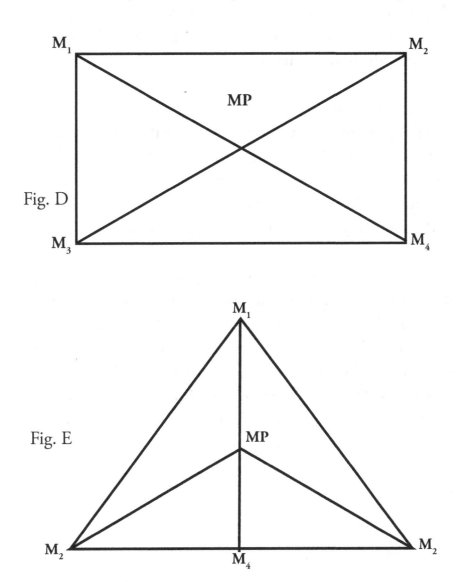

Fig. D

Fig. E

A mathematical, metric look with a philosophical understanding of the diagrams above reveals that mathematical principles are the only factors that can be used in providing mathematical theorems, which must exist as the determinant policy of philosophy. This aspect is why the significance of the chapter is borne out of the fact that the mind of philosophy is equally the soul of mathematics, and the spirit of philosophy is equally the heart of mathematics.

In reality, with objective and consummate concept, mathematics is real with its principles and theorems. Philosophy is equally real with a dose of achieving purity and perfection. That is why both are harmonious and electrifyingly driven in creating inventions using their functions, particularly those that are at the proximity of world star and World Wide Web.

The dynamic functions of philosophical mathematics are defined as the centre points of Internet hibernation, including living on the real-time concept of PHIL/W = M/W = M/M = M/M = M/M = M/M = M/M, which in explanation is revealed on the philosophy and Internet incubator of all inventions. This work exists as what is beyond mathematics and what is above philosophy, continuously and progressively; see the diagrams and expressions below.

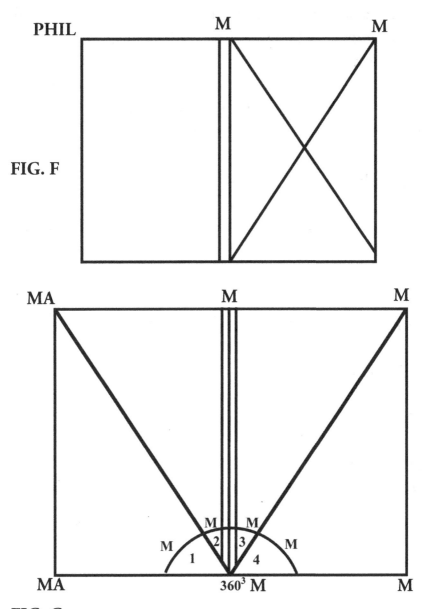

FIG. F

FIG. G

A look at the drawings above reveals the component of philosophical mathematical elements with their scientific existence and co-existence, which must be appreciated as the real anatomy of the practical functions, figures, and formats popularly known as the mathematical frequency of philosophy.

$$\frac{PHIIL}{MA} = F1 + F1 + F1 = \frac{PHIL}{MA} = \frac{F + F + F}{F + F + F}$$

Conversely

$$\frac{MA}{PHIL} = F = F = F$$

$$\therefore F = M + M + M + M$$

while

$$PHIL = F + M + F$$

$$MA \quad \textbf{M} \quad \textbf{F} \quad \textbf{M}$$

$$\frac{PHIL}{MA} \alpha \frac{MA}{PHIL}$$

CHAPTER 3

WHAT ARE THE OBJECTIVE PURPOSES OF THESE POLICIES?

A look at the functions of philosophical mathematics reveals that it maintains a parallel and compact dynamism between its principles and policies. This is why the principles and policies of philosophical mathematics are practically interwoven in producing an acceptable theorem. These policies must be utilised at the same framework where the principles are utilised. It is here that philosophy is practically driven with principles and policies that are in synchronous agreement with the anatomy of mathematical dynamism.

It is unfortunate to reveal that human evolution and civilization are not practically driven on the objectivity of philosophical mathematics. This attitude has greatly reduced the thinking of science and the philosophy of technology, which is responsible for all the errors material science has given to the un-globalized universe. It must be stated that every institution is driven with policies that can be defined as its tactful framework and that is responsible for the achievement of organized goals using the mathematical kit contained in these principles.

When it is stated that philosophy and mathematics are erratic, rational, and complex with un-amazing facts and figures, then it is evidence that there is a bridge in the functions of their polices and principles. This is the situation that empowered the anatomy of science and technology, and particularly that of engineering mechanics, in the dynamic use of laws with regular policies and principles.

A work to determine and appreciate with scientific assessment on the purpose of philosophical mathematics and its policies will certainly reveal the following.

- Mathematical policies are defined as its power point, and so philosophical principles are known as the legend of the power of mathematical polices.

- Both work to achieve a common goal.

- Both principles and policies, symbolically represented, are best defined as the framework by which facts and figures can be honoured and appreciated.

- They are the enabling institutions that are responsible for the investigation of research method, data processing, and statistical analysis, including being the institution that will certainly provide the needed scientific and technological ladder by which the economies of the world will galvanize the activity of the human race into a globalized format.

- Both principles and policies are best defined as the structure of science, the body of technology, and the anatomy of the universal terrain—including being the enabling force that directs the affairs of humanity.

- The objective purpose of philosophical mathematics in relation to the use and application of its policies is in synchronous agreement with the thinking of core mathematics and core science.

A look at the workings of methodology, polymeric engineering, aeronautic and space engineering, and everything contained in magnetism and quantum science reveals that its purpose and mathematical policy is to ensure that facts and figures are ascertained and harmonized in order to ensure that the purposes of policies and principles are tactfully united with the thinking of balance. For example, if $A \times B = AB$ and $AB \times CD = ABCD$, it could be represented that $AB = P$ and $CD = P$, which can conversely stand mathematically as $AB/P = CD/P$. This statement can equally be shown in the following tabular, which reveals $P + P$ (which represents principles and policies) should equally read $AB + CD$.

A look at the dynamism of philosophical mathematical principles will agree that its policies can be utilised as the objectives of scientific thinking of all that has to do with technical functions and practical relativity. This is why the Pythagorean theorem at all levels is mathematically driven on the philosophy of principles and policies. It is in the opinion of science, as contained in my work *The First Principle of*

Science, to reveal that scientific cosmogony and technological empowerment are neither feasible nor realistic if science and technology are not practically and tactfully driven on the best utilisation of policies and principles, ensuring that they are carefully amortized on the natural law of balance. That balance must include the philosophical laws of purity and perfection that understand and underscore the thinking of mathematics as a universal asset, with which the institution of facts and figures can be galvanized in ensuring that the principles and policies of these illumined sciences are utilised to create and protect a united province, particularly that which is in consonance with the thinking of science and technology. This reasoning is why the purpose of the policies is to ensure that the eminent difficulties in understanding these subjects are reduced and corrected so that the human race will appreciate that philosophy and mathematics are the dynamism of the one-world purpose.

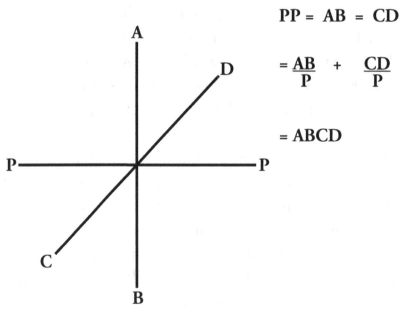

$$PP = AB = CD$$

$$= \frac{AB}{P} + \frac{CD}{P}$$

$$= ABCD$$

CHAPTER 4

WHAT ARE THE DIFFICULTIES IN USING THE PRINCIPLES AND POLICIES OF MATHEMATICAL PHILOSOPHY?

The academia, the research institutions, the world of commerce and industry, and the mathematical and philosophical segments of politics, administration, and government—including the social classes—are finding it immeasurably difficult to use the principles and policies of mathematical philosophy. These difficulties have greatly reduced human dynamism, including the progress of civilization in evolution with a view to attacking the growth of science and the purpose of technology, which are the logo of the people. The mysterious understanding of these policies has made the world reject and neglect the ominous wisdom that is contained in philosophy as the mother of science, knowledge, and mathematics

It is unfortunate to observe that material science has ingloriously dehumanized the objective utilisation of principles and policies. That high level of neglect has greatly reduced investment in the areas of research method, technical intention, scientific thinking and appreciation, and best understanding in the areas of geometry, arithmetic and aerial studies, geophysics and geochemistry, logic and meta-logic, aeronautic technology, and physics production with engineering know-how. This is why it is important to overhaul, appreciate, and understand the electrifying workings of philosophical mathematics' principles and policies. These difficulties have ingloriously orchestrated the incubations that are visibly seen in the areas of computer studies, information technology, mass communication, and infrastructural setup with a high level of sociological anarchy. This is why all efforts must be made to reduce these un-sacrosanct philosophical trends.

As chemistry and physics propel and project the thinking of organic and inorganic estates, including all that is contained in matter, philosophical mathematics is in harmonious agreement with the great minds in ensuring that there is a high level of growth in the areas of knowledge, particularly the core and harder sciences, technology, sociology and philosophy. It must be stated that these difficulties have remained monumental defeats that are attacking the human race at all corners of our ecosystem. Considering that the aim and purpose of philosophy is to achieve purity and perfection, and to ensure that the universe is governed with balance, mathematics as a figurative bank of objective cornerstone always graces the scientific harmonization of philosophical thoughts in ensuring that creation is luminous, purposeful, and dynamically driven. It is here that most of the difficulties can be itemized as follows.

- Lack of objective, balanced, and universal education.

- Inability of man to understand that because life is mathematical, our endeavours are equally philosophical.

- Universal instability, which is greatly directed with high level of man's inhumanity to man.

- The negligence of man in understanding that he is the centre point of mathematics, including being the anatomy of philosophy.

- Inability of the past eras to hand over the light to the well informed.

- Illicit transactions that did not encourage the areas of mathematics, philosophy, science, and technology.

- Lack of data, including universal formats, that will help all and sundry to appreciate that philosophical mathematics is the logic of the universe.

- Constant changes in scientific principles and policies, including facts and figures that are greatly infected with viruses.

- Inability of our civilization to respect and adhere strictly to the wonderful and illumined position of Thomas Edison, who greatly opinionated during his era that material science will certainly lure the world into an anarchy if technology does not follow or adhere strictly to the principles and policies of philosophy, mathematics, logic, and meta-logic, including what is considered as the mind of the Oriental wizards.

It is in the thinking of this work, and particularly of this chapter, to explain that philosophy and mathematics are interwoven and interconnected. This is why both are contained in the anatomy of Internet knowledge, including being the positive and objective stigma that is serving the human race as a vanguard.

It is here that I am greatly concerned about the mistakes and errors the principles and practises of material science are developing. A look at the way and manner Galileo, Albert Einstein, Euclid, Isaac Newton, Michael Faraday, and the mysteries that are portrayed by the philosophical space brothers reveals that all that is being contained in computers with electrical electronics is not greatly in agreement with the opinion of philosophical mathematics. This work can be defined as an advancement that examines the work of Stroud, K. K. Clerk, Durell, and Laurel of the immortal memories. In my opinion as a scientist, there is no separation between the principles and policies that are contained in mathematics with the anatomy of philosophy. This belief is why a great understanding of the two must certainly and arithmetically reveal that $P/M = P/P$, and conversely that $P/P = P/M$, and that P/M in the ratio of P/P will certainly give $P + P = M/P = P/M$.

In the wonderful and illumined understanding of geometric technicality, philosophical mathematics can equally be in agreement with principles and policies. It must be stated that P/M can equally help us to appreciate that in philosophy and mathematics, principles and policies are best understood as the frameworks in which what is contained in their anatomy can be assessed, appreciated, and utilised when man desires to know that every human endeavour is cardinally configured in the objectivity of philosophical mathematics. This reasoning is why the difference can only be removed when there is a united understanding that philosophy and mathematics are the symbolic institutions that advance our course, knowledge, and wisdom, because it is technically and naturally nurtured. This chapter is compact in ensuring that man understands himself as the fulcrum of philosophical mathematics.

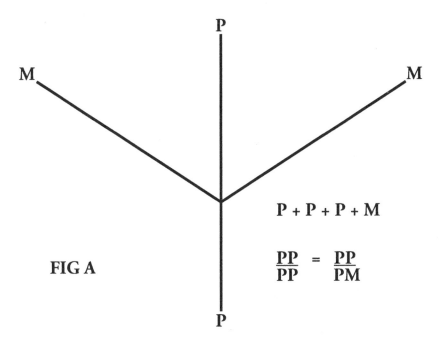

In the above drawing, it is scientifically established that the structure, content, and technology (which are all shown in a right-angle triangle) are configured in the thinking of philosophical mathematics. This is why a triangle, a square, a circle, a quadrant, a cross, a pyramid, and more are technically structured and balanced relative to philosophical mathematics, which is the reason why Pythagoras and Euclid eloquently stated that the structure of the human anatomy is scientifically configured on the balanced and simultaneous relativity of mathematics and philosophy.

CHAPTER 5

WHY ARE HUMAN ENDEAVOURS PURPOSE-DRIVEN BY THE OBJECTIVE FUNCTIONS OF PRINCIPLES AND POLICIES OF PHILOSOPHICAL MATHEMATICS, WHICH ARE THE ENABLING ORIGIN OF RESEARCH METHODOLOGY?

Since the dawn of consciousness, human endeavour is scientifically philosophic and technically mathematical. This is why their actions are practically rooted in the use and application of research methodology. Science gives the world a material cosmogony, and the objective function of philosophical mathematics is to partner with the creations of science in order to create an enduring civilization. Unfortunately, a lot of lapses are being created in the process of achieving this result. The consequences of these lapses have greatly stunted growth, development, and empowerment; this is why there is an urgent need to introduce what can be designated as a philosophical mathematical blueprint.

A scientific look at the objective functions of the principles and practises of mathematics and philosophical ethics will certainly reveal that research methodology can be analysed as the anatomy of creative inventions, including being the logical way by which philosophy achieves a clandestine goal.

The position of Aristotle in projecting and propagating logic as a consummate extension of philosophy was practically talented in instituting a mystified opinion that will help every human endeavour to be instituted in the wisdom of philosophy, including appreciating the wisdom of mathematics. At this point, the chapter will define what the endeavours, purposes, and anatomies of philosophy are; how it has achieved the purposes of purity and perfection; and how mathematics is configured in ensuring that objective calculations are practically certified as logical facts by which every human endeavour is practically certified.

In this context, "endeavour" means every objective act that has a technical and practical meaning. Endeavour can equally be defined as every action that is purpose-driven with objective methodology. It is here that the anatomy of an endeavour is purpose-driven with technical and scientific creativity, which involves research methodology and working to harmonize competence with objective dynamism.

If one looks at the definition of endeavour, one will certainly appreciate how every human action is purpose-driven with the objective functions and the principles and practises of philosophical mathematics, which aims at ensuring that there is always a practical action in every reaction, including a scientific reaction.

The way and manner that Peter Goras, who is popularly known as Pythagoras, exhibited a mystical endeavour which has not been equalled in the areas of mathematics reveals that he can best be defined as another rock—or better put, a figurative and mathematical fountain thrust. It is unfortunate that a lot of his mathematical creations, which where borne out of his dynamism in philosophic engineering, were directed to Euclid. That callous mistake could not stand the test of time, because Pythagoras lived a mathematically mystical life, which greatly endeared him to scholars, professors, and renowned institutions.

A work to determine his mathematical endeavour, which is frankly rational in creations, must certainly appreciate that the world is driven with the coefficient amalgamation of logic and mathematics. In this respect, human endeavour is purpose-driven with objective functions of the principles and policies of philosophical mathematics, which is the balanced origin of research methodology because of the following reasons.

- Every endeavour of man is a product of research practicality.

- Human endeavour at any point in time is both mathematical and philosophical.

- Since the dawn of consciousness, research methodology partners mathematically with the principles and policies of Philosophy.

- Every human endeavour is objective in providing satisfactory utility.

- Scientific endeavour is demand-driven with facts and figures. This is why it partners with the province of research methodology.

- Life is mathematical, and so every human endeavour is equally logical.

- At any point in time, creation can never sever its partnership with the use and application of objective functions of philosophy, mathematics, logic, and research methodology. This fact is why they are all configured as the balanced anatomy of objective technology.

- Human endeavour creates satisfaction with a lot of consequences; most of them are avarice, while a lot have produced monumental incubi.

- Globalization at any point in time works to harness the diversified human endeavour, by which there could be an enhancement in the objectivity concepts of millennium developmental goal.

- The purpose of every endeavour is to uplift human standard, which is in synchronous agreement with the visions of objective mathematics.

- As the computer age is tactfully and technically jet-driven with multiple facts and figures, it works to harness all that is contained in the objective dynamism of human endeavour.

- Practical technology is the logo of the people, and so human endeavour can be defined as the welfare of the society.

It is important to explain these points with facts and figures, because philosophical mathematics is essential in the employment of figurative concepts, including utilising these concepts to discover and ascertain the province of population and how the orchestrated problems that are involved in these endeavours can be amicably resolved. This is what informed my technical dynamism in looking beyond the thinking of philosophy and above the reason of mathematics. As research is scientific, methodology is technical. This is why the enabling institution of philosophical mathematics is achieving and harmonizing the material effects and disadvantages, which are associated with development, human growth, and civilization. It is in this format that philosophical mathematics is always there and is designated as a scientific watchdog.

CHAPTER 6

A Scientific and Technical Look at the Creations of Material Endeavours, Which Work Against Philosophical Mathematics

One of the critically inglorious creations of material endeavours is the lack of employment of philosophical mathematics, which is balanced in purpose, reasonable in appreciation, and pragmatic in input and output. This is why orphic philosophy opinionated that philosophical mathematics will provide the needed data by which the consequences of material creation will be dissolved.

The creations of material science that work against philosophical mathematics include the following.

- Lack of purpose-driven principles and policies.

- Lack of objective methodology, which is research-driven in empowerment and protection.

- Inability of material science to provide a scientific and technical locus, which can be utilised as a technological logo.

- Constant changes in principles, practises, and myopic understanding of the use and application of research methodology.

- Creation of multifarious departments, which do not utilise the illumined wisdom of philosophical mathematics.

- Myopic understanding of material laws, and the inability of material science to be purpose-driven with the use and application of these laws.

- Propagating the concept of unorganized data and statistics, and working to restructure the existing laws.

- Inability of material science to organize and understand the structure of the following scientific and technological sectors:

  1. Petroleum and petrochemical engineering.

  2. Polymer and textile engineering.

  3. Geophysics and geo-chemical engineering.

  4. Astronomical technology with its numerology.

  5. Statistics with the use and application of data and phenomenon.

6. Behaviours that relate to astral biology, physics, biology of the spirit, and astral and astrological insight into the spiritual lights of man, to name but a few.

Material science does not work to appreciate the understanding that is contained in the science of a bottom-line approach and that which is designated as desk and field appraisal. This fact is why material science encounters lots of monumental difficulties: because its philosophy neglects the illumined wisdom of philosophical mathematics. Pythagoras critically examined the creations of material science with mathematics, but it was in the province of Aristotle to scientifically express that meta-science, in its creations, has created a scientific melancholy by the usage of substandard materials, including the manufacturing of atomic weapons, which work to destabilize the universal economy. Judging from what is happening in our chaotic civilization, material science can be designated to have created the following problems.

- Lack of human relationship.

- Unorganized standards that do not produce permanent principles and policies.

- Inability of material science to practically depend on the balanced wisdom of philosophical mathematics.

- Material science does not understand the phenomenon and philosophy of research methodology.

- What is considered as scientific research methodology can be defined as an understanding of moribund phenomenon.

- Constant changes in formulas, formats, data, and statistics are responsible for the inglorious creations of material science.

- Inability of material science to understand that technology, like logic, extends its opinion from the illumined vision of science.

These points scientifically reveal that stunted growth, which has become the culture of all human endeavours and sectors, practically originated from the mistakes of material science. This is why there must be honest endeavours to approach these problems with the use of philosophical mathematics. In the philosophy of Handrel Orwell, philosophy hibernation and the errors of material science, when compared with their benefits, appear to be lopsided in form, format, and statistics, as well as in human understanding.

The rigorous and orchestrated methods by which material science arrives at a common purpose can be defined as another, more vast economy of human endeavour that is against the ordination of natural science. It is in the tension of this work to express with modest understanding that the creations of material science must not be rejected—but neither must they be taken hook, line, and sinker. They must be well scanned in order to appreciate whether its contexts are in conformity with the illumined thinking of philosophical mathematics.

As the world is in a octave of understanding, biorhythm and bio-phenomenal in existence, human endeavour must be purpose-driven in the feature and science of practical concepts and practical thoughts with illumined thinking. It is in the mandate and ordination of this work to explain with a modest approach that man is a supreme fountain of philosophical mathematics, where actions and reactions are essential in ensuring that creation is created, promoted, and propagated. That includes being provided with the honest resources that abound in the chemistry of philosophy, which should be translated in creation. My opinion rests squarely on the wisdom of the Oriental apostolate, which was cautious in ensuring that the world was naturally driven with equity factors, forces, and understanding. That is why this chapter understands that material science is still at its starting point.

CHAPTER 7

WHY IS THE WORLD MATERIALLY DRIVEN AGAINST BEING PHILOSOPHICALLY DRIVEN, WHICH IS THE OPINION OF MOTHER NATURE?

The chaotic nature of material science is responsible for the dangerous and detrimental pollution through which the world is passing today. This chaos has affected the phase and the psyche of the universe. It is unfortunate that material science, which is the cause, is not making any reprisal to address these deadly consequences.

The immortal words of Professor Charles Fid Moore of the University of California on the dangers of material science—and what the Crawnwell Lectures in neo-metaphysics designated as "material science, the greatest enmity that is attacking the eco-system"—have all it takes to appreciate the dangers of material science.

The world is materially driven against being philosophically piloted because of the following reasons.

- The greatest terrorism, which originated from the barbarian invasion of the 18th and 19th centuries.

- The destruction of the College of Alexandria, both in Chris and Crotona, which was edified as the Museum of Philosophical Mathematical Knowledge.

- The consequences of trade imbalance, which worked against the harmonious principles of rhythmic, balanced interchange.

- The errors of trade by barter, which can be said to be one of the deadly errors in the history of man.

- The consequences of the world wars, which cannot be neglected.

- Imperialism and colonialism practically reduced right thinking, action, and balanced transactions to a materially driven system.

- The errors of industrialization, with its localization, can equally be designated as dangerous polyester in the annals of history.

- The high negligence of the laws of nature, particularly by man and the province of the uniformed, is a dangerous situation.

- The practise of cheating in commerce, trade, and sociocultural transactions must equally share in the blame.

- The high level of humiliation to the teachers of truth who brought the gospel of love is designated as a threat to the understanding of the brotherhood principles.

- The dehumanization of man by the invention and introduction of the machine age, which was purpose-driven in grinding man through the gears of machine, can best be said to be an era of human and material massacre.

- The way and manner the human race neglected the province of philosophy, including the use and application of philosophical laws, must be defined and designated as dangerous laws to the province of Atlantis, which was later discovered and directed by the wisdom of Plato, who is known all over the world as a philosophical and logical man.

- The invention of space science with its emeritus technology must be seen as the greatest technique that demystifies the thinking of man.

- Lack of human knowledge that was Platonic in nature, including astrology, which can be defined as the high cost of human negligence.

Mathematically stated, the universe is philosophically driven with facts and figures; this is why the world is fallen on the dictates and directives of philosophy. A look at the content of Plato's metaphysics has not appreciated the world as a metaphysical conclave, which is being directed by the mandate of philosophy. The wisdom of rational thinkers practically reveals that the world cannot know peace and harmony, and cannot understand the will of the law of rhythmic balanced interchange, because man is still at the starting point of his thought. We are facing the greatest enmity, which is being orchestrated by material science, material philosophy, material logic, and material technology. It is here that Walter Russell said that the greatest errors of material science have remained in its calculated attempts to lock the creator out of his creations. A look at the mathematical wisdom of this statement, with its technological eulogy, reveals the imbalance is the making of commerce and industry. This is why philosophical mathematics is practical and purpose-driven: to ensure that the errors of material science must be revisited in the use and application of objective technology. That technology is itself dynamic, practical, and pragmatic in ensuring that the law of balance governs the affairs of man through the use and application of international philosophy with international logic.

For the record, it must be stated that material science is at the zenith of its growth; this is why advanced countries are already stagnated. It must be understood that philosophy assiduously aims to achieve the desire of perfection and purity while it is empathic in ensuring that facts and figures are practically driven in its wisdom.

There is an Internet between philosophy and mathematics, between mathematics and logic. It is here I make the following scientific and philosophical hypotheses, which will be used in the future of human endeavour as mathematical hypotheses.

- All triangles are philosophically driven by the mathematical rhythm of facts and figures.

- Material science is stagnated both in action and reaction because it is not philosophically driven in producing a permanent course.

- Material scientists all over the world appreciate that actions can be defined as being and becoming a robot; this is why most of the creations are infected with scientific and technological viruses.

- Euclid was practically egocentric on the achievement of Pythagoras, and so material science is equally robotic without a basis on the balanced wisdom of natural science, particularly philosophy.

- Globalization at the level of material creation can be defined as one angled triangle.

- The computer age, which is materially driven, will certainly produce a scientific incubus that will affect the province of commerce and industry, because the human race is still developing at the level of the fifth kingdom.

- The desire of the millennium developmental goal, if not philosophically driven, must certainly hit the rock in the future of our material civilization.

The need of explaining these points is borne out of the fact that material science has created a monumental divinity that will be difficult to be united, simply because a look at chemical and petrochemical engineering will certainly understand that the two are almost the same. The need to use this as a case study is practical and important. The way and manner that material science has demystified the operational mechanism of the world calls for the human race to develop a philosophical blueprint that will be utilised in addressing the enmity with which material science has ingloriously infected into the world. That enmity has remained one of the consequences that affect growth and objective civilization. Emperor Acinto of Egypt, who is designated as one of the Twelve World Teachers, opinionated that material science had reduced the world into a moribund semi-hamlet. The royal prince of knowledge was stating that the human race was creating a critical amalgamation that originated from the province of material science.

It is absolute and cohesive to state philosophically that material science should be watched with the practical rules and regulations of philosophical mathematics. This is why this noble science of knowledge is of help to achieve a balanced standard, which is in consonance and in synchronization with the laws of Mother Nature, and which is essential in the achievement of purity and perfection that have remained the mystical essence and natural ecstasy of the dynamic functions of philosophical mathematics.

CHAPTER 8

THE URGENT NECESSITY OF REDUCING MATERIAL INCURSION INTO THE AFFAIRS OF THE UNIVERSE

It must be philosophically explained that the achievement of a universal cosmogony is not possible if the occupation of material incursion is not practically and scientifically reduced to a minimal level. Presently this is one of the greatest problems of the jet age, including the purpose-driven globalization. Edison was figurative and eloquent regarding this issue. That is why the eminent and scholastic words of Aristotle, who is known as a philosophical logical guru, must be strictly adhered to. The un-globalized usage of the one-world family and its one purpose is unnecessarily scientific. A look at the constitution and institution of philosophical mathematics must appreciate that the urgent necessity of reducing material incursion in the affairs of the universe should be understood as a pragmatic wakeup call.

Going by the rules of man, including the practical knowledge of the astronomical principle into his spiritual transaction, reveals the following concepts.

- Man the unknown.

- Man who has refused to understand and conquer himself.

- Man the maker of great thinkers who does not work in tune with the infinite to appreciate his philosophical relativity.

- Man who is known as an immortal seed sower and the un-seeding of all seeds.

- Man who has refused to be the true image of his maker but the ephemeral image of material world.

Philosophy was introduced into the annals of Mother Nature in order to perfect the thinking of man and to be in line with the ecstasy of alchemy. Unfortunately, it was ingloriously swallowed up and attracted by the un-philosophical nature of material science, which works most of the time to reduce the evolution of man to his spiritual civilization. This ugly trend practically works against the illumined thinking of philosophy, including the figurative calculation of mathematics, which conjoins philosophy and mathematics as an enabling, fraternal science.

The urgent necessity of reducing material incursions in the affairs of the universe must be achieved in order to reposition the universe as a philosophical, mystical province that is harmonious with the thinking and anatomy of Mother Nature. Nature practically and perfectly drives the emotions of man into a lot of diversified transactions, and most of them are a lot of myopic Internet viruses. Consequently the following points must be directed and noted.

- The reduction of material incursion will certainly hold the key in actualizing the vision of one-world family.

- It will help to achieve a time-frame philosophical globalization.

- It will help to reduce the manufacturing of dangerous armaments, including other weapons of wars.

- Terrorism and militancy will be reduced to a minimum level.

- Human trafficking and hijacking will become a thing of the past, simply because creation will think of a new dawn that will propel balanced thinking with harmonization.

- It will help to reposition the universe as a dependable alchemy of immortal conclave.

- The concept of unity in diversity will be accepted as a method that will benefit all and sundry in realizing that the hurt of one man is a universal hurt to all men.

- It will attack and tackle the dangerous segregation of gender constitution.

- It will help to achieve the desire of the millennium developmental goal, which is purpose-driven with appreciating the desires and efforts of the Internet age.

- It will promote permanent creation of great ideas to replace ephemeral and material creations.

- Science and technology will be intertwined in the mutual relativity of ensuring that all are purpose-driven with unity.

- Constant changes of material principles will be reduced, and technological principles will be greatly enhanced.

- Edison's understanding of technology will galvanize with the thinking of philosophical mathematics in providing data, statistics, and research methodology by which the onus task of constant changes with hibernated illusions will be relegated to the background.

It is important to explain the urgent necessity of reducing material incursion. The actual fraternity of philosophical mathematics seeks to achieve protection, provision, purity, and perfection. That is why this chapter is purpose-driven with all embodiment of providing a philosophical blueprint, which will help all and sundry in tackling the dangerous incursions of material science. Those incursions have greatly eluded the natural wisdom of man and the universe. It is naturally on the tedium of this orbit to explain without bias that man is still in search of a reality that is contained in the realism of philosophical mathematics, which in earnest is universally driven in serving him and his ecosystem with the best objectivity of scientific balance and technological purpose.

CHAPTER 9

How Has Man Related and Utilised the Protection and Perfection of These Principles and Theorems?

Since the dawn of consciousness, man has manufactured and mitigated lots of errors in the origins of calculus, including the driving force of the purist science designated as philosophy. These errors have remained his greatest problems because changing and renewing his thoughts in order to be driven in the luminous thinking of philosophy has erroneously dwarfed his creations beyond the level of the fifth kingdom man.

This critical pollution of ideas necessitated the likes of Socrates, Plato, Pluto, Aristotle, Peter Goras, and others to design and dedicate themselves to be original and purpose-driven with the realism of philosophy. They had an honest truth about mathematics, enabling powers of logic, aesthetics, metaphysics, meta-logic, and astral understanding. These eminent scholars were strong-willed in the wisdom of numerology and astrology. Thanks be to heavens for their deep understanding of creating a philosophical province that was purpose-driven with the understanding of principles, theorems, policies, and practises that designated Pythagoras to remain a mathematical idol, including being a philosophical millicent; his era introduced culture, tradition, morals, and ethics.

It is on record that after their era, mankind started its degeneration to the lowest rank of philosophy and material engineering. This was a critical melancholy, including being a cancer in the annals of universal understanding, technical development, engineering propagation, and creation. That degeneration included the demystification of mathematics, which comprises algebra, statistics and further mathematics. All these were seen as calculated attempts to bring man to his present melancholic state. A look at the workings of mathematical theorems with their principles will certainly appreciate philosophy to wisdom, wisdom to reasoning, reasoning to metaphysics, and metaphysics to mysticism. Above all, philosophy and mathematics work to achieve a scientific and technical tedium that is purpose-driven in ensuring that knowledge is propagated and employed as a way and means of understanding the networking of science, including the international relativity of technology.

Unfortunately, the availability of mathematical principles and policies were greatly rejected by the province of material science, which is encumbered with a high level of technological myopism, which does not relate well with the alchemy of philosophical mathematics. A look at the wonders that are going to be carried in part 5, which is going to present Pythagoras as a front-liner who greatly initiated Euclid and Albert Einstein, will prove that what is practically seen now as mathematical principles and policies tactfully negate the constitution of engineering mathematics, engineering science, and engineering philosophy. This is why a look at how Mother Nature reengineers her rudimental creation will certainly appreciate that man has fallen short of glory from the thinking of mathematical methods.

It is here that the protection and perfection of these theorems are narrowed, and most of them appear to be unscientific and unrelated. That is why material science has greatly created lots of viruses in the annals of human invasion and occupation, which must be tactfully corrected if the human race is to avoid falling into a dangerous technological oblivion. For example, pure mathematics believes strictly that $1 + 1 = 2$. But logically, mathematics argues that $3 - 1 = 1$ and $1 + 1 = 2$. Conversely, $4 - 3 = 1$ and $2 - 1 = 1$. In the scientific opinion of truth, natural mathematics is figurative, constitutional, and dynamic. This is why it is the only science that is institutionalized in producing facts and figures with the utilisation of technical and scientific empathy.

In the genealogy of philosophy, it can be explained that $1 + 4 - 1 = 4$ and that $5 - 2 + 0 = 3$. Conversely, $1 + 1 + 1 - 3 = 0$. In logic, it could equally be seen that 0 and 1 can drive the realism of the above configuration. This is why there is a harmonious relativity of philosophy as a combined millicential science.

A look at the explanation and anatomy of every angle reveals that all angles at any horizon are scientifically right-angled either at 45°, 60°, 90°, or 180°.

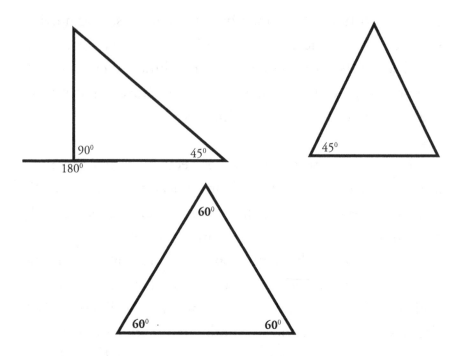

A look at the structure of these triangles reveals that the book is purpose-driven with the infection and affection of the thinking of Pythagoras, Euclid, Albert Einstein, H. A. Clement, W. W. Yurel, and S. A. Smith. Mathematical principles and policies work harmoniously to achieve a purpose-driven theorem which, when further investigated, has remained the power block by which engineering mechanics, chemistry, and astrology are used in explaining what is known as a figurative constitution of balanced calculus.

It is here that the simplest engineering mathematics designated as statistics ensures that philosophy is utilised in research anatomy. A look at the constitution of metallurgy, with its obvious facts and figures, reveals that there is nothing like technology without the enabling constitution of science.

This revelation is why the world is practically diversified in ideas, knowledge, and wisdom.

Philosophical mathematics' principles and theorems can be understood when the human race appreciates and understands the logic of the four cardinal points, which explain that in the north there is wisdom, in the south there is mathematical knowledge, and in the west there is objective reasoning. North, south, and west look at the unequalled wisdom of the east, where it is stated that the wise men found their shelter, propagated, and harmonized their idea—which was responsible for the foremost creation of a philosophical and mathematical pyramid that has remained the greatest structure. The pyramid illumined those designated as the thinkers of the unknown world who bequeathed knowledge to humanity, including wisdom, which came at the proximity of the blessed angels.

The blessed philosophers reengineered this great knowledge as the sacred teachings of all ages, which contain the protection and perfection of philosophy and include the principles and theorems of mathematics.

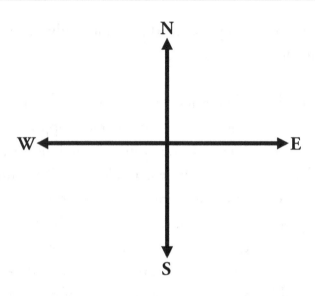

It is mathematically proven that N = S = E = W, S = N = W = E, W = N = S = E, and E = N = S = W. Therefore, N/S = E/W and N/W = W/S = S/E = E/N. It must be stated that the four cardinal points maintain an angle of 360°, which is both converse, rotational, and proportional with facts and figures. That is why it has forever remained the dynamic and tedium of philosophical nucleus, which when further investigated reveals that the four cardinal points are the tedium of the universe, including being the scientific rotor of the world by which all discussions on the subject of globalization can be achieved. This is why mathematical principles and theorems have always fallen on the structure of the four cardinal points with a view to ascertaining that its astral creation is indeed the scientific anatomy of man as understood by the illumined illuminates who originated with the honest understanding of mystical and philosophical mathematics.

CHAPTER 10

AN OVERVIEW ON THE PHILOSOPHICAL MATHEMATICS OF PART 4

A look at the functionalities and compartmentalization of philosophical mathematics' principles, policies, and theorems will understand that the theorems and principles are practically intertwined with the honest reason and logic that are embedded in calculus. By compartmentalization, we must understand the segments, the levels, and the aim these principles and theorems focus to achieve. It is unfortunate to observe that modern philosophers and mathematicians are not objectively driven by the dynamics of mathematics. This ugly situation calls for a lot of technical and scientific restructuring, including ensuring the fact that philosophy, at any point in time, does not segregate its functions from the natural relativity of its oneness with mathematics.

A look at the content and functions of the Internet system, with its figurative concepts of globalization, reveals that it is purpose-driven with forces and principles of natural mathematics and natural philosophy. This aspect is why all that is achieved by the province of electrical and electronic engineering has a lot of the input of magnetic philosophy and dynamic mathematics. From the octave reasoning of mathematics, we can appreciate that the human reasoning cannot function without relating its objectives with the anatomy of core philosophy, which is purpose-driven with the ethical understanding of logic, calculus, further mathematics, algebra, geometry, and qualitative and quantitative reasoning, with a view to achieving what is contained in verbal dynamics with an implicit understanding.

It is unfortunate to observe that the human race has not appreciated that the world is mathematically driven, whereas the universe is configured in the anatomy of logic, which always achieves its desire through the use and application of calculus, magnetic mathematics, quantitative mathematics, and statistical mathematics. This is why the province of electrical and electronic engineering, mechanical engineering, instrumentation, aeronautics, geosciences and geophysics, quantity surveying, aerial studies, and more are practically galvanized in the formats of philosophical mathematics. This is why the achievement of human endeavour is essential to the reasonability of philosophy to mathematics.

S. H. Smith, who is known and designated as our modern father of mathematics, opinionated strongly that philosophy functions practically and technically with the use and application of mathematics, particularly that which is embedded with core mathematical dynamism. It is unfortunate to observe that our modern thinkers could not understand the opinion of Smith.

A work to examine the functionality and technicality of mathematical principles and theorems, including how it relates with strategic understanding, must observe the figurative contents of philosophy, logic, and verbal reasoning. This is why all honest efforts must be made in order to ascertain, understand, and appreciate that the world is practically and technically driven with the objective of philosophical mathematics.

The opinion of the idol of mathematics, Pythagoras, whose wisdom affected Euclid and Albert Einstein, must be utilised as a scientific VSAT simply because he was destined at the proximity of inspiration to give the world the following.

- The mathematical proportion of a right-angled triangle.

- The functions and functionalities of mathematical principles.

- The quantum and diameter of mathematical theorems.

- The reasonability of how these principles and theorems can be intertwined, including how and why these theorems and principles cannot function without the figurative dynamism of philosophy.

The origin of philosophical mathematics is the origin of Mother Nature, which embraces man, the universe, and the one-world purpose, including what I call the wisdom that is contained in the views of metaphysical philosophy.

For the record, philosophy is a galvanized tedium of objective reasoning, while mathematics is a functional compartmentalization of all figures, facts, principles, and theorems that can be utilised in buttressing mathematical ideas and concepts. It is in my intertwined understanding to present this overview, which is pivotal in the objectivity of philosophical mathematics as a databank, a functional mathematician, and an illumined philosopher whose ideas are pragmatic and creative.

Part 4 and the overview can be defined and designated as the views of honest thinkers, including the opinion of those who are greatly reasonable, technical, scientific, and philosophic. It also includes being possessed in giving the world what is known and acceptable as the way of truth, which lies in the anatomy of philosophical mathematics.

The Oriental mathematical philosophers were natural in producing a banking format, including propagating the course of philosophical mathematics, which will greatly utilise the principles, policies, and theorems—as well as what can be considered as asphalt mathematical facts, which drive the opinion of luminous thinkers to be purpose-driven in creating a one-world purpose with a universal philosophy. This is why both mathematics and philosophy cherish the understanding of science and technology.

A look at the functions of trigonometry, including the usage of log tables with sine and cosine formulas, reveals that we can have a balanced universe with a balanced economy when philosophical mathematics is given leave to correct the errors of material science. Those errors include the dangers that are contained in technology, which in my opinion is triggering disaster, including other chaotic and related enmities that are stunting the growth of the human race. It is unfortunate to register that science and technology at their material postulations have not borrowed much from philosophy, which is natural in action, reaction, functionality, and segments.

When Dr. Walter Russell opinionated that the greatest errors of science lie in locking the creator out of his creations, he was only educating the apostolate of his province to appreciate that man's actions and transactions cannot function without the use and application of the dynamic functions of philosophical mathematics. Consequently, it is in the view of this work to produce what can be defined as a regularized mathematical thinking, including ensuring that philosophy (which is the maker of great thinkers) is being utilised in strengthening the world as a globalized village and one-world purpose. That earnest effort should be purpose-driven in the engineering and technicalities of globalization.

For the record, this overview is scientifically denoted, designed, and appreciated from the bank of the author as being in the becoming of a philosophical mathematics tedium. Its simultaneous relativity will help the human race to appreciate that the universe is a mathematical Sandra, whereas the one-world purpose is a mathematical sandalwood which, when understood, will make all and sundry know that the world is a fulcrum of philosophy and that the universe is understood at the anatomy of mathematics. Mathematics exists at the opinion of Mother Nature, whose actions and reactions are intertwined with the dynamic functions of philosophical mathematics.

PART 5

PYTHAGORAS'
PHILOSOPHICAL MATHEMATICS

"There is geometry in the humming of the strings, there is music in the spacing of the spheres."

—Pythagoras

INTRODUCTION

In many problems in which right-angled triangles are involved, the theorem of Pythagoras is necessary. This illumined theorem states that the square on the hypotenuse of a right-angled triangle is equal to the sum of the squares on the other two sides. Hence if two sides of a right-angled triangle are known, then the third side can be calculated.

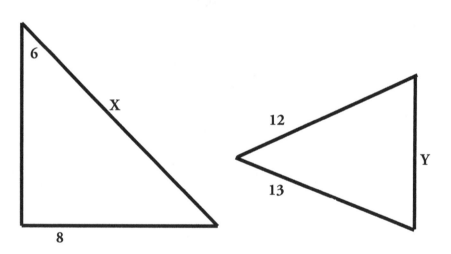

$X^2 = 6^2 + 8^2 = 36 + 64 = 100$

$\therefore X = 10$

$13^2 = 12^2 + Y^2$

$\therefore Y^2 = 13^2 - 12^2 = 169 - 144 = 25$

$$\dot{.} \ . \ Y = 5$$

$$\text{or} \ \ Y^2 = (13 + 12)(13 - 12) = 25 \times 1 = 25$$

$$\dot{.} \ . \ Y = 5$$

It must be stated that this introduction shall explain much about the philosophical mathematics of Pythagoras, as well as explain how it finally formed the nucleus of the universal mathematical anatomy. The chapter will actually and abstractly introduce one of the highly gifted philosophical mathematicians who existed at the era of the Oriental wizards: Pythagoras was known, designated, adored, and honoured. A work in philosophical mathematics that does not bring him to the foreground cannot be considered in the reality of philosophy, because he was a mystical mathematician, a metaphysical philosopher, an acclaimed logician, and an illumined apostolate.

The University of Crotona is nobly designated to this great Orient, whose contributions to the problems of calculus and philosophy were honoured with a universal chair at the proximity of these dynamic postulations. It must be borne in mind that the opinion, which can best be defined as a philosophical epigram, that lies in introducing his philosophical mathematics was borne out of his mind that most of the contributions of mathematics extolled the likes of Euclid and Albert Einstein. His origination in the dynamic understanding of philosophy equally extolled such noble humans like Emmanuel Kant, Aristotle, and Max Miller.

This is why Plato was convinced that he had an illumined successor who would carry out the cross of knowledge and the wisdom of philosophy, despite human upheavals and social tempests.

Pythagoras was a Grecian from a humble family, and the strength of his parents was tilted towards ensuring that they produced a philosophical illuminate. Goras' father was a teacher by occupation, and his mother was a farmer. Both were Grecians from a humble origin.

The initiation of Pythagoras was at the proximity of Plato the blessed, who was among the Twelve World Teachers; Plato was initiated by Socrates, who is designated as the immortal teacher of Greece. He was greatly dedicated to his illumined teacher and was adept with humility to understand him, ensuring that Plato mystified and translated his mystical knowledge into him.

The history of calculations—particularly triangles and angles, mathematical fractions, simultaneous equations, and remainder theorems—is incomplete without adequate recognition shown to Pythagoras. He was the first human since the dawn of consciousness that came in contact with Mathematics, the goddess of calculation and wisdom of scientific reasoning. This is why it is important to carefully examine Pythagoras, both the content of what he is and in what he is becoming.

Goras' knowledge, particularly in the areas of philosophy and mathematics, was acclaimed to be the best among humans. He used his gifts without melancholy and regrets. It was stated that he was at the proximity to discover the science of numerology, including the mathematics of logic, which is an abstract science; the mathematics of astrology; the mathematics of astral physics; and the mathematics of astral plane.

It is unfortunate that the barbarians invaded Crotona and destroyed most of his creations and concepts, which in return reduced knowledge and humanity beyond the fifth kingdom. The barbarians were also the beast that destroyed the College of Alexandria and other institutions that were known as the zenith of knowledge of that time.

Pythagoras believed that $1 \times 2 = 2$. He further buttressed this point that $1 + 1 = 2$ while $2 \times 1 = 2$ and $3 - 1 = 2$. This celestial scholar equally opinionated that $A + B = A + B$ and that $AB \times CD = ABCD$. It must be stated that just as Mozart was known as the mystical gadget of music, Pythagoras was known as the mystic of calculations, figures, logic, and balanced reasoning. The universal seat of wisdom honoured him with an illumined chair as Pythagoras the immortal, mystical mathematician.

It is in the wisdom of this introduction to bring to the forefront most of the qualities of Pythagoras, which include humility, sincerity, Internet wisdom with compact understanding, and the ability to reason with electrifying thinking. This is why at a certain age in his life, the philosophical mathematician was designated as Goras, the Peter of knowledge.

Pythagoras should be extolled as the foremost philosophical mathematician who can be designated as the founder of truth, of philosophy, and of metaphysics, including being defined as the universal polyester whose creations are objective with good conscience.

It is in the interest of this book to ensure readers know that Pythagoras was not a spirit, as people believed he was, but a noble and rational soul who was permanently sowed in the illumined soul of the angels of mathematics, philosophy, logic, science, and technology. Pythagoras was willed to have a lot of followers, disciples, apostles, and humans that believed he was a philosophical crescent from the luminous province of mathematics. This introduction cannot be completed if it is not boldly written that he kept a permanent seat and chair in the hall of fame designated for immortals, rational souls, and great thinkers. His greatness as a man of immeasurable knowledge is practically foretold in one of my books, *The Philosophy of the Oriental Wizards*. In this work, Pythagoras is designated as a wizard of immortality and a prominent philosopher whose grace reengineered noble souls to look at him as the foremost abstract thinker, including being a comprehensive Benedict of mathematics, Statistics and Databank. This is why he will live forever: because his creations will endure and extol him as one of those philosophical mathematics, having extolled and designated him as the knob of rational thinking with abstract creativity.

Below is a tabular view of some his mathematical concepts.

Some Selected Pythagorean Triples

a	b	c	a	b	c	a	b	c
3	4	5	6	8	10	20	21	29
5	12	13	8	15	17	28	45	53
7	24	25	10	24	26	33	56	65
9	40	41	12	35	37	36	77	85
11	60	61	14	48	50	39	80	89
13	84	85	16	127	129	48	55	73
15	112	113	18	80	82	65	72	97
17	144	145	20	99	101	56	90	106
19	180	181	22	120	122	60	91	109
21	220	221	24	143	145	44	117	125
23	264	265	26	168	170	51	140	149
25	312	313	28	195	197	88	105	137
27	364	365	32	255	257	85	132	157
29	420	421	36	323	325	57	176	185
31	480	481	40	399	401	95	168	193
33	544	545	44	483	485	119	120	169

All multiples of these triples also form Pythagorean triples.

CHAPTER 1

WHAT ARE PYTHAGORAS' CONTRIBUTIONS TO THESE EMINENT AND SCIENTIFIC PROBLEMS?

It is important to philosophically examine the contributions of Pythagoras in the areas of mathematics, calculus, quantum science, magnetic thinking, and Internet postulations. We must examine how and why these eminent contributions have remained immortal challenges to our present civilization.

The Grecian community adopted Pythagoras as a rational soul. This ancient province designated him as a mathematical Buddha who is known to give the world a philosophical Buddhism. In Western countries and among the mathematical Amazons, he is remembered as the seed of calculation, algebra, geometry, and statistics. This is why every era must recognize him as a philosophical mathematical bank. It must be noted that Pythagoras is respected for being a mathematical nun and a philosophical emengard. This is why his scientific contributions are ever welcomed and remembered as the origin of mathematical technology.

Eminent scholars like Euclid, Albert Einstein, Aristotle, Emmanuel Kant, St Augustine, and St Francis, can always attest spiritually to the fact that Pythagoras was a living mathematical anatomy whose thoughts and contributions, including his illumined knowledge, can be appreciated by souls who are mystically driven.

At this point, we must present a logical epigram that exists at the proximity of Aristotle in favour of Pythagoras, which reads thus.

- If any man should be honoured for possessing Internet knowledge of Mother Nature, it cannot be any other person than immortal Pythagoras.

- The wisdom of Pythagoras is beyond words, thoughts, and ordinary understanding. This is why I strongly designate him as a rational soul.

- Mathematical thoughts, with its anatomy, were first explained to the world of science by Pythagoras. This is why he is known as the seat of balanced science.

- Logic, which was explained in detail to the world of philosophy, was infected at my province by the eminent and scholastic wizardry of Pythagoras.

- Philosophy, which is known as the mother of all living sciences, including being the fulcrum of technology, was mathematically bequeathed to the world of knowledge by the deep thinking of the consummate, rational soul designated as Pythagoras.

A look at these eminent epigrams of Aristotle will certainly reveal that Pythagoras was the prince of philosophical mathematics who paid the cosmic price of its possession by ensuring that posterity gains from his illumined knowledge through scientific impact by the establishment of the University of Mathematics, including the Pythagoras Institute in Crotona.

After thorough examination of Pythagoras' contributions, I was affected by inspiration to present the following philosophical mathematics eulogies as addenda to the immortal contributions of Aristotle, who was practically educated and initiated by the highly illumined philosophical mystic.

- Pythagoras had a careful and consummate patience with the world of wisdom; this is why the province of philosophical mathematics designated him as the supreme possessor and harvester of supreme knowledge.

- Pythagoras can always be remembered for his service of love to God, humanity, and the province of knowledge. This is why I thereby authenticate respect with abstract possession that his illumined views on mathematical philosophy can be defined as the structure and anatomy of every triangle.

- Given that he was a messiah of philosophy and mathematics, it is equally honoured that he became an immortal idol who gave his all to ensure that posterity does not lack.

Otherwise stated, "for their tomorrow we give our today" was a philosophical Oriental maxim that was celestially granted at the proximity of Pythagoras. Any man that is not infected with philosophical mathematics cannot be called a human. Such a man is not fit to live, and in him are found all the structures of illiteracy, which are more expensive than education.

It is here that Pythagoras has forever remained a philosophical, fine ray, including being a mathematical pathfinder. His contributions to this scholastic province include:

- The establishment of a mathematical, philosophic, democratic province.

- The establishment of the University of Philosophy, including the Institute of Mathematics.

- The development of ethical codes, which have to present philosophy as the supreme and immortal science.

- The improvement of people at the province of Crotona, including developing apostles and disciples who were greatly infected by his highly illumined wisdom.

- Pythagoras was the foremost mathematical thinker who discovered the anatomy of the triangles and quadrants. The concept of the properties of right-angled triangles was also at his honest proximity.

- Pythagoras was the mathematical hero who opinionated logically that all triangles that are contained in isosceles triangles are equal and balanced with all mathematical constituents.

- Mathematical fraternization was a symbolic arithmetic, which he introduced to the province of reasoning, logic, and meta-logic. This is why Pythagoras was an expert in metaphysics, numerology, astrology, and psychic reasoning with empathic chromatization—hence all over the world his mystical mathematics is greatly symbolic in the world of science and objective reasoning.

Pythagoras utilised his wisdom in medicine. From the illumined words of Andrew, the fulcrum and founder of modern anatomy, Pythagoras is known as the wizard of wizards, particularly in psychic dictation, analysis, and octave reasoning with construction.

The eminent philosophical scholar in the person of Manly P. Hall, who is the living founder of the Philosophical Society in California, opinionated that the write-ups that are contained in the list of Atlantis, including the discovery of the New World, cannot be fully explained without bringing in Pythagoras and in all the pages with human evolution and mathematical development. He concluded that the world of philosophy, mathematics, engineering technology, quantum science, and magnetic discoveries (with its propagation) were all structured in Pythagoras' deep thinking and engineering dynamism.

It is lucid with mathematical opinion that Pythagoras is the father of the sacred teachings of all ages. I feel it is important to explain these facts to the scholastic world, to the might of ages, to the world of commerce and industry, and to the province of philosophical and Internet thinkers, because a man is not worth his name until he has evolved and developed to give the world a new thinking in the areas of science and philosophy. Pythagoras' contributions to the development of the world's knowledge, particularly in philosophical mathematics, have earned him a glorious chair in the universal hall of fame. This is why his contributions will always be echoed in the souls of posterity, including the minds of modern men who are yet not vast in the knowledge of philosophy, mathematics, logic, epistemology, science, aesthetics, theurgy, ethics, and morals.

The universe is greatly asking for the likes of Pythagoras. Those people are best designated and defined as the anatomy of rational souls whose thoughts and lifestyles are always piloting the affairs of the Internet universe.

CHAPTER 2

What Are the Philosophical Relationship of Pythagoras and Plato, and How Did They Use These Dynamisms to Improve the World of Mathematics?

The life and time of any genius must be purpose-driven with a practical and spiritual adaptation to an illumined adept. The case of Pythagoras, his history, his life and time, and his achievements must remain un-philosophical without due credit and respect to Plato. Apart from being educated and trained for the realism of his destiny by Plato, Pythagoras was initiated at his proximity, which can be dedicated by a philosophic maxim that reveals that great ones are called practically for the assignment which they are destined to fulfil.

It must be explained that the contact of Pythagoras with Plato started a new dawn. The illumined wizardry of Plato at any point in time affected Pythagoras widely and positively; this is why Plato was greatly humbled to initiate him as one of his apostolate who would continue the apostolic foundation of his ministry when he left the mortal body.

In the areas of mathematics and calculus, Pythagoras was more illumined. He was equally there at the areas of philosophy, morals, ethics, and mystics, as well as in converting inspiration to a business necessity.

Plato was refined and was referred to with logical inference as an Amazon of wizardry. Consequently the pious relationship that existed between him and Pythagoras started a new life with a new dawn, which included the following.

- Both philosophically worked to establish the institute, whereby the University of Alexandria was meritoriously credited to their achievement.

- Their positive and philosophic lives were instrumental in changing the character of their eras, particularly the Oriental province.

- Both were foremost philosophical emengards who were responsible for developing philosophical codes, technological prowess, and scientific dynamism. They also introduced the science of astrology and numerology in different schools of higher learning in Greece and most Asian countries.

- They were the philosophical geniuses that started a new dawn in the thinking of science as being all about nature, the cosmos, and how man can be involved as a competent image of his creator.

- Both collectively established a school of thought at Croton, which was responsible for training humans to be, bear, and become philosophically in tune with the purpose of Mother Nature.

- Pythagoras and Plato were the foremost philosophical geniuses that understood the law of rhythmic, balanced interchange. They buttressed how this law could be used for honest and commercial transactions.

- They established an administrative college of thought that was responsible for training politicians in the use and application of philosophical ethics to better the lives of the community.

- It was the logical thinking of the two that empowered philosophy to be known and called a science that is purpose-driven with the wisdom of the rational souls.

- They were responsible for the empowerment of Aristotle, who started from where these eminent scholars left off.

- The mathematical wizardry of Pythagoras affected Aristotle to appreciate logic both as a mathematical databank and a serious science that is purpose-driven with inference, reference, convention, and conviction.

- Both worked to establish statistics as a branch of mathematics that is purpose-driven with formats, figures, and data, including how it could be utilised in buttressing concepts.

- The philosophical dynamisms of the two were responsible for the establishment of the will of science and methodology, the institution of strategic management, and the university of philosophical psychology. This is why history designated them as immortals.

- Pythagoras was single-handedly responsible for the introduction of the metric system and industrial mathematics with its eminent psychology, which has contributed to the works of the cosmic province.

- Pythagoras was a scientific initiate in the best understanding of astronomy, astrology, numerology, and space science. This is why most of his contributions were responsible for affecting Euclid and Albert Einstein.

- Pythagoras was defined and designated as the mathematical mystery of all ages.

Thus Plato, in his epigram to the world about his philosophic friend, designated Pythagoras the best of men and the truth of figures. An honest appreciation of this fact reveals that the relationship between Pythagoras and Plato greatly produced tedium of achievement, including ennobling human beings to understand the dynamism of life.

A look at the fortunes of this great man, who was philosophically known as the Millard of his time, reveals that he is best defined as a man of mission and of eminent scholastic wisdom, particularly in the areas of mathematics and philosophy. This is why the honour that the universe has ever credited to Pythagoras will always challenge all eras, particularly all humans, in knowing that every great man is a great art, and every great art is symbolic, creative, and luminous. It is here that the relationship of Pythagoras and Plato originated the mathematical anatomy, which is best equated as P + P = P + P, which represents Plato and Pythagoras.

CHAPTER 3

WHY IS PYTHAGORAS KNOWN AS THE PRINCE OF PHILOSOPHICAL MATHEMATICS AND THE ANGELUS OF SCIENTIFIC AND MATHEMATICAL THEOREMS?

It must be explained that Pythagoras is known all over the world as the prince of philosophy, the mind of logic, the anatomy of aesthetic postulations, and the Orion fort of mathematical wizardry. A Grecian adept who is highly illumined and celebrated all over the world as Orion bequeathed a lot of mystical thought and wisdom to Pythagoras. This is why his knowledge has remained vast in the annals of universal history.

A work to determine with scientific explanation why and how Pythagoras was known as the prince of philosophy and the Angelus of philosophical mathematics must understand the following.

- He is an ennobled blessing who worked for the best.

- He is known as a philosophical initiate who metamorphosed gloriously into being the foremost founder of philosophical mathematics.

- His dynamism in the areas of logic and creative and constructive concepts cements him as being in the light of mathematical understanding.

- His achievements in the areas of transformation of his province and transmutation of great ideas consummated him as a monumental saint.

- In the Grecian province, he represented himself with an immortal chair.

- In the universal hall of fame, Pythagoras gave the world what is known as triangular theorems, principles, policies, and practises. This is why he is known as the originator of the isosceles triangle theorem, which includes being known as the monumental discovery of right-angled triangles.

- His unequalled knowledge of equality and inequality of letters strongly established him as the fulcrum fort of figures.

- His knowledge of numerology philosophically designated him as the Orion of his era.

- His vastness in the knowledge of astrology and planetary bodies glorified him as an immortal pyramid.

- His contributions to uplifting the Rosicrucian Amorc, including other sacred teachers, ennobled him as a monumental teacher of the universe.

- His opinion about the rising sun, including the wonders of the stars, designated him as the enmitied structure of the planetary bodies.

- His glorious knowledge of the astral world, including transcendental mechanism, gave him a universal chair as the Orion metaphysician.

- The way and manner Pythagoras utilised mathematical wizardry to mystify the province of light prompted his era to call him the idol of illumination.

- His chastity to live an odd life without a woman designated him as the prince of purity and perfection, which represents $P = P = P$ or $P + P + P = 3P$.

- His wider knowledge of the universe, particularly the cosmos, eulogized him as the mysterious mathematical matador, ably represented as $M + M + M = 3M$ or $M \times M \times M = M^3$.

- His knowledge of himself, which was a philosophical beacon, made him be a man of the people and a soul that is galvanized in the soul of all souls.

- His thinking about the equality of all establishes him as a mystical soul whose ideas about creation are purpose-driven with high ideas.

- His opinion about algebra, geometry, LCM, as well as the invention of logarithms, triangles, squares, and parallelograms, honoured him as the wizard of trios.

- His high level of understanding about simultaneous equations, remainder theorem, Euclid postulations, and the physics of Albert Einstein cemented his reputation as a mystical mathematician.

- As a prince of knowledge, Pythagoras ensured with all human ability that creation would understand the purpose of life and the meaning of death. Certainly he was the first person that designated death as a transformation or a call to glory.

According to Manly Palmer Hall, who was the founder of the Philosophical Research Society, the immortal consummation of Atlantis (which was at the proximity of Plato and Lord Bacon) cannot be a complete work without due mention of· the eminent philosophical contributions of Pythagoras, whom it was said lived in that region before the evolution of human consciousness.

The ennobled soul of philosophy in the person of Pythagoras was instrumental to the creation of an idealistic social order, which started from the eras of Plato and Karl Max. This is why the next chapter is going to examine the metaphysics of Pythagoras, including how he was ennobled as a rational soul. It is unfortunate to observe that our material philosophers and scientists have not been able to appreciate the immortal contributions of Pythagoras and how he was elevated to the post of being the prince of philosophy.

Pythagoras was a great soul, an immortal monument, a consummate and philosophical mystic, and a mathematical genius whose thoughts and lifestyle will continue to affect all those who yearn to investigate the mind of this matador. This chapter is humbly denoted to the investigation of how he was ennobled and designated as the prince of knowledge and the wizard of calculus, including being honoured with universal respect as the mathematical scientist whose contributions will continue to provoke the legacies of posterity. This is why the world is tactfully and scientifically humbled, because of his eminent and scholastic dynamism.

CHAPTER 4

WHAT IS THE MATHEMATICAL METAPHYSICS OF PYTHAGORAS?

$2 \times 2 = 4$; $4 - 2 = 2$; $2 - 1 = 1$; $1 + 4 = 5$; $5 - 3 = 2$; $2 + 2 = 4$.

I need to explain that mathematics is greatly intertwined with metaphysics. The first man in creation that explained this logical philosophy was Pythagoras. This is why my illumined delight in giving humanity this work is practically borne out of my knowledge of mathematical metaphysics.

It is important to explain that the mathematical metaphysics of Plato includes the following.

- His purist knowledge of letters.

- His perfect knowledge of formats.

- His inspired knowledge of sine and cosine formulas.

- His illumined understanding of creative construction.

- His illumined understanding of the anatomy of triangles, including other sequential concepts.

- His wizardry in the institution of simultaneous equation, particularly that which deals with the Pythagoras simultaneous equation.

- In the areas of metaphysics, Pythagoras was a gifted lamp holder who was interested in passing the light to the new era. He was able to give others light, illumination, and betterment, including using his illumined knowledge of metaphysics to improve his era.

A look at the mathematical formula that started this chapter will certainly understand that Pythagoras was pure and perfect in the use of figures. This is why he could multiply the figure 1 to denote millions. In the mystery of what he stands for, his understanding of perfection was purpose-driven with the lifestyle he lived. He was possessed with unequalled infectious humility; this is why he was designated as the philosophical Angelus.

To the ordinary man, Pythagoras was a benefactor, to the illumined he was known as a Benedict, and to his fellow mathematical mystics he was a Buddhist scholar. A look at the life and time of his metaphysical mathematics reveals that his purity of heart, purity of purpose, and charitable life—including the way and manner he galvanized his thinking into creative practicality—was beyond genius. That is why his physical body greatly changed into a metaphysical metamorphosis.

In presenting what can be considered an acceptable treatise for the mathematics of his metaphysics, Plato's position never differed from Pythagoras' understanding of the Crawnwell Lectures, the Cannon Laws, the Orphic Logic, and the Sophia and Theosophical Thoughts. This is why Pythagoras can best be defined and designated as the Peter of mathematical dynamism.

It is unfortunate that most of his unequalled contributions were lost to the barbarians that massacred Crotona. But let us accept that Pythagoras was a monad of his time, including being a mathematical metaphysician of our modern technology, whose ideas are gifted with consummate concepts. Manly Hall defined Pythagoras as an illumined mathematical dictionary, including being an encyclopaedia of metaphysics. I hereby designate him as the illumined Britannica of mathematical metaphysics, mathematical philosophy, mathematical logic, mathematical technology, mathematical science, and mathematical thoughts. Crotona also defined Pythagoras with respect as the luminous soul of dynamic conscience.

At this point philosophy, psychology, sociology, anthropology, theology, theosophy, Hinduism, Islamism, Christianity, metaphysics, and mysticism were the greatest teachers of Pythagoras the immortal companion. The mathematical metaphysics of Pythagoras was greatly borne out of his conviction that nobility is like genius, which is inherent in every man. This is why he worked spiritually in extolling his mathematical metaphysics, which is known as a dignified virtue that was instrumental in exposing him with perfect exaltation. It is here that the world appreciates his contributions, his purpose-driven life, and his scholastic engineering as one who is philosophically balanced to be named the soul of mathematical metaphysics. He's also the prince of philosophy, whose contributions to the dynamism of the one-world purpose will challenge all and sundry in knowing that the creator is the fulcrum foundation of mathematical metaphysics.

CHAPTER 5

WHAT CAN BE DEFINED AS PYTHAGORAS' PHILOSOPHICAL LEGACIES TO MANKIND, AND WHY ARE THESE LEGACIES TRANSLATED AS THE ORIGIN OF PHILOSOPHICAL MATHEMATICS?

Pythagoras, who is known as a man of mathematical philosophy with indomitable dynamics and original legacies, must always be remembered for giving the world a chair in the origin and reality of philosophical mathematics. He established an original mathematical and philosophical province with the purpose-driven creativity of logic, metaphysics, meta-science, meta-mathematics, and meta-technology. This is why a work to practically explain his philosophical legacies must be in synchronous and dynamic agreement with the reality of his personal conviction of life.

It is unfortunate that our present era appears to have neglected, rejected, and practically degenerated in the philosophical ideas of Pythagoras. For instance, our modern era is not technically illumined in the objectivity of mathematics as a pathfinder to the true knowledge of philosophy. This is why the province of further mathematics has greatly degenerated beyond words and thoughts. The philosophical eulogy that Pythagoras worked to bequeath to humanity is not being used; this is why there is a thwarted agenda from the uninformed that works against the natural agenda of Mother Nature. Pythagoras believed in the monumental use and application of the agenda of Mother Nature, which was tactfully coherent as being the institution of his legacies. Most of his legacies were practically drawn from the life and times of Plato and Socrates. It must be stated that Pythagoras, even though understood as a rational soul, was practically and scientifically souled in the soul of the universe. In this respect, he worked to achieve the ideas of philosophy, which include:

- Ability to achieve a purpose-driven ideology.

- Determined to achieve perfection at all times, despite the challenges of the chaotic and incoherent universe.

- Determined to achieve purity, particularly its effects as a driven force. This is why philosophy at any point in time is known as the mother of all.

- Ability to accept its onus task as being the mother of science. This is why Manly P. Hall, in his work entitled *Pathways of Philosophy*, designated that philosophy is always the seat of knowledge and the greatness of all powers. This illumined thinker opinionated strongly that Pythagoras can best be defined as the first principle of philosophy, including being the foundry of logic.

- Philosophy works to establish a purpose-driven concept, which must utilise the objective of Mother Nature in explanation, agreement, argument, and relativity. This is why most of the philosophical legacies of Pythagoras are always defined as the original roots of philosophical dynamism.

- Philosophy is not stylistic per se, and so Pythagoras was not equally stylistic in action and reaction. This is why the province of philosophical mathematics honoured his thoughts and his life for the achievements of his constructive philosophical creation.

- As one created with many leads of rational thoughts, he was tactfully honest both in the use of words and letters. This is why the protection of his philosophical perfection was foremost in his agenda.

It must be stated that earth life requires a purpose-driven life, with philosophical ideas and mathematical dynamism. The legacies of Pythagoras can best be defined as the immortal Christ of mathematical knowledge which in earnest was sequential in the reality of orderly pattern that the rational souls have known, recognized, and respected as the foundry of a mathematical databank. Even though the uninformed, particularly the barbarians, destroyed a lot of Pythagoras' legacies at Crotona, it must be stated that his philosophical exception, particularly to the world of objective and balanced reasoning, cannot be wiped out simply because he has not had a substitute in the areas of mathematical analysis, engineering mathematics formats, philosophical engineering, and logical engineering. The donation to Aristotle as the prince of philosophy was at the consent of Pythagoras, who is known all over the world as the philosophical kit and the mathematical magician. His thoughts on objective triangles, particularly those that have to deal with right-angle triangles must be remembered, honoured, and cherished.

This is why objectively Pythagoras is best defined in the minds of our new generation, particularly in the areas of creative construction, mathematical engineering, meta-logic, numerology, astrology, thought transference, creative arts, and the formation of right-angle triangles. In this respect, these legacies are translated as the origin of philosophical mathematics because of the following reasons.

- The reality of the concepts that are involved.

- The reasonability of the ideas that are being imparted.

- The dynamism and the synchronous agreement of his reasoning, particularly in the areas of philosophy and mathematics.

- The miracle of his translation of ideas to objective conviction.

- His Platonic and mathematical way of explaining points with algebraic equations, including the application of his geometric forts.

- The reasonability of his creativity in ensuring that philosophical constructions agrees with the database, which is always full of facts, figures, and features. This is why Pythagoras is known as the foundry fort of F^4, which means $F = F = F = F$ or conversely $F/F = F/F$. It can equally mean $F^2 = F^2 = (F \times F) + (F \times F)$.

- Notationally, Pythagoras is honoured and denoted as a factory of letters, a doctor of figures, and an institution of commerce, including a philosophical partner to mathematical proof. This is why this province designated him as the philosophical origin of all mathematical proofs.

It must be stated that any work on Pythagoras that is not correcting the use and application of mathematical data does not originate in the reality of his ideology.

In the next chapter, which is going to look at the mathematical logic of Pythagoras, a soul inspired like him must certainly leave a consummate and coherent legacy to posterity. I can fondly be designated as originating from the roots of Pythagoras, and I am taking up this challenge to give the world a philosophical and organized mathematical Britannica that is orderly written with inspired ideas and philosophical concepts that will help all eras in knowing and appreciating that as the world is mathematically driven, the universe is equally philosophically compacted in objective technology. This is why this work exists practically and at the proximity of the anatomy of philosophical mathematics.

CHAPTER 6

WHAT IS THE MATHEMATICAL LOGIC OF PYTHAGORAS, AND WHY WAS THIS LOGIC THE ANATOMY OF HIS ORIGINALITY?

As a rational soul, Pythagoras of Crotona had a mathematical logic; this is why he lived a different life when compared with his contemporaries. The originality of his mathematical logic empowered him to be philosophic, honest, and an institution of dynamic and objective expression. His educational understanding and achievements, including his lifestyle, were purpose-driven in technical and mathematical logic. A scientific look at his achievements, which came between Plato and Marxism, reveals that the originality of his mathematical logic was unequalled if compared with the achievement of his contemporaries.

Apart from being a wizard of philosophy, Pythagoras was a devoted logician of mathematical thinking. This devotion is why he was able to give the world the following ideas.

- The mathematics of ratio.

- The mathematics of fractions.

- The logical mathematics of formula.

- The mathematics of facts, figures, statistics, data, structure, and analysis with methods.

- The mathematics of log tables.

- The mathematics of the use and application of LCM.

- The logic of the mathematics of geometry.

- The logic and mathematics of algebra.

- The logic and mathematics of fraction.

- The logic and mathematics of engineering mechanism, particularly in the areas of quantum and electro-mechanical dynamism with its engineering philosophy.

- The logical and the use of table of values.

- The logical mathematics of equations.

What can best be defined as the mathematical logic of Pythagoras is practically essential regarding his deep thinking in the areas of angled triangles, equal triangles, and isosceles triangles. This is why his mathematical logic is original, natural, and platonic in nature, structure, expression, expansion, and reasoning. It is here that the originality of his thinking technically formed the anatomy of his logical mathematics.

Immeasurable contributions of Pythagoras to the areas of philosophy and science, particularly in mathematical logic, can best be summed in the following: $A + A + A = 3A$ and $A - A - A = O$ and $A \times A \times A = A^3$. In this arrangement, Pythagoras is of the opinion that A can logically be used to express the anatomy of mathematics in reference, inference, conviction, and convention. He believed that $2 \times 2 = 4$, while $4 \times 2 = 8$ and $8 \times 2 = 16$. This is why $16/2 = 4 \times 2$, $16/8 = 2 \times 1$, and $2 \times 2 \times 2 \times 2 = 8 \times 2 = 16$.

A look at the structural method of Pythagoras' multiplication principle reveals the proof in his belief of the use and application of mathematical logic. The originality of his formats was in empathic confirmation with the reality of letters and opinion of formats. It is here that the world of mathematics is always incomplete without the wizardry and contributions of Pythagoras, who lived his life as a mathematical logician.

As logic is technical and dynamic in the use and application of conventional concepts, there is a need to explain that human endeavour, at any point in time, is incomplete if one does not fall back to the use and application of mathematical logic, including the use of philosophical mathematics as a way forward.

Pythagoras utilised his mathematical logic to create a pathway for the province of philosophy, metaphysics, meta-logic, science, and technology. It is in conformity with the radius of mathematical thought that Pythagoras was able to realize the originality of his roots, having been initiated by Plato as a mathematical emengard. This initiation was why he was greatly ennobled in the areas of mathematical designs and constructions. His belief was that the world existed on the formulation of letters; this is why all his creations were relative and creative with the wide dimension of letter A, including how Figure 1 can always be utilised in the ebonite expression of his logical thoughts.

At this point, I must explain that logic and mathematics are interwoven and intertwined. Each is scientifically essential to practical independence. This is why a work to examine the mathematical logic of Pythagoras must always appreciate that he was greatly rational in the use of words and with the involvement of letters. Mathematics has remained the logical kit by which the world is driven. The universe is logically formatted, and it was Pythagoras' intention to reveal that man's actions and reactions must be tilted towards the direction of universal ideas, which will help him to appreciate, dialogue, and compact his thinking into the originality of natural logic.

It is on this note that Pythagoras' methodology with mathematical compartmentalization inspired the likes of Aristotle and Euclid, so that the world will always live at the tedium of logical lubricants, including being driven with the anatomy of logical diversification. This situation is why he strongly opinionated that man requires a mathematical regeneration, including a logical uplifting which, when properly galvanized, will help one to understand that creation is purely and perfectly the originality of philosophical mathematics, including being the anatomy of logical expression. This idea will forever live with him and with creation as the mathematical logic of his thoughts, including the originality of his anatomy, which the academia and research fellows (including the province of mathematics) has not made honest effort to investigate with practical resources as to how and why Pythagoras is named the idol of mathematical thinking with logical postulations.

CHAPTER 7

Why Is Pythagoras' Mathematical Logic a Point of Philosophical Reference?

Pythagoras was greatly endowed with the use and application of logical mathematics, logical wisdom, logical dynamism, and logical engineering. This is why he was able to perform and propagate the use and application of philosophical reference as a logical measure of values. Our universe is mathematically driven with the use and application of logical reference, which in earnest has remained the opinion of all particularly on issues that have to be argued upon—for instance, the way and manner Pythagoras utilised the wonders of mathematical logic in the following areas: pure and applied mathematics, pure and applied science, pure and applied wisdom, pure and applied logic, pure and applied philosophy, philosophy and logic, philosophy and meta-logic. Those areas were a great achievement in the advancement of human knowledge.

His driving force was essential in achieving the desire of philosophical purity and perfection, including all references and influences that have to deal with mathematical perfection with logical standards. Unfortunately his primitive and un-illumined era could not understand or appreciate his contributions. He was defined as the messiah of pure mathematical logic and of pure philosophical understanding. If any man is defined as the messiah of mathematical knowledge, including being an idol of calculus, it is Pythagoras.

It is on record that his creations and achievements, particularly in the areas of science and technology, were instrumental in taking the Grecian world to the pinnacle of knowledge. His mathematical knowledge was greatly recognized and inferred with philosophical reference as a point of understanding of issues, arguments, and dialogues, particularly those that had to deal with mathematical logic, mathematical philosophy, and meta-mathematics, including meta-logic.

A look at his contributions will certainly eulogize him as the prince of electrifying thinking, the hero of unequalled knowledge, and the idol and messiah of deep thinking. This is why a work to examine the impact of philosophical mathematics in the area of general knowledge must certainly bring Pythagoras to the fore.

The enabling way he was trying to establish Institutions of philosophical learning and colleges of mathematical teaching, including the university of science, philosophy, knowledge, and wisdom, was instrumental to his being noted for his arts of wisdom.

At this point, his mathematical logic was a point of philosophical reference because of the following reasons:

- He utilised mathematical logic to advance the course of philosophical reference.

- He utilised mathematical logic to create awareness of what philosophy is, particularly in the province of Crotona.

- His initiation at the proximity of Plato helped him to understand the purpose of life, including utilising mathematical laws in appreciating the philosophy of logic and the logic of philosophy.

- He was able to attract noble minds from far and near simply because his past knowledge in calculus, geometry, quantum science, and statistics was honest in the reality of philosophical thinking.

- His monumental inspiration was drawn from the celestial bank of mathematical logic; this is why his ideas about life and creation were philosophically driven with empathic reference.

- As a philosophical province, he welcomed living a rational life, including accepting his role as a servant with the use and application of mystified philosophical mathematics.

- As a philosophical crown, he was able to affect the life of others through the use and application of mathematical logic, philosophical logic, scientific logic, and technical know-how. This is why he is remembered as the prince of mighty knowledge.

- Understanding the blessedness of his roots, Pythagoras was able to discover his mission on earth and its purpose. This understanding empowered him to tackle obnoxious circumstances beyond words and thoughts.

- Logically, this rare gem was not cruel, and he never wanted to owe anybody. This is why his mathematical life was purpose-driven in achieving the concept of philosophical purity and perfection.

- As one that is greatly endowed with wisdom, he rejected pride and neglected all that had to do with ideas. This monumental idol of wisdom always believed that religion, culture, and tradition were all part and parcel of human existence. This is why he was the foremost philosopher to opinionate that in any effort of man, prayer is necessary.

Pythagoras believed in a life of quietude with patience, perseverance, and contentment. His life was not loud, but his actions spoke loudly for him. His mathematical logic was a point of philosophical reference, and that this is why he is remembered for his service of philosophical love.

CHAPTER 8

HOW DID HE USE THIS POWER TO DEVELOP A META-LOGIC PROVINCE?

Logic is the course of human endeavour that has to deal with advanced reason, which can philosophically be translated into meta-logic dynamism. The thoughts of Pythagoras were essential in the reality of this power. His realization of the use and application of meta-logical philosophy can be denoted as being the part of the rational souls. This is why his colleagues were greatly biased about him—his province was confused about where this idol of human knowledge could be placed.

Pythagoras was an orchestrated being in the areas of science and technology, logic, mathematics, psychology, geometry and algebra, calculus, and statistics. He was greatly known for being a panelled and impeccable mathematician. Respectfully, he was able to use this power to develop a meta-logic prowess because of the following reasons.

- Honesty of purpose with balanced reasoning.

- Being able to live a purpose-driven life from the point of practical empathy.

- Appreciating the desire of philosophy as the science that works to achieve purity and perfection.

- Recognizing that man is an infinite matrix from the province of mathematical kingdom.

- Recognizing without sentiments that meta-logic holds the key in advancing the spiritual man.

- Appreciating that life which is lived outside the dictates of philosophy is vanity upon vanity.

- Rejecting greatly all that has to do with falsehood and false teaching.

- Ornamenting himself as the messiah of pure reasoning, which includes mathematics, science, and technology.

- Resisting the views of others, particularly from the point of honest philosophical reasoning.

- Recognizing that every man is an institution of higher learning.

- Understanding that life is a game of two-way thinking.

- Appreciating that philosophy holds the key by which our intention to discover the wide province of Internet wisdom can be gained.

- Pythagoras was a compendium of ideas, a monument of knowledge, and a legacy of consolidated reasoning.

A work on philosophical mathematics must represent Pythagoras as a human being who is possessed with wisdom of the triangle, wisdom of three letters, and wisdom of the angled triangle. In the north he is at 90 degrees, and in the east and west he is also fully represented at 90 degrees. A look at this phenomenal ratio by which Pythagoras is being represented will certainly mean that he is a man of many visions, including being the philosopher of many versions. As the world of science respects her luminous knowledge, the mathematical know-how of mathematics contributed to the growth of quantum science, magnetic thinking, and information technology.

It is recorded in the annals of world history that this philosophical emengard must be honoured by the well informed, adored by would-be philosophers, and highly respected by philosophy and calculus.

Science and technology designated Pythagoras as the supreme master of objective principle. This is why there is a need to explain the anatomy of his meta-logic development, which finally gave rise to the development of this province.

The generic development and history of creation has shown that the messiah and the saviour are not being respected in their homeland. The illumined Pythagoras was a victim of this natural omen, but in earnest, it did not bother him because he was a philosophical iroko, a scientific shelter, a mathematical oracle, and a technological idol whose mission on earth was to establish a province that is purpose-driven with meta-logical ideas, philosophical understanding, mathematics, and scientific prowess. His thoughts for human life were practically cued in achieving perfection and purity, which is in confirmation with advanced knowledge for which philosophical mathematics works to harness, uplift, protect, and appreciate.

It is in the desire of Pythagoras and his colleagues to uplift human standards through the use and application of meta-logical mathematics, which in earnest is in synchronous agreement with the empowerment of all and sundry through the use and application of philosophical reasoning. Only a few in creation have worked to achieve this, of which Pythagoras, the idol of mathematics, stands tallest and most esteemed.

CHAPTER 9

WHAT ARE THE OBJECTIVE LESSONS OF PYTHAGORAS' LIFE IN REFERENCE TO PHILOSOPHICAL MATHEMATICS?

I should explain that the purpose-driven concept of philosophical mathematics is to analyse, ascertain, and buttress with cross-examination the objectives that make life what it is. It is a lesson that will make one appreciate the objective dynamism of Pythagoras, and it must certainly produce the following.

- The reality of the lesson.

- The reason for the lesson

- The being of the lesson.

- The becoming of the lesson.

- The certainty of the lesson.

- The importance of the lesson.

- The character of the lesson.

- The philosophy of the lesson.

- The language of the lesson.

- The phenomenon of the lesson.

- The science of the lesson, including all that is galvanized in explaining that the cosmos is compact and purpose-driven with a relevant objective.

Pythagoras illuminated his life philosophically from the point of honest knowledge. This institution designated as Pythagoras can best be defined in the originality of his creation as a co-creator, a co-monad, and a co-Leonard. This is why the province of posterity will forever remember the lessons that he impacted to the annals of human history purposefully.

It is here that the creations and inventions of Pythagoras are to be recognized in the following areas.

- Pythagoras as a philosopher.

- Pythagoras as a psychologist.

- Pythagoras as a metaphysician.

- Pythagoras as an initiate.

- Pythagoras as a Rosicrucian from the consummate bank of Ross cross.

- Pythagoras as a mathematician.

- Pythagoras an astrologer and numerologist.

- Pythagoras as a scientist.

- Pythagoras as a technologist.

It must be mentioned that the life and time of this great philosopher was greatly cultured in developing a province that will be free from academic, social, administrative, economic, and industrial poverty. He greatly worked for the upliftment of human dignity. This is why the province of philosophical mathematics donated what can be designated as a life chair to the immortal memory and mathematical credit of Pythagoras. The human race has a lot to learn from his unequalled objective lessons. A work to analyse and understand the magnanimity of these lessons will certainly extol the life and time of this millicent, including one who is practically angled at the righteous angle of a triangle, which in reality is synchronous with the detects of mathematics and philosophy.

It is unfortunate that our present era is practically dwindled, technically dwarfed, scientifically doomed, and technologically irrelevant and irreplaceable in the objectivity of the life and time of Pythagoras. This is the situation that has created a culture of Armageddon, particularly in the educational, administrative, environmental, and social sectors with a coefficient demystification of philosophy and mathematics, which in return has placed the knowledge of the youths at an incoherent sea. Our modern civilization and leaders have not reasoned to find out why the academic and academia are lacking in the full knowledge of Pythagoras, particularly as it affects his illumined contributions to the world of science, philosophy, technology, and logic.

A great research conducted among the academic institutions universally has revealed that the provinces of mathematics, logic, science, and technology are greatly enmitied with uninformed thoughts and myopic lessons that are creating viruses with monumental deluge. The way and manner the material teachers of mathematics explain mathematical principles, theorems, and formats, is un-Pythagorean, both in nature and in action. This is why the urgent necessity of making a reversal is the only thing that will save the human race from abundantly becoming a mathematical and calculative robot.

A work to analyse, examine, and appreciate what can be a configured, transmitted, and transcendental analysis of Pythagoras' lessons will appreciate that man has been revolving without growth and knowledge. This is why we cannot be able to place a benchmark between what is defined as philosophical chemistry and logical alchemy. We cannot determine how mathematics can be qualitative and quantitative in analyzing facts and figures using the radius and radii of the thinking of Pythagoras, which is systematic, pragmatic, and luminous.

It is here that this chapter is greatly interested in what can be defined as a mathematical data point that exposes the life of Pythagoras as a wizard to his province, a genius to modern man, and a metric mathematician to all eras. The dynamic functions of philosophical mathematics is galvanized with philosophical crystallization that Pythagoras is mystified as the idol of facts and figures. That includes being the ornamented scientist whose lessons will explain how and why the human race can prosper using the wide ecstasy of Pythagoras' knowledge, which is luminous in the best objective of the lessons used and configured in the anatomy and structure of philosophical mathematics.

CHAPTER 10

AN OVERVIEW IN FAVOUR OF PART 5

Pythagoras was a teacher of truth, a master of knowledge, a philosophical windfall, and a splendid human with titanic structures. This character greatly affected his lifestyle. It is in the thinking of this overview to present a man that is mathematical in action, triangular in reasoning, quadratic in argument and presentation, simultaneous in logic, and fractional and realistic in relations. This is why this overview is simultaneous in presenting Pythagoras in the world of his own, in others' world, and in the world of unusual understanding.

It is unfortunate that our present era appears to be materially driven. The un-philosophic transactions are presenting what can be defined in this context as a hibernated virus that works against one's creations, actions, and transactions. This is why one needs to have a break in the present, uncreative world. A look at the simultaneous agenda of Pythagoras' actions and what he left for posterity will certainly appreciate the following.

- The validity of his logic as mathematical principles and policies.

- The vastness of his philosophy in reasoning, actions, relations, and transactions, with a view to appreciating why and how he was honoured and adored as the prince of philosophical mathematics.

- The validity of the inference and reference of his philosophical opinion.

- The enabling power that is contained in his meta-logic province.

- The way and manner he was initiated, and how he utilised the lessons of his initiation in forming what is a mathematical apparatus with philosophical apostolate.

- Why and how he is designated as a rational soul among his contemporaries.

- His contributions, particularly in the field of meta-mathematics, meta-philosophy, meta-religion, and meta-science.

It is important to examine and explain how he was strongly admonished and advocated as the monumental saint of his era, a man that is mystical and mystified in the business of transmutation of iron into chemical, of alchemy into forms and formats. This is why the history of polymatric dynamism (at any level) and quantum science (with magnetic understanding) are absolutely incomplete without reference to the mystical ingenuity of Pythagoras, both in projecting philosophy and in propagating mathematics.

It is unfortunate that our scientists are not greatly focused either in reason or in action. The way and manner the life of Pythagoras affected other philosophical emengards such as Aristotle, Emmanuel Kant, Max Miller, and Karl Max is evidence to show that the province of philosophical mathematics is tactfully secured and protected by the single effort of one who is denoted for this peculiar assignment.

A concrete study of Plato's metaphysics from his era to the Renaissance has composed a mathematical symphony that is in synchronous agreement with the desire of Mother Nature. The tedium of this overview is practically in harmony with the aspects needed to produce what is esteemed and recognized as a purpose-driven treatise. It is in the absolute recognition of his great contributions that the world is forging ahead, particularly as a mathematical kit, a logical anatomy, a philosophical province, and a relative entity.

Pythagoras, who is known and mystified as a realism in angled triangles, including being a reality in composite analytical mathematics, is hereby donated to the one-world family for appreciation, particularly to the well informed who appreciate that the best we can do is to be aware of all inspired ideas as metaphysical action plans. It is through those plans that all is contained in the nectar and hectare of Mother Nature, and it can be accessed and invented using what can be accepted as perfect and rational laws, which helped Pythagoras to practically and scientifically dominate the world of balanced reasoning as a mathematical millicent and a philosophical adept. His ideas will ever be remembered by posterity, particularly in the worlds of science and philosophy.

This overview stands high and tall in remembrance of Pythagoras and his aborigines as monumental Orients who were supreme gifts to the world of compound and absolute knowledge.

Pythagoras Quotes

"Do not say a little in many words but a great deal in a few."

 "Friends are as companions on a journey, who ought to aid each other to persevere in the road to a happier life."

"Above the cloud with its shadow is the star with its light. Above all things reverence thyself."

"Strength of mind rests in sobriety; for this keeps your reason unclouded by passion."

"Concern should drive us into action and not into a depression. No man is free who cannot control himself."

"Rest satisfied with doing well, and leave others to talk of you as they will."

"A thought is an idea in transit."

"Silence is better than unmeaning words."

❖

"Anger begins in folly, and ends in repentance."

PART 6

PHILOSOPHICAL MATHEMATICS AS RATIONAL AND ABSTRACT SCIENCE

"There are things which seem incredible to most men who have not studied mathematics."

—Aristotle

INTRODUCTION

Since the dawn of consciousness, humanity has always looked upon philosophy and mathematics as a rational science and an endeavour that is the fulcrum on issues which cannot be gained directly. It is unfortunate to say that the jet age, which was formerly celebrated in California in 1979, has not changed the thinking of humanity. The appearance of material computers, Internet usage, and more has not produced a trailblazer in the original belief and understanding of humanity in respect to philosophy as a rational science, and the abstract nature of mathematics with its quantious explanation. This ugly situation has affected the entire universal educational growth, intelligence, commerce, and industry. The political sector, with its administrative mechanism, has equally suffered the vision effect of this myopic understanding.

Philosophy and mathematics are not abstract in nature—they are not even rational. But it must be stated that their electrifying unity is greatly imbibed in presenting and uniting them as an aesthetic and cosmogonical science. Philosophical mathematics is the umbrella of human dynamism; it is the dynamic engine of all luminous thoughts. This is why I opine with absolute conviction that philosophy is the maker of great thinkers.

I should explain practically that the way and manner civilization has handled the affairs of philosophy and mathematics is practically responsible for the negative and stagnated effects of human dynamism, particularly in the areas of philosophy, mathematics, science, engineering, technology, and medicine. It must be stated that philosophical mathematics is all embracing with a lot of purpose-driven agenda.

The purpose of this introduction is to regenerate and remove the un-philosophical cataracts that have greatly clothed humanity because of wrong notions about the purpose of life, the dignity and magnanimity of logic, and the way one should hold honest transactions. This is why the world has had fewer honest and illumined mathematical philosophers whose lifestyles, including their creations and contributions, have remained unequalled in the annals of human history.

Consequently, the rationality of philosophy, including the abstract nature of mathematics, is long overdue to be detached from the thinking of humanity. It may be explained that science is philosophically driven, whereas technology is mathematically driven. This is why I am presenting a universal formula and format of the above statements using the following signs.

$$S/P = T/M, \text{ or } S/P + T/M = ST/PM.$$

It is important at this point to use what can be configured into technological formulas, which will certainly open a new horizon for the appraisal and appreciation of scholars in knowing that philosophy, which is galvanized to be the mother of science and knowledge, originally conceived the thinking of mathematics. A work to explain the relative and scientific nature of philosophy and mathematics will always appreciate that the world is philosophically driven and the universe is mathematically operating with the use and application of scientific logic, technical dynamism, abstract esteem, and rational thinking. This is why the world is counselled by the province of philosophical mathematics to remain at the sea and opium of human dermatology. At this point, it is agreeable and visible in the face of human endeavours that the anatomy of human inventions, at any point in time, cannot be detached as being incubated and driven in the realistic dynamism of philosophical mathematics.

As opinionated by the Oriental wizards, whose lives and times created a province of abstract and rational thinking, the human province must be charged and challenged in appreciating what philosophy, in its originality, is including in explaining the deep thinking of mathematics. This is why we need an apparatus of the magnanimity of philosophical mathematics, including using what we can access to create a scientific balance, a technological incubator, a political province, and an engineering format.

It has remained in the opinion of meta-logic and meta-science that philosophical mathematics must be utilised in formatting concepts and ideas that can be purpose-driven with scientific relativity. At any point in time, science cannot produce a globalize community, a united province, and an acceptable data if there is no practical involvement of philosophical mathematics, which is naturally known to be the hub and framework of science.

As opinionated by Manly P. Hall in his classical work *Astrological Keywords,* this honest, balanced, and mathematical thinker was able to reveal that there is no abstract or rational constituents in the ability of philosophical mathematics in the reasoning of philosophy and in the psyche and system of mathematics. It is here that he gave the world what I may define as a philosophical key through which the universe can tap and access the wonders of mathematics, including the podium of applied mathematics.

The human race is advancing in material technicality, but it is not advancing in natural and illumined philosophy. This is why there is imbalance in human relations. The way and manner that scientific laws, principles, and policies keep changing is evidence showing that man has materially understood the dignity and the electrifying functions of philosophical mathematics as a short pen, which is vaster than long memory.

The purpose of this introduction is to open up a vista of knowledge with a tedium of wisdom which, when understood, will help humanity in appreciating the sine and cosine formulas in which we can use S/P and T/M to achieve, create, understand, and appreciate.

There is a need to regenerate the original thinking and belief of the universe that philosophical mathematics is both a rational and abstract science that can only be appreciated and proved by those whose characters, transactions, and understanding of life are cued up with inspired understanding. Every human being must appreciate that he is a rational soul as well as an abstract entity, which is not separate from the universal workings of mathematics.

It is unfortunate that our thoughts, including our myopic education, have not been able to launch us into the great adventure, ornament, ecstasy, and province where it is said true that all knowledge abounds. For the record, philosophical mathematics is the seat of wisdom and is equally the luminous work of knowledge. This is why the well informed have always designated it as holding the key by which we can connect the varsity of Internet unity with its dynamic and pragmatic knowledge, which is free for all to access, accept, and understand. It is here that the open connection is hereby effected for human upliftment, either as a rational science or an abstract knowledge.

CHAPTER 1

WHAT IS ABSTRACT SCIENCE?

I should explain the ancient belief that philosophy and mathematics are wider sciences that are abstract. It is important to give a dictionary definition of **abstract** and **rational** for the use of scholars, institutions, research fellows, and the general universe. Our definition is going to be both a reference and in inference from some sources like *Webster's Dictionary*.

This work defines the word "abstract" as "to draw from or to separate". It equally stated that abstract stands for "thoughts from any particular instance not concrete".

I define the word "rational" to mean the following: "based on or derived from reasoning; able to reason including in mathematics; designating and/or quantity expressible with a radical sign as an integer or as a coefficient of an integer". A look at this definition will certainly appreciate that philosophical mathematics can best be defined as an integral science with a coefficient integer of balanced reasoning.

In my own understanding, the word abstract can best be defined as:

- Drawing an inference from the reasonability of existing information.

- Appreciating the anatomy of reality from logical realism.

- Existing in the originality of reality without mixture or pollution. This is why such elements like water, the desert, the mountains, the clouds, the stars, the moon, and the sun can philosophically be defined to exist in the perfect anatomy of perfection.

Reasonably stated, the word abstract is best defined as *a coherent substance or material object not concrete in one format but united in existence, in understanding, in expression, and in explanation.* This is why the dynamic functions of philosophical mathematics can best be seen as an abstract science that exists in the originality of its own.

The principle of originality in this context is coherent in the use and application of policies, powers, purity, and perfection, which are dynamic standards of abstract sciences, of which philosophy stands tall as an amalgamated canopy.

In essence, the philosophy of abstract science, which philosophical mathematics stands to explain, can best be defined in the following methods.

- Being a word to the wise.

- Being a light in the desert.

- Being a concretized and illumined anatomy.

- Being philosophical in reasoning.

- Being objective, dynamic, and regenerating.

This is why a work to determine the definition of the word "**abstract**" as it relates to philosophical mathematics must not fail to appreciate that $A2^1$, $B3^2$, $C6^4$, $D9^6$, which can be difficult to analyse, explain, and understand simply because it does not exist in a standard mathematical method. In this respect, abstract science has to do with reasoning objectively in the tedium of philosophy, the language of mathematics, and the reasoning of technology. This is why philosophy is sequential for the use and application of abstract wisdom.

The diagram below will explain my originality of the definition of the word "abstract" in order to explain how and why philosophy is the abstract science that warehouses and empowers knowledge with electrifying wisdom. This diagram, which is not going to represent the thinking of mathematics, will certainly be utilised by the province of philosophy, mathematics, and transcendental technology in explaining and empowering humanity to know that the universe exists in the anatomy of abstract omniscient, omnipresent, omni-reality and omni-being, which exists in omni-realism.

It is important to explain that the general abstract science that originated philosophy was indeed the podium of all reasoning, which man has not been able to understand, behold, and appreciate. This is why the universe has made a tympanic strive in establishing a world province into a philosophical purpose. In this respect, abstract science can best be defined in the anatomy and structure of a full format that designated reality in the originality of inspiration as follows.

- Abstract science is the science and knowledge of man in the entire cosmos.

- It is the tedium of the philosophical being and becoming.

- It is the imperial of rational analysis with statistical method.

- Abstract science can best be defined as the philosophy of meta-logic.

- It can equally be defined as the originality of technical structure of developmental technology.

- I define it as an idealistic democracy with idealistic reasoning.

- It can equally be defined as a simultaneous mathematical dynamism.

- Abstract science can be defined as the ration of truth that expands and appreciates the electrifying power of unity.

- It is the tedium of the honest knowledge that must be backed up with earnest and purpose-driven wisdom.

- It is not a material polyester, and neither can it be defined as the alchemy of material transmutation.

- It can be defined as the philosophical science of light, including being the chemistry of the numeral.

- Abstract science can best be defined as the Oscar of input and output mechanism.

- It is known and understood as the maker and the originality of all things.

- It can be explained as the natural polyester of polymeric anatomy by which all the institutions of aesthetic dynamics are gained with scientific appreciation.

A look at this diagram will certainly let the reader appreciate this work as being scientifically and mathematically rooted in the understanding of what abstract science is all about. This is why and how an adventure into this province must be seen: as being in the pathway of philosophy, including becoming the Oriental journey of deeper life.

The concept of a journey in abstract knowledge is cancelled with the use and application of all originality, which is scientifically enabling, both in knowingness and appreciation.

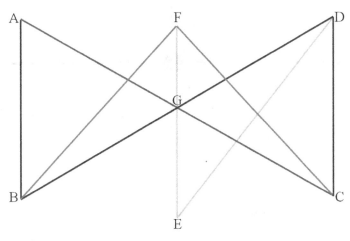

Fig. A

ΔABF = ΔDCE with angle G perpendicular to both angles ΔABF and ΔDCE.
Ratio-wise, AB = FE = DC and AC = BD
∴EG = FG and GE = GF

The philosophical mathematical drawing above shows an input of abstract thinking that can be used to explain and appreciate the rational of the reality of original science, with the tedium of natural technology and with both info-mechanics and info-dynamics reasoning.

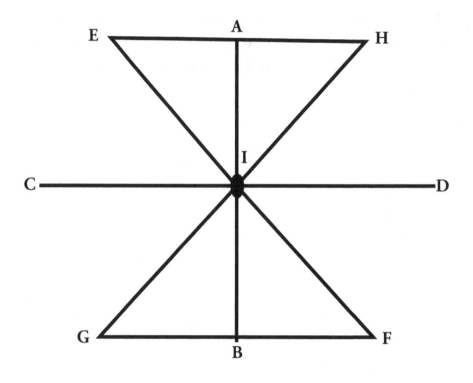

Fig. B

AEH = BGF

ECG = HDF

AIB = ECG = HDF

AH = HD = DF = FB = BG =

GC = CE = EA

EA = CI = AH = DI

∴CI = GB = DI = FB

A look at the scientific explanation of the full, philosophical creations will certainly make one appreciate that mathematics is real in originality, whereas philosophy is realistic in reasoning. This is why the universe has greatly remained a podium of abstract knowledge, which is forever recycling ideas with the use and application of material enabling, material, and meta-informatics enabling. This is also why the urgent necessity of explaining this chapter as both a practical and astute energizer is borne out of the fact that the thinking of humanity is not meta-logic in the original appreciation of abstract science—which is queued up in the enabling philosophy of abstract rhythm, and which is transcendental and transmutational. The use and application of the wider community that abstract science as best defined and understood from the living angle of mathematical trigonometry, with the philosophical input and output of balanced reasoning.

CHAPTER 2

WHY IS IT SEEN AND KNOWN AS SUCH BY THE COLLEGE OF THE UNINFORMED?

Epictetus said—

The first business of philosophy is to part from self-deceit. In this analogy, he estimates and explains this fact as follows.

- A man who overestimates himself will underestimate his world.

- To be humble is to admit the greatness of the universe.

- Out of humility arises the capacity for understanding.

Aristotle said—

Philosophy is the science which considers Truth.
In this analogy he states-
Though Pilate's question remains unanswered, it is the opinion of the wise; it is not unanswerable. Philosophers know that there is but one way to discover Truth and is to become Truth. Philosophy is the science of becoming.

Lavater says—*True philosophy is that which makes us be ourselves and to all about us, better.*
He further explains-

There is no merit in wisdom; there is no reward in knowledge, there is no comfort in faith, unless these things manifest outwardly, subduing the violence of action and bringing us to a harmless mode of existence.

Voltaire opines that—*The discovery of what is true and the practise of that which is good, are the two most important objects of philosophy.*

He stated further—
Thinking is not merely an exercising of the mind. It is the directing of the mind. Only such as have organized thought to the accomplishment of some actual good are worthy to be called the wise.

Londos says,

A true philosopher is beyond the reach of fortune.

To be truly wise it is necessary so to love wisdom that there is no place left in the mind for anxieties concerning the temporal state. He who lives in desire for plenty or in fear for loss has no right to call himself a philosopher.

Larmatine: *Philosophy is the rational expression of genius. In this regard he expresses:*

We may define genius as special attitude but when special aptitude is directed to the most important of all efforts, the perfection of self, it is termed philosophy.

A philosopher is a genius who has discovered the most perfect use of his abilities.

Jouber says,

Whence? Whither? Why? How?—these questions cover all philosophy.

Whence—God. Whither—to God. Why—Law. How-Wisdom.

From the Infinite to the Infinite we must proceed. The why of life is known only to the Maker. But from philosophy we learn how to fit ourselves for final identity with Cause.

Carlyle states—*The philosopher is he to whom the highest has descended, and the lowest has mounted up; who is the equal and kindly brother to all.*

He explains—

Between heaven and earth stands the wise man. His earthly part has been raised to its highest perfection the instrument of a divine purpose. The higher part, the soul itself, has become tolerant of the limitations of the body and wise in its own weakness, is tolerant of the limitations of all other things.

A systematic appreciation of these philosophical epilogues, which can be etched in deep thought with mathematical methodology, must certainly answer this chapter's question. Philosophical mathematics is not rational per se, but it is logically dependent on the inference. This is why life is mathematical and logical, and why the science of honest logic is practised with methodology.

In this respect, philosophical mathematics works to seek expansion of honest knowledge, including understanding the enabling power of driven purpose. It is unfortunate that from the dawn of consciousness, humanity has enmitied a lot of abysmal and un-philosophical thoughts in this province of pure and perfect reasoning, with the hope to decentralize the anatomy of philosophy—which indeed is the maker of great thinkers, of which man remains symbolic and protected. The concepts of rationality and abstraction are naturally incoherent and are practically the making of man. Science is both cosmic and comic, and philosophy is practically glorious with the mathematical enabling of the cosmos, by which the tedium of honest understanding that makes the wise can be gained.

At this point, we must not fail to defray these ancient eulogies, which have not helped humanity in promoting permanent and prospective concepts:

- Philosophy is essential in monumental foundation. This is why it formatted mathematics as a partner in thoughts, in expression, in understanding, and in principles and polices, existing with modesty and empathic methodology.

- Philosophical mathematics is the mandate and modus of life.

- If it is rational, it can be argued that the rationality of philosophical mathematics is in line with universal intakes and outtakes. This is why its dynamic feature can better be accepted and explained as the synergy of honest life.

- As the first principle of thoughts and the foremost policy of physics, philosophical mathematics is in the knowledge of empowering creation to be purpose-driven.

- As a mathematical universe, which has a futuristic solvent, it can best be defined as a mathematical genealogy.

It is important to explain this not for the uninformed but for the well informed, simply because philosophical mathematics has remained the dynamic technology of creation and the engineering of the soul. Every effort must be made by all and sundry to appreciate and uphold the tedium of this explanation, because philosophy exists purely and practically on the anatomy of inspiration. This is why such noble and rational souls like Plato and Pythagoras must be appreciated as being in the becoming of philosophical mathematics and becoming in the being of philosophy.

For the record, it must be explained that philosophy exists dynamically with its mathematical concepts as leading the bull by reasoning, and reasoning to protect the bull. This is why a work that will explain this chapter must be esteemed in the objectivity of practical realism, and why every noble creation is best defined and designated as the product of philosophical ingenuity.

CHAPTER 3

WHAT IS THE LOGICAL REASONING OF THIS ARGUMENT?

It is important to explain that out of objective conviction, one establishes facts and figures. Philosophical mathematics is logically argumentative and practically subjecting all facts and figures to critical examination. In explaining what is the logical realism of the argument on how and why, the college of the uninformed treats philosophical mathematics as both an abstract and a rational science. It can only be explained using the institution of critical logic, which must be synchronous with the ennoblement of scientific and mental logic.

In this respect, philosophical mathematics is realistic in sustaining its province as a reality that is dependent in realism. It is explained that the logical argument can always be dated on facts and figures using the following numbers as a representative institution: 0, 1, 2, 3, 4, 5, 6, 7.

Logically, the above letters can be progressively reversed to 0, 14, 15, 16, 17, 18, 19, 20, 21. In this respect, we must appreciate the phenomenon and the expression of philosophical mathematics tactfully and scientifically with the use and application of progressive mathematical concepts. Four important elements are usually at work with each other.

- The multiplication principle.

- The addition principle.

- The subtraction method.

- The equation format.

We can equally appreciate the logical argument of these facts with the use and application of sine and cosine. This is why the use and application of compound proportion and compound configuration can equally be used as logical bases to buttress this argument. The concept of the realism of this logic must remain an issue and will certainly bother the universities, research institutes, and colleges of the academia.

It is important to explain in the tedium of this argument that this inconsistency is capable of tearing ideas apart. This argument is equally capable of providing two parallels that will be difficult to meet. For instance, the abstract understanding of philosophical mathematics during Plato's metaphysics was not in sequential pattern with its appreciation during the era of Max Miller. This is why a work to logically argue on the realism of this argument must certainly achieve those elements, which will help the colleges of philosophy and the universities of mathematics further examine what can best be defined as to how and why the ethical world has greatly appreciated the province of philosophical mathematics.

For instance, it is established that Plato's metaphysics was scientifically, morally, and spiritually driven in developing a philosophical mathematics province, which was greatly responsible for providing a basis for such eminent scholars like Pythagoras, Aristotle, Emmanuel Kant, Oscar, and Einstein. His influence in the lives of these people greatly propagated the engineering of philosophical mathematics, including its dynamics, applications, and appreciations.

The time has come when there should be a departure from the indecent understanding of philosophical mathematics to a decent and purpose-driven understanding of this science. In this respect, there is contravention about the realism of this argument by the well informed, the uninformed, and the would-be informed—including what I have, in this context, designed to be the global argument on the logic of this realism. Mentally and morally, philosophical mathematics is looking for facts and figures. This is why a work to align it to the province of material science has always produced limited ideas. The argument can only be explained with philosophical mathematics, and with the application of the following.

- The first principle of science.

- The first principle of philosophy.

- The policies and practises of mathematical dynamism.

- The enabling realism of the Twelve World Teachers.

- The sacred teachings of all ages.

- The secrets of light.

- The quantum grail, which is essential in the light of truth.

- The dynamism of science.

- The philosophy of technology.

- The philosophy of beauty.

- The metaphysics of mathematics.

- The meta-logic concept of universal configuration.

- The natural laws of transmutation.

- The universal laws of "Man, know thyself".

- The use and application of different codes of ethics.

A look at the quantum and tedium of the wordings of these ethics will certainly buttress a permanent and co-creative argument on the logical realism of philosophical mathematics as a rational science or an abstract endeavour.

The twenty-first century aborigines must be well prepared for the challenges of the universe, of which their only solutions can come when they are practically regenerated in the use and application of philosophical mathematics. For instance, the tedium of inquiry at the province of California by the arrival of the jet age was practically mechanical and scientifically electro-mechanic. The practical issue remains that all these endeavours were greatly geared towards achieving a mathematical civilization. Invariably, this is not the opinion of philosophical mathematics—its opinion rests practically in the quadrants of the following.

- Establishing true philosophy, which is beyond the reach of fortune.

- Energizing a practical philosophy, which is in agreement with the eyes of the soul.

- Ensuring that philosophy circumnavigates the human nature in order to be purpose-driven.

- Establishing philosophical mathematics as the rational expression of genius.

- Ensuring that philosophical mathematics is utilised in achieving the positive functioning of the mind.

- Ensuring that philosophical mathematics becomes the choice, the act, the law, and the strategic purpose of life.

- Using philosophical mathematics to discover what is known as the value and volume of truth, including the practise of that which is good, and which in realism forms the basis for the honest living with philosophical mathematics transactions.

In this respect, man must always appreciate that the essence of this argument, which is both practical and logical in realism, is to explain that to some people, philosophical mathematics appears to be an abstract science, whereas to others it is a rational endeavour. But to the uninformed, it is the way, the application, and the occupation, including being the spiritual endeavour of genius.

I provide this maxim to help humanity arise.

Until a man appreciates himself as an ornamented philosophical asset, his transactions and creations will continue to be chaotic with multiple and infamous deceits.

CHAPTER 4

HOW CAN THE MATHEMATICAL REALITY OF THE ARGUMENT BE ESTABLISHED?

Philosophical mathematics is a coherent and balanced science that is practically driven on the general concept of mathematical logic, technical logic, and extensive and expansive philosophy. The nature of this science prompted Pythagoras to philosophically caption this system as the enclosure of all postulations. That is why this argument has bothered the academia and the institution of mathematical philosophy, including all the colleges of practical and rational science.

It is in the tedium of this argument to coherently explain that there is always an argumentative reality that drives the fragmented units of philosophy with mathematics. This is why a look at what can be defined as mathematical equation always maintains a natural and philosophic emblem with the logical workings of philosophy—simply because no mathematical equation can be explained without a scientific involvement of philosophy. This is why human endeavour is philosophically aquatic and logically simultaneous. The practical understanding of this strenuous logic will make one to understand that philosophical mathematics maintains a benchmark between the borderlines of conventional science, including being the broader base of natural science.

It must be stated that the reality of this argument cannot be accepted without involving the following philosophical mechanism, which indeed remains a modest and logical method.

- What is in the embodiment of the law of mathematics?

- What is the scientific enclosure of logic as a practical, methodological science?

- What is the origin of the misconception of this reality if any?

- Have these misconceptions produced any format or standard by which further investigations can be realized in mathematical realism?

- How can we appreciate philosophical mathematics as a pure science that originated in pure reasoning?

- Why is purity essential in the best understanding of philosophical mathematics?

- If philosophical mathematics can be defined as a scientific emengard, what can we designate as the locus of its illumination?

- If it is philosophic in lintiousness, how can we appreciate the anatomy of its broad line concept?

- How and why was it designated as the meta-logic of Plato's mysticism?

- What is the rationale behind its Platonism as a practical thinker whose ideas are technically realistic in producing conducive and coherent results?

- Why is the academia fragmented with abysmal disagreement on the philosophy of philosophical mathematics?

- How and why is it recognized by the well informed as the imperator of scientific globalization with universal cosmogony?

- What can be defined as the target of the logic, including the titanic reality of the argument?

- From the eras of Plato to Pythagoras, which produced a purest civilization of logical renaissance, how and why did Karl Marx develop a technical, practical, logical, and enabling philosophy of mathematical tedium?

One should appreciate that these arguments must continue to live with creation, simply because the unity of philosophical mathematics is scientifically essential in establishing the realism that is contained in the best galvanized postulation of mathematics, philosophy, and logic. This is why I have the following as the simultaneous conviction of the argument, which is contained in the fraternization and fragmentation of the reality and can best be understood as a standard compartmentalization of data, figures, statistics, calculus, mechanics, mechanism, and magnetism with quantum mathematical theorems. It is here that the following should be established.

- That every triangle is philosophically classified as mathematics in producing an abstract technicality.

- That the quadratic equations are naturally and philosophically quantious.

- That its locus, which can be termed as the anatomy of Ohmity of all, is mathematically explained using the abstract nature of logical philosophy.

For the records, Ohmity is a standard and scientific explanation of data that is strenuous and abstract to understand. It is the concept of mathematical Ohmity which led Pythagoras to appreciate and define his uninformed era as being and becoming at the ecstasy of metals and hoodlums from a dependable understanding of the argument.

It must be established that every endeavour—which includes research methodology, management and technology, engineering production with production technologies, and physics electronic—cannot be appreciated without knowing the structure of its realism, the dynamism of its reality, and the anatomy of its philosophical mathematics. That includes what can be explained as using the scientific approach to appreciate the galvanized fragmentation, which in essence is purpose-driven in scientific and practical encasement.

The urgent necessity of providing what can be seen by the advanced soul as the rationale of a realistic argument is borne out of the fact that the world is mathematically driven with logical fact and factors. The universe is practically at the tedium of being involved, and it revolves due to the honest wisdom of mathematics. This is why the following has remained the honest way of pure philosophical mathematics.

- The principles of simultaneous equations.

- The methodology of the quadratic equation.

- The logic of triangular theorems.

- The driving force of the remainder theorem.

- The philosophy of linear equations

The use and application of circle theorems, which is mostly explained with the inverse and converse systems of right angled triangles, can be patterned and positioned in the following methodologies: 45°, 90°, and 360°.

Philosophical mathematics is simultaneous as a rational science, an abstract endeavour, and a knowledge that is the philosophy of the informed.

In my illumined understanding of this work, I designated the reality and the conviction, as the living, oblong orb of philosophical mathematics—which is the formula, format, method, science, and engineering mentality that inventors and inventions involve with the technical utilisation in creating standard and acceptable mathematical products.

CHAPTER 5

PLATO AND PYTHAGORAS ARGUED THAT PHILOSOPHICAL MATHEMATICS IS NOT ABSTRACT BUT A KINGDOM OF KNOWLEDGE, WHICH IS MEANT FOR THE WELL INFORMED. WHY WAS THIS OPINION FORMED?

Platonism, as recorded in the ledger of the wizards, has remained the focused thought and development in Pythagoras' metaphysics of mathematics. Philosophy is not abstract but is presented as a collective and colossal wisdom of knowledge, which can best be defined as the monumental power of the well informed, of which Plato and Pythagoras were greatly known. Uplifting the experience of their era slightly differed with our modern and contemporary experience. This fact is why the history of the thought of Plato has remained a classic diction of knowledge, particularly in the province of the academy.

A lot of abstract and configured opinions were responsible for their philosophical argument. The following can be defined as abstract benchmarks.

- Plato, as an illumined philosophical metaphysician, opinionated that nothing is abstract except man, whose actions and reactions appear to be titanic in abstract realism.

- Pythagoras seconded this opinion by insisting that abstract philosophy cannot be the way of honest souls. This is why he redialled the meta-philosophical thought of Plato: in order to synchronize with the philosophical thoughts of his own Oriental horizon.

- Both gurus were philosophical idealists whose ideas were spread and prompted the creation of a united philosophical universe.

- Their arguments were accepted to buttress and configure all colossal and disunited thought of humans in their era. This is why their idealistic reasoning was balanced with the wisdom and thoughts of philosophy.

- Both men can best be defined as epoch epistemological realists who were inspired in the best objectively of philosophical illumination to detach, segment, and moderate the affairs of humans with the use and application of ethical and scientific realism.

- Plato was an initiate of the hermetic and holistic principles. Pythagoras was a mathematical initiate who was greatly attached and driven with the configured thoughts of the Kabala.

- Both believed in the reality of unity in diversity, including looking at life from the ontology of metaphysical provinces.

- Plato believed in the gospel of the divine journey. Pythagoras believed in the powers that configured these journeys as monumental planetary easterners.

- Both were illumined in the best understanding of the Zodiac, astrological science, and astronomy. This is why Plato has remained configured in relating the best knowledge of palmistry scientology.

- Both existed in the era that was dignified in human history as the period of objective reasoning, scientific enabling, and technological impact. This is why they were able to utilise philosophy both as a science and as an enabling art.

- They were illumined occult masters who understood that man could be ennobled with the use and application of the hidden forces, which included the unseen powers and the unseen realities. This is why they argued greatly that occult would forever remain the only authority, which would change and recharge the thinking of man as an illumined creative millicent.

- Plato is known as one of the Twelve World Teachers, and Pythagoras is appreciated universally as a demigod of mathematical engineering and figurative dynamism within the laboratory provinces. This is why I define him as the databank of universal statistics.

Mathematically speaking, Pythagoras is practically illumined as the Oriental wizard whose knowledge of mathematics has remained unequalled. In this respect, both illumined thinkers and actors presented the universe as cardinal crescents of triangled esteem, including being in an estate that is horizontally cycled and recycled by the philosophy and knowledge of known and unknown powers. It is here that philosophical dialogue has remained ensured and assured in the best opinion of balanced objectivity.

Mathematics is configured in letters and alphabets, but philosophy is mystified in the best utilisation of words, thoughts, medium, maxims, eulogies, and epigrams. This difference is why contemporary issues have continued the ways of determining what can be found on the databank of philosophical dynamism.

The locus of the cardinal, which has remained a figurative argument, is that Plato and Pythagoras are coordinators and co-operators. This is why they were able to influence the likes of George Hegel, Emmanuel Kant, and Aristotle. Religious fathers like St. Augustine, who has invariably remained the doctor of the church; St. Thomas Aquinas, who was known for his orchestrated and tympanic logical argument; St. Francis, who was designated as a philosophical homily; and such other metaphysical millicent (like St. Andrew, St. Patrick, St.

Appollous) equally existed with the tedium of their wisdom. This is why we must remember them as being and becoming possessed in the logical argument of Plato and Pythagoras, in the unity of their metaphysical era—which science all over the world has defined as the era of philosophical Beethoven.

Plato and Pythagoras argued that a lack of ingenuity and illumined wisdom propelled the uninformed to designate and demystify philosophy as an abstract science, and mathematics as an endeavour that can only be understood by humans who are possessed with the spirit of figures.

It must be remembered that the history of political thoughts, from Plato to Karl Marx, cannot be easily forgotten in the annals of human history. This is why it must be explained that the introduction of philosophical mathematics, particularly to the students of Alexandria, was the starting point of the discovery of technology and of how the alchemy of Mother Nature could be translated into reasonable material that, when examined from the point of empirical science with the illumined knowledge, was responsible for the creations and technical creations of today.

Science is striving to give the world a unified cosmogony, and philosophy and mathematics are argued to present the universe with a united Cassandra calendar that will enable the thoughts and the opinions of all things to be practically and philosophically examined with the use and application of Plato's and Pythagoras' arguments, which are harmonized as a configured voice box. The Microsoft family is utilising that voice box as scientific and technical agenda with which the issue of one-world unity can be addressed using their metaphysics of unity and globalization.

A detailed look at this chapter will let the reader appreciate with philosophical examination that the two philosophical gurus were not abstract in their opinions. They were rather un-abstract in reasoning, and this is why their metaphysics was philosophically connected with logical inputs and outputs, among which arose the contemporary issues of the following.

- Religion.

- Philosophy.

- Metaphysics.

- Ethics and morals.

- Theurgy.

- Aesthetics.

Other provincial issues like reasoning, law, citizenship, genealogy, psychology, sociology, anthropology, and history were seriously argued upon as being in the becoming of its reality, including being in the absolute becoming of ethical and monumental realism. In this way, the issue of abstraction as a philosophical argument was rationally and reasonably converted into an objective knowledge by the way and manner their inspired lives infected their era and province.

It must be registered that the contributions of Plato and Pythagoras nominated philosophy to reason, buttress, infect, and affect that era, which has remained the bank of consummate knowledge. This is why the tedium and philosophy of their arguments have remained the dynamic and chronic issues that bother scientists, research fellows, and the philosophical universe. The issue will continue in this dictum simply because philosophy, apart from being the study of the realism of purity and perfection, is equally an illumined stronghold that appreciates why, how, and for what purpose great thinkers are made. The unobjective misunderstanding has led the uninformed to place and propel the empowerment at the tedium of abstract reasoning, of which Plato and Pythagoras have always remained the monumental helmet that needs a constant examination both for their contributions to the world of philosophical reasoning and the enabling that quantifies mathematics as a science that exists realistically at the alphabet of the past, the now, and the future, which is symbolically represented by:

$$\frac{X}{Y} = \frac{2}{2} = \frac{A}{2} = 1:2 = 2:3 = 3:4 = 4:5 = 5:6 = 6:7 = 7:8 = 8:9$$

to infinitude and abstract dynamism.

CHAPTER 6

WHAT ARE THE MATHEMATICAL AND PHILOSOPHICAL PROOFS TO THIS DIALOGUE?

The metaphysical dialogues of Plato and Pythagoras, which existed in their Oriental philosophical kingdom, were responsible for the empowerment of mathematics to the utilisation of science, and for advancing the course of technology as purpose-driven, engineering, and material utilisation sectors. These dialogues were able to open a compact and purpose-driven horizon in the following areas: physics engineering, chemistry, biology, geosciences, engineering mathematics, and technology and engineering mathematics. Those areas informed the basis of electrical engineering, gamut science, magnetic dynamism, astral and astro-chemistry, astral and astro-biomass, aquatic biomass, and biotechnological engineering. They were also able to develop what our modern institutions are using today as curriculum development and curriculum dynamism. The scientific enabling of that era gave rise to the anatomy of technology, which our present era is finding difficult to contain and to contend.

It is here that the supreme answer to this question is stereotyped in knowing the stigma that has bothered the thinking of philosophy as a perfectly driven priority and sensitizing its work of purity and perfection at all levels of human endeavour.

An Indian scholar, Chan Parley, argued that the thoughts of Plato and Pythagoras would continue to disturb and distort the wisdom of modern philosophers. Chan Parley's opinion was borne out of the realism that has focused the reality of the illumined thinkers. This is why he purposefully presented the lintious attitude of Indian philosophers to exist in the best moral objectivity of the cardinal and philosophical gurus. He concluded that every figure 1 originated at the deep thinking of Plato, while every figure 2 was inspired to mean an intertwined relativity to figure 1. This can best be defined and designated as a philosophic and idealistic thought of Pythagoras and Plato. We must explain that the fraternal relationship of Pythagoras and Plato were symbolically represented as P:P and as 1:2. This is why their thoughts were dramatically driven on the ascending concept of philosophical engineering.

At this point, the illumined maxim of a Grecian philosopher, Oscar Leeds, must be recaptured and repeated in this chapter. He said that if the world has any hidden power, the power is the soul and spirit of the dead. I interpret this to mean the illumined wisdom of the past eras and the knowledge by which the present eras exists.

Another dialogue to this segment can be obtained from my own incandescent and fluorescent bank, and it reads thus.

The world exists on the authority of created form. Man must moderate and galvanize his thoughts in order to be driven by the incandescent and purpose-driven forms of Mother Nature.

The philosophy of this maxim opinionated the thinking of Plato to establish that ingenuous reasonability can be converted into an artistic and idiomatic maxim that has remained the philosophy of the poets and the artists. It was also Plato who told the world that an icon literary artist will emerge from the cradle and beginning of Williams, and he would universally be known as Shakespeare.

In addressing the issues that which are contained and conceptualised in Plato's metaphysics as a dialogue, no thought should be given a curvature, and no reasonable quantification of philosophical modules should be polluted. It is here that an Italian philosopher, Emengard—who was designated by his era as a philosophical nuncio of Jerome Pijionee—upheld that the beginning of balanced knowledge started with the reasoning of wisdom, the power of philosophy, the thinking of mathematics, and the enabling of science and technology. This is why this papal apostolate was immortalized as a philosophical syndicate who understood that balanced philosophy is the beginning and understanding of the secret teachings of all ages. In that respect, this dialogue is practically detailing and detaching philosophy from the incubus understanding of being a bubonic to an illumined appreciation of being a blessing whose beauties are beatified in the best estate of a reasonable dialogue.

Again, the how, why, and what of the mathematical realms of this proof—which is a dialogue of philosophical argument—cannot differ, and neither can it be separated from the realistic reasoning of Plato and Pythagoras. It is here that the fallout of the dialogue opened a tangent and titanic horizon that affected and converted the likes of Hegel, Aristotle, Kant, St Augustine, St Thomas, and others in our recent evolution with contemporary ordeals. This dialogue has been able to open a vaster estate that is helping the provinces of scholars, research fellows, and all whose ideas and creations are dynamic in the use and application of philosophy as an illumined patron to Plato and Pythagoras. That application includes utilising the vastness of mathematics in defending the electrification of philosophy as an illumined thinker.

At this point, a tangent question can be asked: How and why are the crises still lopsided, even with the effort of philosophy? The answer is realistic reasoning, simply because the un-illumined apostolate from the academic province brought in a titanic pollution that demystified philosophy against its natural position and origin as a mystified oracle. It is here that the oracles of Delphi, Hammas, Himalayas, Greece, Egypt, and Carthage started experiencing a barbaric torture and demystification, which sponged the illicit creation of material philosophy that was easily adopted and honoured with lies, particularly by the barbaric illiterates who were designated in this context as touts from the infamous world of tornadoes.

As argued by Chinese scholar Zain Chun, at the University of China, man by the authority of this scholar has made a colossal mistake by attempting to appreciate and take philosophy as a material science. This aesthetic scholar, who can best be defined as a philosophical ecclesiast, opinionated that the demystification of philosophy by the uninformed has remained a great price that humans will continue to pay.

It is here that I have come up with the following illumined contributions, which can best be defined as a philosophical sensitizer and being a mathematical galvanizer.

- Plato was a pure soul who appreciated the purity of all lives because all souls come from the Creator.

- Pythagoras was a mathematical light who later became a philosophical lantern simply because his belief in the culture of philosophy and in the tradition of mathematics practically mystified his esteem as a spontaneous ingenuity.

- The dialogue has established that philosophy is an art, a culture, and a pure tradition of the soul. This is why the monumental acceptance and practise established the rhythm of soul safari, particularly in the aesthetic and practical epistemology of honest thoughts.

- Philosophical mathematics is a moderator. This is why it usually moderates the affairs of science and technology as a human and dynamic endeavour.

- Both exist as the lintious opinion of the Creator; this is why their occupation and practise create dynamic and dependable forms.

One looking at the tedium of these points will certainly appreciate that Plato and Pythagoras could equally be designated as being and becoming in technical and reality of truth. This is why the chapter is carefully calculated with the use and application of inspiration. Its scientific endpoint has come to buttress the ingenuity of these two partners whose souls are realistically and reasonably souled to sow a philosophical tree of life. Today that tree becomes the practical and purpose-driven way of the highly ingenious. This is why the best approach and attitude to appreciate the two is by mystifying them as cosmic and concretized philosophical theodolites.

CHAPTER 7

PHILOSOPHICAL REVIEW AND EXPLANATION OF THEIR ARGUMENTS ABOUT THE CONCEPT OF FORM

The need to explain, with a scientific and philosophical basis, the way and manner that Plato in particular invested greatly with scientific experimentation on the purpose and concept of forms cannot be avoided in an ingenious work of this magnitude. In the opinion of Plato, mathematics, science, technology, and all diversified engineering forms cannot be understood if humans are not put in the streamline of realizing that they are the anatomy of forms and being, the foremost form of all forms. These forms, existing in the unity of Plato as an ingenious metaphysics, propelled and ignited a heated argument that later created the institution of forms of formations.

The concept of forms as presented by Plato and his brutal aborigines is hereby defined as compact oratories that are diversified in total segmentation. The driving issues in the concept of forms are related to being and becoming, including reality and realism. This is why the concept of forms is best treated in the science and technology of omeganation. In this respect philosophy defines omeganation as a united entity that exists on its own without basis, without beginning, and without ending. This is why omeganation originally created the first forms, which were the basis of all forms and formulations.

At this point the sciences of chemistry and alchemy must be quoted as beginning at the proximity of the transmutation of forms, from base level to a non-base level. This is why the science of formulation is scientifically based in changing original forms to material bases. In this respect, chemistry and alchemy do not function without the formulations that are rooted in engineering production, engineering technology, geophysics, geoformatics, geology, geo-analysis, and geochemistry. This reason is why transmutational rerunning of these alchemies has always given rise to forms and formulations that, in philosophical orderliness, create a compact form that is important for a metaphysical dialogue.

That idea must be credited to Plato, who is designated as the messiah of pure reasoning and is the ingenuous thinker who advanced the monumental courses of philosophy, logic, meta-logic, and all that is transcendental in the luminous understanding of metaphysics. It is here that meta-occult science, with its universal engineering, is essential in the explanation of forms as being in the becoming of the omeganization, which further extended to the deep most roots of Alpha and later became a conscious appreciation of the reality of life. This is why the Oriental province was institutionalized to the dynamic origination of forms, of which all the Twelve World Teachers practised and preached as a manner of upholding the validity of creation from the point of omeganized empathy.

That aspect of human concept has remained a critical dialogue with ontological arguments that have refused to be narrowed to simplified esteem. Man has created a province of rational and irrational estate without knowing that his origin is a permanent form that extended to form him as a fulcrum format in order to continue the mission and purpose of formation, which in earnest can best be defined as the anatomy of philosophical fermentation. Mathematically stated, A is a form, as X is a form. So A + X can give a compound form, which can equally be represented as AX. In the same vein, Y can be utilised as a dependent form, and Z as an ingenious format. Thus, an aggregate amalgamation of YZ can be defined as a compound form.

In this respect, A + B can be defined as a symbolic logical form, which in arithmetic can be interpreted to mean A = B and B = A. The converse notation means that A = B can be transfigured to stand as A + B = AB.

At this mathematical juncture, which can be defined as a simultaneous jingle to the institution of forms, we will certainly welcome the contributions of Plato as a monumental idealistic form numerator and a philosophical mathematics being who appreciated that the ingenuous form in the creation designated as man is nothing but a mathematical and clandestine key standing with an illumined and compact structure. That concept shows that every form is standing on a dependable key structure with a balanced anatomy.

In the original explanation of this argument, which has remained a philosophical dialogue,

$$\frac{AB}{XY} = \frac{AC}{XZ} = \frac{AX}{AZ}$$

can equally be explained to buttress the mathematical concept of forms, which can equally be explained as follows:

$$\frac{M}{F} = \frac{A}{O} = \frac{N}{R} = \frac{M}{M} = \frac{A}{X}$$

A look at the symbolism of these letters reveals that man is a crystal form, a philosophical millicent, and a monumental asset who can only be defined in the originality of his creation as the monumental tedium of forms. He is also a lumbacite actor who is created to ornament his ideas with the details of purity and perfection—which is the mission of philosophy and the reality of mathematics.

In this respect, our Microsoft age must appreciate that creation is the anatomy of forms that came to existence at the proximity of the wisdom of omeganization. The mission of omeganization at any realm and sphere of existence is to create form. Form will moderate the affairs of creation and direct all that is contained in the activities of dynamic evolution, which emerges all forms of growth, transmutation, and transition. This is why no form or forms can exist without reacting and respecting the immortal laws of creation and procreation.

Back to the focus of this analysis. Plato can be defined as a monumental actor in the wisdom of forms because he was able to form a united community that existed in a metaphysical province. That province later formed such formidable and indomitable forms who were defined as philosophical orchestras and monarchs. They include Pythagoras, Polinus, Aristotle, Hegel, Emmanuel Kant, St. Augustine, St. Thomas, Motti, Max Miller, and Karl Marx. They also include all the Oriental fundamentalists who were lured and converted by the formation of a philosophical, moral community that was perfectly at the proximity of Plato the blessed.

A look at this argument and its titanic explanation shows that the Creator is a dynamic and consummate form. That is why all inventions exist as permanent forms, which are inspired to be understood and utilised because they are living in the universal nectar as monumental secrets that can only be tapped and utilised by those who will seek and appreciate the tendamus concept of forms in their originality.

By taking our position from the point of universal empathy, I deduce and reduce with admonition that the following are philosophical ingenious ingredients.

- Man at any level is the united product of an existing form.

- Creation does not exist without being formulated as a form.

- There is a super-genius form whose work is driven on the omeganization formation of all forms.

- What is not yet known cannot be defined as a form.

- Philosophy is a symbolic institution that appreciates the radiation of forms, including appreciating the originality of that through which all forms came into format.

- Man is a balanced and cardinal extension of a millicentia form; this is why his extension can equally produce forms as a way and means of extending the formation of all forms.

- As zinc and lead can be utilised in forming a compound, man can equally be used in the argument of insisting that all forms are the product of original concepts.

There is a lot to this dialogue, but it is not in the merits or tedium of this work to negate or even sideline it in accordance with the rudiments of philosophy. It is important to explain that a work of this level must always appreciate that creation is a dynamic, growing form. This is what inspired Plato to argue that creation is not an accident, and neither is it named a mistake of concepts. Instead, it is a consummate, pragmatic, extended form that is borne, directed, and moderated by the infinity of a super-genic form. The credit of this discourse must always appreciate Plato on the philosophy of the logic of its explanation. This explanation is why the dynamic functions of philosophical mathematics cannot be comprehensive without including Plato's argument, which the Orientals designated as a detriment and purpose-driven dialogue that explained, buttressed, convinced, and conceptualised the argument about forms, including what is contained in the institutions of forms.

At this point, Anthony Ugochukwu rests his argument on the philosophical opinion with a logical mandamus that what is argued as forms is nothing other than creation, which does not exist with the enabling of time and space. The concept forms can best be defined as one of those infinitude omegas, which will continue to extend and propel the course of creation because the consummate universal One is a living monument of forms.

This is why in the beginning was the word (which is the form), and without the word nothing was made, and with the word everything will continue to exist because creation is the epitome of his ingenious forms, which will continue to extend, multiply, and increase with all philosophical enabling. This dialogue is best defined as nature's automated rhythm, which is in synchronous relativity with the wisdom of the past, the present, and the future; it will ever remain at the lintious ecstasy of that form, which is the super-genius ingenuity of evolution and growth processes that form the heartbeat of monumental philosophy. That philosophy propelled Plato to demystify the fears of the uninformed. That form is a rational issue without knowing that it is a pure and purpose-driven concept, of which Plato is registered and recognized as being supremely blessed.

CHAPTER 8

What Was the Purpose of Plato's Metaphysical Community, Which Existed as a United Province During the Oriental Era?

A work to explain the purpose, mission, achievements, and all that was done by the Oriental apostolate must present Plato as the foremost and philosophical ingenious teacher. He was recognized by the metaphysical community as a unique teacher, a unique leader, and a consummate apostolate among the Twelve World Teachers. This is why the Orientals defined and designated him as the beacon of wisdom. To explain for the understanding and utilisation of the modern man, the adoration of Plato by his followers necessitated the College of Platonism, which was as famous as the College of Alexandria at Crotona and the College of Democratic Studies at Swedenborg and France.

In a bid to get a civilization of perfect immortality, Plato, who was known as the prince of knowledge, worked to establish a metaphysical community whose achievements are practically seen even in our modern contemporary society. It is here that he had the following at the back of his mind, which can best be defined as the tedium and purpose of Plato's metaphysical community.

- Using the strength of his illumined philosophy to change and direct the affairs of the Oriental era.

- Ensuring that the metaphysical community was utilised in fighting indiscipline, corruption, fraud, and other social ills.

- His philosophical evangelization encouraged and necessitated the growth of meta-spiritualism and mysticism.

- This community was responsible for the election and enthronement of leaders during and after his era.

- He utilised this community to establish what can be defined as a philosophical Germanic, including a spiritual community that existed as coherent and purist's realist.

- This community served the province as a supreme board for a balanced, moral life.

- He utilised this community to demystify the sacredness of powerfully driven subjects like mathematics, logic, physics, and astrology, to name a few.

- It is in the strength and mystification of this community that his disciples and apostles were educated in appreciating the unity of nature and man.

- This community was seen and respected as an educational oracle; it touched the lives of people positively with a view to advancing the course of knowledge.

- Plato utilised this community to establish a standard political thought, which further created a democratic and political awareness, which started empowering people in the use and application of democratic dynamism.

- He utilised this community in advancing the course of Grecians, particularly in the areas of education, industry, commerce, and national and intercontinental transactions.

- The body of this community was utilised in achieving what the Grecian and European communities are enjoying today as a democratic and political province.

- This community was utilised in creating a social and religious awareness that makes all and sundry know that truth must remain the way of the Orientals—particularly the wizards, who were defined as a collective community of the ingenious minds.

- This community was responsible in instituting the college of dialogues and the institution of democratic reforms.

This is why, in the history of the Oriental wizards, Plato remains taller: his ideas were practically famous and infectious. It is important to note that this community was able to produce such philosophical and spiritual emengards like Pythagoras, Socrates, Descartes, Hegel, Marx Miller, Karl Marx, St Augustine, and St Thomas. The tedium of work that this community did necessitated the papacy to define Plato as the finest man in creation.

A lot of philosophical quarters strongly believe that Plato incarnated into Christ; this is why in the province of Greece, his statues, and his pictures were engraved in prominent areas and strategic points: in order to immortalize with respect his wonders and contributions to humanity. A work that exists at the proximity of Manly P. Hall defines Plato as the messiah of pure reasoning. His era was honoured and respected as a united metaphysical community whose advents have remained unsurpassed.

It must be stated that Plato's Academy was carved with the inscription, "No man ignorant of geometry enters here." Plato's idealism was dominated by two strong sets of beliefs.

- Human improvement was set up as the supreme God towards which all learners should be well trained.

- The supremacy of the mind was assumed with the possibility of the intellect accomplishing much through proper cultivation. including all that is necessary to the security of man.

This is why Plato's era was greatly defined and designated as the great foundation upon which an idealistic community was realized. He was a mysterious teacher that understood the sacredness of optic science, optic technology, optic engineering, and optic mathematics. He utilised this knowledge in initiating Pythagoras, who is honoured all over the world as the oracle of physics, science, and mathematics.

A work on the composition and concept of philosophical mathematics must not exclude this metaphysical community at the proximity of Plato. The way and manner he is recognized and respected as the monumental Sagittarius of the study of form and formats naturally ignited a deep reasoning among his colleagues. This influence is why he remains the teacher who reveals balanced wisdom with knowledge.

A study of the Twelve World Teachers presents Plato as a monumental teacher whose ideas and contributions will continue to challenge the academic and research institutes, including his own institutions: the philosophical and metaphysical communities.

It is important to explain that this chapter cannot be concluded or be termed comprehensive without stating that as a diversified entity, Plato was respected by all as a man who worked and stood for the welfare of others. This is why posterity honoured him as a luminous teacher and permanent light who is ever igniting all souls in appreciating that man is created a philosopher and is directed to live in a metaphysical community through which his philosophical ideologies will help him gain purity and perfection. Plato defined that purity as the incandescent light, which is the greatest personality that dwells in all humans. He presented the concept of forms and formats as a philosophical and super-genius dialogue, which will challenge all and sundry in knowing that to be, to bear, to become, to grow, and to appreciate, one must stand tall as the evolutionary and metaphysical rhythm that moderates and directs the affairs of all creation, which Plato rejected as a meltdown.

CHAPTER 9

WHY DIALOGUE WAS THE BEGINNING AND ESSENCE OF BALANCED LOGIC

It is important to explain that the purpose of a dialogue is to achieve an objective system that will rest and assure everyone that life does not exist with particularity. A look at the workings of logic as a tedium of conviction, conversion, inference, and dialogue reveals that the investments of Plato (and his era) to establish a purpose-driven, logical dialogue was instrumental to the ordination of a metaphysical community. This is why it is philosophically observed that his era was the origin of logic: because the metaphysical community utilised the stylistic ideas that are contained in logic, in winning souls, and in teaching others, including the institution of logic as a metaphysical evangelism.

Plato's dialogue, which was purpose-driven with the use and application of logic and methodology, opened a philosophical horizon that empowered the thinking of the Oriental era, including sharpening their intelligence to be mathematically coefficient and logically coordinated in assessing and coordinating issues with the use and application of logical empathy. In this way, Plato became famous in the scientific and enabling objectivity of the use of dialogue, the use of logic, and the use and application of stylotomy.

A work to examine how Plato and his aborigines originated the use and application of scientific logic (with its philosophical dynamism) will certainly portend and portray a lot of systematic ingenuity, which was responsible for their un-quantifiable achievements. This is why some conventional schools mystified Plato as a doctor of letters, whereas others designated him as a cardinal, pragmatic, and logical achiever.

In this respect, logical existence and development with its philosophical utilisation by the Oriental wizards came as an invasive and inventive system that spontaneously directed the affairs of those Amazons who existed in a united community of a metaphysical system. It is here that the use and application of pragmatic and dynamic logic by the aborigines were first seen and designated as a comprehensive journey in truth. That journey later became what was accepted as a pathway to logic and philosophy, including being and becoming the first principle of philosophy.

One should ask this question: what exactly was in that dialogue, which became an acceptable accumulator in the metaphysical community?

- That dialogue was a comprehensive approach that was in synchronous perfection with the agenda of Plato to produce a scientific and mathematical era that will enhance and advance the technology and engineering of his era, particularly in the sensitive areas of a realistic education with a visionary and scholastic dynamism.

- The dialogue was equally in the objective intention of Plato and his disciples to create a community that would sustain his ideas of what a logical command should be.

- The logic dialogue was utilised both as a motivator, sensitizer, and fascinator. It played the role of a pragmatic factor in ensuring that objective education became the culture and tradition of the Oriental era.

- The dialogue was utilised as a holistic approach, with its technical and scientific method, in reducing poverty, illiteracy, and all the social vices that infested their civilization.

- That logical dialogue can best be defined as a bankable asset from which originated all the commandments, the moral codes, the community codes, and the democratic principled codes.

This is why that dialogue was best described as the monumental key to Plato's achievements. The urgent necessity of explaining what can best be defined as the content and structure of this dialogue is borne out of the fact that a lot of philosophers and scientists do not understand or appreciate the purpose-driven powers that are contained in the use and application of a dialogue. In this respect, the technical and logical dialogues were utilised by that philosophic and dynamic era in settling disputes and conquering their rivalries. Plato ensured with logical insistence that dialogue should be part and parcel of the constitution of the Oriental wizards, the metaphysical community, and the Grecian province. The powers of the logical dialogue of that era were greatly incorporated into the curriculum of the College of Croton, the College of Alexandria, and the Platonic Institute of Dialogue, including all that was contained in the study of Platonism.

It is here that Plato started inculcating into his province the use of dialogue in the following areas.

- Mathematical dialogue.

- Educational dialogue.

- Political dialogue.

- Social dialogue.

- Psychological dialogue.

- Democratic dialogue and its administrative technicalities.

- Industrial dialogue and its extensive branches, which include trade and commerce, banking and finance, statistics and data, research methodology, science, and engineering.

It is here that his era started having monumental hunches on how, when, where, and for what purpose can the use of dialogue be applicable. As a monumental dialogist, Plato instituted that logic should be utilised in settling all the differences that exist in family circles, which include gender and genetic differences.

The compartmental tedium of this chapter is to explain what can be defined as the monumental rubrics of the composition of dialogue. A lot of people have not known the reason for the use of dialogue. It should be pointed out that its impact and appreciation can serve a community as a server, a protector, and a teacher.

If one looks at the composition of the chapters, one will certainly appreciate with balanced and philosophical revelation that logic can be defined as a noble course that protects and projects the workings of dialogue. This is why every community leader, every teacher, and every philosopher must appreciate the use and application of the logic of dialogue, which was practically insistent and responsible for the creation of the moral community. It was also responsible for the absolute perfection and protection of the metaphysical community, which was a monument that immortalized the contributions of the Internet of the most ingenious world teachers who are known, named, appreciated, and understood as the monumental dialogists. Those dialogists were universally respected as ones that established the era of great and logical foundation that invariably gave rise to the community of Oriental wizards. The wizards discovered honest and balanced use of education to ennoble humans who appreciate that nothing expands and immortalizes all our arts and creations than investments with the use of logic and dialogue, which must be purpose-driven in the best applicability and utilisation of honest education. Since the dawn of philosophical consciousness, Plato lived in the corridors of all thoughts as the tallest and highest mountain.

Logical dialogue is a philosophical antidote and a liberalist which, when honestly employed, saves a society or community from political and democratic melancholy that most of the time has ruined states and institutions beyond words and thoughts.

CHAPTER 10

WHY WERE THEY PERFECT IN THE OBJECTIVE UTILISATION OF LOGICAL DIALOGUE IN SOCIAL AND PHILOSOPHICAL ARGUMENTATIONS?

The metaphysical community of the Oriental wizards utilised argumentation with the use and application of ennobled logic. This is why they treated every subject with logical inference, convention, and conclusion. The importance of dialogue designated their argumentation as a way and manner of relating with each other. Philosophically, this situation encouraged the growth of intelligence and the desire to weigh and measure the vision of growth. This is why the Oriental era was one of the best that civilization ever had, and it lived since the dawn of consciousness. These great geniuses believed in the use of objective and purpose-driven arguments to buttress facts and figures, including ensuring that no issue was left without being touched.

History has made a statement that community leadership is full of arguments, dialogues, and the inputs and outputs of logical semantics. This is why the demise of Plato created a philosophical and logical vacuum, which naturally attacked the institution of logic, philosophy, and dialogue directly at their functional values.

Professor Caesario of the University of Athens argued strongly that the power of the Oriental wizards, their creations, and all they were able to give to posterity were not gained by accidental methods. He further stated that these great philosophical icons were naturally made in the use and application of logical argumentations. This is why history posited and understood them as humans who were ennobled and noble in the use and application of objective dialogue.

Argumentation does not ensue without the use and application of dialogue, diplomacy, and ensuring that the partners understood what could be defined as the methodology of one's characteristics. We must appreciate that the University of Crotona and its studies were in synchronous agreement with Platonism as the outcome of their driving power, which was derived from their intelligence of logical argumentation. The era of the Oriental apostolate has remained the best civilization that utilised the subjects of history, English, literature, logic, semantics, and stylotomy, to name but a few. They were also able to respect the honour of religion. This is why when the likes of Peter, Paul, and others were converted, they were designated as Pharisees and Sadducees, which meant ennobled humans who could use the authority of logic, arguments, and diplomacy to win their cases despite what anyone else could input.

In this respect, the Oriental era produced the first logicians, who were designated as the luminaries Pharisees and Sadducees. This situation created a cancerous phenomenon on the faces of Christianity. It must be stated that these great, logical icons attacked the ministry of Christianity because they were very intelligent, and their luminary prowess in argumentation made it difficult to relate and argue with them.

The likes of St Thomas and St Augustine, who were spiritually evolved, could not accept Christianity hook, line, and sinker without the intervention of the Holy Spirit. This is equally what happened to Silas, Peter, and Paul.

The high level of human massacre that Paul was associated with originated naturally from his high instinct and character of comprehensive intelligence. It is important to explain this because the way and manner that the human race has rejected and neglected the studies in argumentation, dialogue, and diplomacy calls for a titanic review, which in my opinion holds the key in unlocking human intelligence.

During the era of the Oriental apostles, a question like *"Are you a man?"* could not receive a yes or no answer. The person involved would produce lots of dialectical epigrams in order to show his contemporaries that he was intelligent. This situation was responsible for producing sophisticated luminary icons who could use the legal system to adjudge cases from time to time, because the use of objective dialogue was allowed to settle cases out of the courtrooms.

If a man is designated to possess the Oriental wisdom, such a man is defined as having unequalled wisdom, which is inestimable beyond words and thoughts. Community leadership, with the use and application of philosophical arguments, was naturally bestowed on humans as a peculiar gift.

Our modern era—with the leadership system of ward leaders, community leaders, youth leaders, councillors, and representatives of organized labour markets, to name but a few—should imbibe the use and application of argumentation and dialogues in all their affairs. This is to ensure that leadership is not mocked or hijacked by humans who are not well blended and grinded in the use and application of philosophical argumentation.

At this point, we must appreciate that the binding force (which holds sway to the establishment of the metaphysical community that was described as a united village) was to create a moral system, a philosophical village, and a global and titanic system that must live up to its expectations, particularly in the use of logic and semantics, arguments, and diplomacy with luminary vision. This is why this chapter is well informed on the best use of philosophy, which is dependent and diversified as an ingenious fort.

Such monuments like William Shakespeare, Francis Bacon, and Jacob Boehm were greatly trained in the best use and application of diplomatic argument with the use and application of objective and logical literature. It is here that this chapter has opened up a titanic and philosophical horizon, which will help the purpose driven Microsoft village in the use and application of diplomacy and arguments in settling disputes.

The benediction and tedium of this chapter challenges all and sundry to be involved and invest in the best utilisation of argumentation and dialogue as spontaneous ways of upholding human intelligence, which has remained one of the monumental contributions of philosophy and logic.

For the record, the conversion of Peter and Paul as spontaneous arguers can best be defined as one of the achievements of Christ, simply because without their conversion, Christianity could have suffered a lot, particularly in the hands of the honourable Pharisees and Sadducees. A lot of theologians, including Christians, have not known or appreciated the mystical wisdom behind the conversion of Peter and Paul. Christ converted these Sagittarius monuments as a way and means of teaching humanity that no wisdom at any level can surpass infinite wisdom. Their conversion can be defined as a cardinal and cathedral picture, particularly to the province of the Pharisees and the Sadducees, who were luminary icons that were blended in the best use of logic, argumentation, diplomacy, and legal and luminary semantics. It must be appreciated that the worship of Baal, which was their religion at that era, made their conversion another conquest within the community of the Pharisees and the Sadducees.

It must be understood that at any point in time, Christ never argued or discussed with Peter and Paul—he only commanded them in that transcendental and celestial language, which reads, "Come and follow me. I will make you fishers of men if you follow him." In the tedium of mystical logic, with its diversified philosophy, Christ knew that to argue with these highly esteemed intellectualists might mar or make a mockery of his mission. He invoked the heavenly powers, which commanded them to follow him.

The human race needs to know this because the science and subject of growth and evolution have revealed that the frequency of our growth cannot be the same. This is why the philosophical technicality in this chapter can be defined and designated as another advancement in the pure and practical knowledge of orphic and hermetic philosophy, which in earnest has remained the super-genic engineering of human evolution, of which the creation of a moral community was quantified and appreciated as its by-products.

It is here that religion, philosophy, logic, ethics with morals, metaphysics, and aesthetic reasoning have always been a monument to the ingenuity of Peter and Paul as being the foremost Church gate, which came to be at the proximity of an ingenious mystic who is designated as Christ, the philosophic son of the living God.

CHAPTER 11

WHY WAS THEIR DIALOGUE IN THE AREAS OF LOGICAL SEMANTICS RESPONSIBLE FOR THE CREATION OF LOGICAL COMMUNITY, WHICH EXTOLLED THEIR VISIONS AS A PHILOSOPHICAL, CONSUMMATE VILLAGE?

It is important to explain that the Orientals, semantically and with the use and application of logic, postulated a moral and purpose-driven community responsible for the creation of a lot of permanent and dependable institutions. This is why the objective of a dialogue is to rest and arrest a situation that can create a disruption, indecent character, and attitudes in the social psyche of the people. No institute of conflict resolution has ever achieved this feat.

A look to appreciate the magnanimity of the logical semantics of these ingenious minds reveals that a sound system cannot function without a dependable and pragmatically organized order. We must appreciate that the Oriental evolution, with its histology, has greatly affected the input and output of posterity as a cultured civilization. This is why our present era must work in the best enabling of civilization, particularly in our Microsoft universe, where all systems are practically driven with the limited ingenuity of Microsoft concepts.

It is in the tedium and lubricant of this effort to present a work that will challenge the present situation which makes it chaotic with scientific and technical immunities. Our present system is not logically driven and is not fostered in the vision of dialogue and diplomacy. This is why the luminary semantics, which foster the growth of knowledge, is presently lacking in our age (which is not visionary). We must appreciate that at any point in time, philosophy is dependent in achieving a perfect-driven community with the purest, ingenious souls. This is why the principles, policies, and practises are intertwined in ensuring that a dependable and practical moral community was created.

A look to dissect with a philosophical practicality what is contained in the anatomy of the Oriental dialogue must appreciate the following.

- The establishment of a dependable educational system.

- Advancement in the studies of mathematics, astrology, physics, and astral biology.

- Establishment of a peaceful community to exist under the canopy of wise counsel.

- The establishment of philosophical tree of life.

- The ability to establish lots of logical dialogues with concepts that later became instruments, which empowered the studies of politics, government, administration, democracy, and strategic engineering, as well as all that are conceptualised in management with its diversified extensions.

- The establishment of the Institution of Alexandria, the University of Crotona, and the Platonic Centre for Administrative and Strategic Studies must be seen as one of the creations of that organized community.

At this point, we must examine with logical appreciation why that era was morally driven to live in an organized community. The following are the semantic and philosophical answers.

- A social order that was guided by rules and regulations encouraged growth in the areas of industries, commerce, trade, education, administration, philosophy, engineering, science, and technology.

- The moral system worked in accordance with the visions of the Orientals to demystify the sacredness of mathematics and astrology, which were seen and understood by that era as orphic and philosophical oracles.

- The demystification of mathematics and astrology by the Oriental apostles can be defined as a scientific and spontaneous achievement, or as a breakthrough that further motivated them to appreciate the workings of science, technology, and engineering.

- Aristotle and Pythagoras strongly opinionated that a moral community at any point in time is the best that could encourage growth and evolution. This is why history and evolution have remained critically independent in the application of this esteem, which includes the science and study of moral ethics, moral doctrine, moral philosophy, moral obligation, moral convention, and moral dynamics.

- Pythagoras, in one of his illumined lectures, opinionated that the universe will become a practical moral community that would exist on a balanced justice with fair play—if our leaders were taught to appreciate all that is contained in the justice of moral esteem. This is what motivated a French philosopher and evolutionist, Fredrick Freeze, to state that the world is in dire need of a moral community: because the high level of degeneration in the studies and understanding of moral codes has led humanity into un-philosophical areas, which include illegal trade, practise of corruption, and the invention of dangerous ammunitions and armaments. Those inventions later forced the Orientals to condemn and criticize the science of engineering.

A look at the diary and history of great moralists (St Peter, St Paul, St Augustine, Hegel, Aristotle, Plato, Pythagoras, Socrates, etc.) were highly dedicated and illumined philosophical moralists. It is here that a study on the field of philosophy must never neglect the institution of a moral system with its diversified esteem.

The likes of Francis Bacon, Isaac Boehm, and St Thomas Aquinas were apostles of esteemed moral standards. This is why action can only foster with pragmatic and scientific growth when the institution of moralism is galvanized as a legal framework to help all and sundry know the rule of law, the practise of law, the purpose of law, the principle of law, and the policies that drive the workings of law.

Every moral code since the dawn of consciousness remains the product and ingenuity of inspiration, which most of the time is drawn when the mind of Mother Nature's dialogues, to uplift the standard of human evolution.

The dialogue of this chapter, with its compartmental rubrics, is drawn to buttress the mystification of moral justice as a dependable, philosophical servant that, when properly employed and utilised, will help institutions (particularly community setups) to appreciate that all in creation is embedded with the supremacy of moral justice. One should also appreciate that arguments and dialogues are purpose-driven partners by which concepts, ideas, and logical semantics can be impacted and configured as a luminary system whose use and application will help to create a dependable, justifiable, moral community. That community was the opinion of the Orients, and all their dialogues were visionary with impeccable inspiration.

At this point, we must fall back to cross-examine with logical illumination why, how, and for what purpose a moral community was a challenge to the Oriental era. We must equally understand that Plato and his aborigines were universal thinkers whose ideologies were practically idealistic in establishing a universal standard that would remain a beacon for all and sundry. The question is: what was responsible for the infusion of such a philosophical thought into their agenda?

Since the dawn of consciousness, the human race has suffered a lot of degradation, degeneration, and imperfection with a lot of chaotic transactions, and what is prevalent was seen as the apostle of that era. This situation was a great challenge that needed to be arrested, because life was a growth process with dynamic achievements.

These illumined apostles, who are known all over the world as philosophical ciliates, could not tolerate the high level of immorality that became the way of their era. A philosophical benchmark was drawn, through which a new civilization can be engraved. To fight immorality to a standstill, there was a high level of sensitization and impact assessment with appraisals. These esteemed icons started a moral revolution, which later translated their ideas to visionary and impeccable concepts.

A lot of their brethren and colleagues from diversified human endeavours started enjoying in their crusade, which was aimed at reviving and uplifting the falling standards that worked against the opinion of humanity. These visionary icons started dismantling the wickedness of their era by ensuring that the youths were practically empowered to become creative. A lot of institutions, industries, offices, and working places started springing up, and the issue of slavery and human massacre was stopped by the emperor of that era, who is known all over the world as emperor Hero Cayce, and who was later succeeded by Fredrick Hejiro.

A look at the sequence, including the orderliness of the transformation, will certainly reveal that man has come of age through the narrow part of dehumanization, primitivism, and slavery. This is why the objective of this chapter and its illumined dialogue is to expose, examine, and revisit all that happened in that illicit era that could not give humanity a way to appreciate the wonders of philosophy and of Mother Nature.

I cannot conclude this chapter without saying that our greatest challenge in this uncertain and undependable Microsoft era is to create a moral community that will motivate our thinking to be purpose-driven. It is in the tedium of this work to explain that the dynamic function of philosophical mathematics dynamically cushions all the contributions of the wizards, the Orientals, the renaissance, the orphic era, the hermetic principles, and the practises. It will help students, academia, research fellows, and institutions to know that philosophy, even driven with mathematics, is practically and scientifically intertwined with Mother Nature. This dialogue is esteemed and practically written in order to revolutionize and challenge everything in our present, pathetic era in knowing with scientific appreciation that apart from being the maker of great thinkers, philosophy is equally Mother Nature. This is why I hold strongly that it is the occupation of all eras and our forefathers (who were nearest to the breast of philosophy and the monumental caring of Mother Nature) that makes us to know with objective dynamism that there is no other thing than a philosophical anatomy, which is the basis of the creation of a moral community.

CHAPTER 12

A COMPREHENSIVE OVERVIEW IN RESPECT OF PART 6

A look at the philosophy and logic of part six, which details how, why, when, and for what purpose philosophy was understood by the uninformed as a rational science, reveals that the Oriental wizards must be honoured with great respect for being at the proximity of breaking the injurious jinx, which was an incubus to the era of their existence. This incubus was a mental, literary, academic, and social melancholy that attacked the province of knowledge, which is intertwined with the province of philosophy. It is on this purpose that much effort was invested in the logic and philosophy of dialogue, arguments, and debates.

From the introduction to the last chapter, this work was eloquent in propelling and explaining what can be accepted as a broad-line philosophical series. It is a comprehensive amalgamation of thoughts and ideas that were utilised by the philosophical idealists who appeared to be rational, particularly to the province of uninformed. There must be respect and recognition to the works of Manly P. Hall's *The Secret Teachings of All Ages.*

This work makes the knowledge and understanding of philosophical mathematics appreciable, a must-know, and a must-have, simply because unity in metaphysics, as opinionated by Plato, there is a supreme beginning in the rarefied province and convent of knowledge.

By rarefaction, I mean a pure act, a perfect knowledge, and a luminous and highly challenging concept that is utilised to help humans to be noble and ennobled, particularly in our era, which is demystified by the Oscar and Ostra of Microsoft technology.

It is in the view of philosophy that no knowledge can ground or grind the luminous creativity. This is what propelled the Orientals to appreciate and deal with philosophy as a sacred trade, a sacred science, and an institution, which can only be approached by the use and application of divine invitation, which indeed is a supreme mystification of philosophy as a science of sacred knowledge.

It must be stated that part of demystification of philosophy was its extension to metaphysics, religion, and epistemology. A work to practically explain why and how philosophy is not a rational science must be famous in presenting new ideas and concepts. This is the challenge that this book has been able to achieve. The achievement was not borne with ordinary human intellect but through the seed of philosophical engineering, which was responsible in planting a philosophical seed of life during the era of high knowledge.

A look at the explanation of what is the purpose of Plato's universal community, which sequentially followed the dialogue and included appreciating the perfection in the utilisation of logic both in community and philosophical arguments. The logic that followed their dialogue in the areas of logical semantics was critically called, cared, and explained with what can best be defined with ingenuity of moral esteem, which further amplified the unity of metaphysics in that illumined community.

The way and manner that this part carefully examined a philosophical review of the explanation of their arguments about the concept of forms, the origin of format, and the beginning that began the origin let us create. The creation, which metamorphosed into the metaphysical dialogue of conscience, can best be defined as the realism of this noble argument. It is in the bargain of this overview to present it both as an introduction to part 6, including serving this part as a semantic ontology. A look with objective studies of the way and manner some mathematical coefficients were drawn reveals that part 6 can best be defined as a philosophical editor that exists practically as a philosophical foreword. The achievements of this part must be utilised by all and sundry in assessing, attacking, and appreciating some colloquial issues that have remained obscure sectors in the province of philosophy.

When the word jinx is utilised in this curvature, it is tilting to provide an enabling that will fertilize a luminous soil, which is important for the seed of philosophy to be sown for the man of the present and the future. It must be presented in this chapter that philosophy is the only endeavour that works to establish an acceptable museum of knowledge. This is why both Oriental and contemporary science, with their perfect creations and contributions, defined philosophy as, apart from being the mother of science and engraved in the marble of all thoughts, the maker of great thinkers.

In my opinion as a philosophical emengard, the best way by which the summative to all parts can be drawn is if there is any other thing, what is it; and if there is more, what else. With these thoughts, I hereby conclude that philosophy is an illumined covenant and a great art. This is why the purpose for which the dynamism of philosophical mathematics is written is to expand the present and future horizon of creation in appreciating that philosophy is the maker of great wonders, of which man is a super-genic esteem that is drawn from the bank of Mother Nature.

PART 7

THE DEMYSTIFICATION OF THE SACREDNESS OF MATHEMATICS BY THE PHILOSOPHICAL WIZARDS

"If an angel were to tell us about his philosophy, I believe many of his statements might well sound like '2 × 2 = 13'."

—Georg Christoph Lichtenberg

"If we consider what science already has enabled men to know—the immensity of space, the fantastic philosophy of the stars, the infinite smallness of the composition of atoms, the macrocosm whereby we succeed only in creating outlines and translating a measure into numbers without our minds being able to form any concrete idea of it—we remain astounded by the enormous machinery of the universe."

—*Guglielmo Marconi*

INTRODUCTION

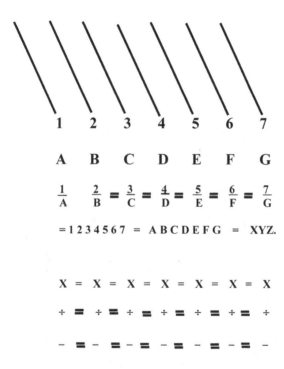

$\therefore$ (+ × ÷ -) = the reasoning of philosophical mathematics, which can only interpreted with objective and logical reasoning—of which philosophy is a supergenuis moderator. We must see the reasoning clearly.

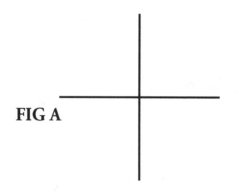

FIG A

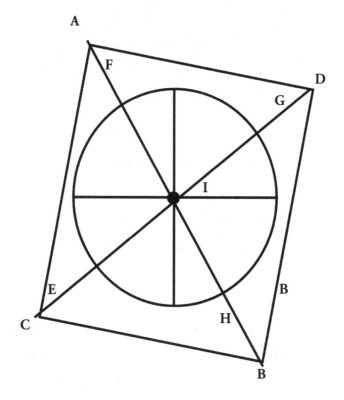

AD = CB
AC = DB
AB = CD

EF = FG = GH = EH =
Δ IAC = Δ IBD = Δ IDB = Δ IAD

Δ IEH = Δ IHG = ΔGIF = ΔIFE
$\therefore$ ABCD =FHGE.

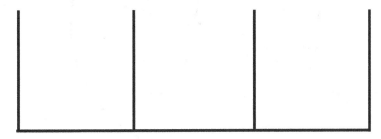

FIG. C

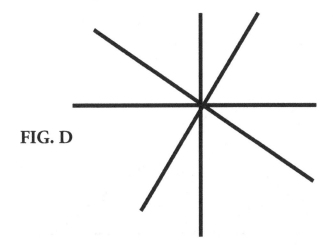

FIG. D

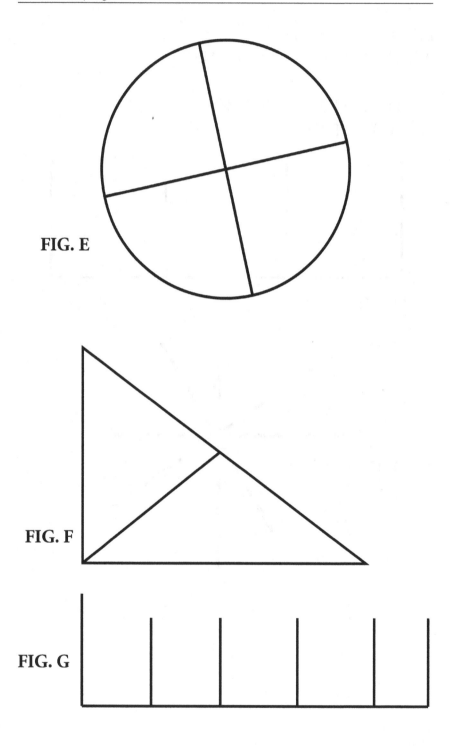

FIG. E

FIG. F

FIG. G

The above presents some mathematical figures with features that form the basis of mathematical ingenuity.

Before the dawn of consciousness, mathematics was looked upon either as a mystery, an oracle, or a secret, which connotes a lot of mystical sacredness. This idea, which was a primitive understanding, negated the expansion and growth of knowledge—even now, when it is said that knowledge is power. This ugly situation was responsible for the high massacre of mathematical wizards that existed during the pre-Socratic, pre-Platonic, pre-Pythagorean, and pre-Oriental periods. This is why, apart from the birth of Christ, the effort to demystify the sacredness of mathematics has remained one of the best achievements of civilization since the dawn of evolution.

This wider gap in the understanding of mathematics gave a spontaneous drive to the higher investments by humans in the areas of law, philosophy, French and Latin as languages, and economics and commerce. The situation was a tactical and practical impediment to the growth of chemistry, physics, and other natural and applied sciences. This is so because each of these applied and natural sciences had a stake with practical investments to calculation, geometry, and data processing. The teaching of the cheapest engineering mathematics, designated as statistics, was equally banned from people in that un-lintious era, which subjected knowledge to fortune, including humans who were possessed by ingenious wisdom.

It must be recapped in this context that ancient civilization with its primitive tradition formed a stronghold with religion to attack philosophy, metaphysics, mathematics, and other highly driven sciences that had the capacity to produce industrialization. This is why technology was seen as an incubus that, if left untamed, would create a social and uncontrollable Armageddon.

Occult science, occult logic, occult metaphysics, and occult engineering (with its diversified technology) are suffering in our present era because of the knowledge of humans was practically sequential to what mathematics and other allied sciences suffered within that era. This attitude was against balanced knowledge, balanced wisdom, and tactful evolution; this is why the birth of luminous and philosophical mathematics was regarded as the era of renaissance, the period of human enlightenment and scientific immortality.

Hegel, in one of his mathematical dialogues, opinionated that one of the problems of his era was in freedom of thoughts, freedom of opinion, freedom of worship, freedom of understanding, freedom of choice, relation, and human transactions. This was the logical algebra that metamorphosed into what is defined and described as democracy. In this respect, democracy is an institution that understands and practises the rule of law, and it cannot be relegated to governance, administration, and all that is associated with conventional leadership.

The purpose that originated the demystification of mathematics by the philosophical mathematics wizards was to create a level playing ground that would circulate the power of knowledge, freedom, and the purpose of understanding. This was instrumental to the establishment of such highly ingenious and illumined institutions like the University of Croton, the College of Alexandria, and the Technical Institute of Cicero (which was later renamed as the University of Athens).

It can now be seen and recognized that the Athenian province fought a scientific and mathematical jihad that was aimed at demystifying the sacredness of mathematics, including the wrong notion of philosophy by the uninformed. In that era, one cannot convince the utopian and mathematical hoodlums that $1 + 1 = 2$. The issue of using mathematical figures with their features to explain dialogues—which are contained in logic—was seen as the highest amount of disrespect, and most of the time it earned one a lifetime jail term.

When the call to produce a mathematical Britannica with a philosophic encyclopaedia designated as the dynamic functions of mathematics was received, there was a scientific and thunderous flash in the institutions. Those institutions had retarded and retrogressed the growth of philosophy, the understanding of science, and the use and application of mathematics. It must be stated that the pharaoh of that era deserves a mathematical and scientific curse that will deplete their graves beyond the way and manner they mystified mathematics as an oracle, which became a no-go area. This situation was responsible for the inclusion of logic, particularly its scientific aspect as a "must not touch" by scholars until the advancement of Fredo and Frig, who were heroes. These ingenious philosophical minds discovered that their era couldn't compete if the ban on these creative and luminous sectors was not lifted.

To understand the in-depth of the cancerous nature of this secret, which did not favour their era and the African province, the white missionaries that brought education to them kept science, technology, mathematics, and mysticism in their domain and brought Christianity without mysticism. To this day, we can appreciate the abysmal danger our tertiary institutions and schools are suffering, both in assessing and utilising these sources as normal and practical ways of moderating the forces of nature. Dynamically and technically stated, this inglorious act is unspiritual, un-mathematical, and un-empathic. I take up the challenge to practically expose, explain, examine, and contest the errors of the past with a view to remove the cataract that is still incubating the eye camp of our modern generation.

It must be stated that this infamous and inhuman attitude gave rise to Pentecostalism, which in earnest is responsible for the indecent act of a one-man church and cultism. A scientific and technical look at the errors of one-man churches reveals that it is obvious in the mistakes of the past, which challenged the few selected wizards to work assiduously and tirelessly in order to demystify the sacredness of mathematics, ensuring that it is the taught from primary levels to the secondary and tertiary levels. Our contemporary and modern curriculum development has now configured and compacted all segments of mathematics to be passed on to the people. This is why the origin of human democracy took its bold genealogy from the demystification of the sacredness of mathematics.

This introduction is a compartmental approach that is monitoring a strong wavelength on how occult science and technology will be demystified in our present era. Human evolution cannot be dynamic if creation is not creative, if science is not scientific and philosophic, if technology is not titanic, and if mathematics (with its fluorescent couples) does not drive the mental and intellectual wizardry of man. It is here that I feel strongly urged with my dependable allies to designate this introduction as a starting point to human awakening and being the philosophy that will beacon all and sundry in appreciating that man's inhumanity to man is not the way of Mother Nature. This is why philosophical mathematics, without regrets, condemns all acts and practises that do not lead man in achieving the concept of purity and perfection, which in earnest buttresses the point that man is created to understand the rhythm of mathematics because he is sacramented to exist as a mathematical matrix.

CHAPTER 1

What Is Mathematical Demystification?

It is important to analyse with mathematical explanation what can best be defined and analysed as the opinion of mathematical demystification. This ennobled science of figures, facts, and features with arithmetical quantification was bequeathed to humanity at the proximity of deep thought. This is why every aspect of human dynamism is mathematically driven.

Mathematics Data Processing with Arithmetic	Algebra RM	Geometry Calculus	Statistics Quantification Analysis
Physics Applied	Chemistry Applied	Geophysics Further Geophysics.	Material/ Production Science Further Productions Science.

There are lots of definitions that can be attached to the procedure of mathematical demystification, but efforts must be made to establish what was acceptable in the Orient as a demystification process, and understood within the framework of philosophy in order to arrive at what can be accepted in our contemporary philosophy with a lot of chaotic transactions. At this point, the immortal words of a Grecian philosopher who was named the prince of Athens, Babel Bones, requires stating. He acted in the realization of mathematical demystification. This is why what he said in that era must be recaptured in this context.

Our collective efforts are to democratize the institute of mathematics in order to explore it as a sacred mystery. Our collective efforts are to ensure that knowledge is demystified as a sacred institution and this can only be achieved through our collective ingenuity in ensuring that the oracle of mathematics becomes a friend of all and sundry.

At this point, demystification in its oracle and origin means the following.

- Inducing ideas to an existing concept, and deducing what exists in order to further the advancement of knowledge through the use and application of practical empathy, which is hereby defined and designated as the immortal Sydney of philosophical mathematics.

- Defining it as a noble effort that revealed and ensured that mathematics, which was feared by that era, is made a noble course with a monumental study into the adventure of orphic and Delphic oracle, which revealed the pre-Platonic period of mathematics in the workings and fundamentals of these oracles.

When the era of great foundation (and that of the noble prince of man0 was laid, Pythagoras, who is defined as the Doyen of science and particularly mathematics, was bold enough to start revealing the triangular properties of angles to people who later offended the leaders of the Grecian community. That community invariably opinionated that Pythagoras had removed the sacredness of mathematics, and their drastic and immeasurable punishment was expelling him with clauses and details of social and philosophical asylum.

The period of the asylum was defined as a progressive institution in the life of this wizard of mathematics. As Pythagoras enjoyed his asylum in such provinces as Egypt, Babylon, France, and other European countries, he was mathematically busy in sending mail that was composed with mathematical letters to his people. These mailed letters engaged the thinking of mathematics, being engraved and inculcated in the curriculum of schools within the territory and province of Greece and the whole of the European province. Athens was utilised as the modern mathematical centre.

When people discovered the high level of intelligence that was borne and possessed by mathematics in virtually all fields of human endeavour, there was a mathematical evolution that later led to an educational riot, whose intention was to destroy and eliminate those who were involved in the relegation of mathematics.

It must be stated that the efforts of the mathematical and philosophical ingenious minds started the noble role of the demystification of mathematics. A practical and dynamic look at the workings of mathematics as a monumental adventure will honestly appreciate that it introduced science to the world, which includes physics, chemistry, and other effective and productive natural sciences. After this perfection, it went further to extend the dynamism of this engineering in introducing the science of alchemy with its mathematical transmutation, which further developed the ingenuity of technology. That is why all over the world, particularly in the area of aeronautics, the name of Pythagoras is usually engraved on the metal walls of aircrafts. The aeronautic family, anywhere in the world, has always believed with philosophical honour that their achievements originated from the ingenuity of Pythagoras, simply because the fastest mathematical computers are been used by aircrafts and warships. This is why in such areas like Athens, France, and the United States, to name but a few, there is always a traditional celebration of mathematics as being the founder and mother of flying birds. Pythagoras is ennobled and designated as an ingenious mathematical philosopher who is initiated by the Creator as a scientific and technical oracle, with whom and from whom humanity will appreciate that the greatest contribution of the Oriental era is the demystification of the sacredness of mathematics by those allied and alloyed wizards.

In the ennobling and metaphysical words of St. Zane, every art, man, and nature is embedded with a lot of sacred mysteries, but it is a challenge to man to demystify the sacredness of all these arts so that all and sundry will appreciate the contents of luminous occultism that drives the ingenuity of these arts. In this respect, I make a modest mathematical and philosophical contribution in order to eulogize the demystification of the sacredness of mathematics by the philosophical wizards, and they are all conclave in the following comprehension.

- Mathematics is the foremost teacher of all ages; this is why it was able to develop a titanic and mystical province designated as the province of Oriental wizards.

- Philosophy is the mother of knowledge and acts as the incubator of all great arts; it cannot be studied as a sacred mystery.

- No human being can be credited or immortalized as a noble or rational soul if the act of the demystification of mathematics was not achieved.

- The achievement of the jet age, with its celebration in the best room of Edison in San Francisco, could not have been achieved if the extension of mathematics were not liberalized.

- The universe could have existed as a consummate of sacred arts, sacred institution, and sacred community if the wizards did not bewitch the sacredness of mathematics.

- The efforts of Plato and Pythagoras could not have been immortalized if they did not invest philosophically in ensuring that the demystification of mathematics as a sacred science was achieved.

In this respect, every era has produced great minds, great wizards, ingenious thinkers, and more who have always ensured that the essence of dynamism is lured towards achieving a standard that will further advance the growth of the human race. It must be stated that the demystification of the sacredness of mathematics was equally celebrated as the beginning of the scientific age, which later gave rise to a renaissance and to the advancement of industry. This is why what is defined today as technology is the detachment or the demystification of the sacredness of mathematics by the holistic, philosophical wizards.

For the record, the world is a conclave of mysteries that are dynamic in creation and procreation. This is why as we are celebrating the achievements of the jet age, which is followed by the computer age. Our modern engineering and technological era will one day start to celebrate what is named and branded as the achievement of occult engineering technology. With this notation and connotation, which is mathematically driven, it is hereby sowed as a permanent seed of philosophy that the human race remains ennobled and balanced with the modest and humble demystification of the sacredness of mathematics. This is what quantified this work to exist as a Britannica of philosophical mathematics, including being a living encyclopaedia that will help all and sundry in appreciating that the universe is a conclave and concave of philosophical mathematics propagation.

CHAPTER 2

What Were the Challenging Visions That Gave Rise to the Demystification of Mathematics?

A	**I**	**Q**	**1**
B	**J**	**R**	**2**
C	**K**	**S**	**3**
D	**L**	**T**	**4**
E	**M**	**U**	**5**
F	**N**	**V**	**6**
G	**O**	**W**	**7**
H	**P**	**XYZ**	**8**

It must be explained that the challenging visions, which gave rise to the demystification of philosophical mathematics, arose as a factor that could not be contained or appreciated during the pre-historical era of the Oriental apostolate. This era witnessed and experienced the highest level of untold sufferings, which was intertwined with poverty of conscience, knowledge, and character. The use and application of the orphic and Delphic oracles, which gave an inglorious rise to human degeneration, could best be explained in this context as a state of life that called for a change. This change was necessary because philosophy was not utilised as a way and means of objective renewing; neither was logic and diplomacy involved in the settlement of discourses and disputes.

A look at the above drawing will certainly prove mathematics and ascertain scientifically that A, when represented with 1, can equally lead to B, which must certainly be represented by 2. The geometric progressions of ascending and descending orders of the mathematical incubator were other visions that gave rise to the discovery of engineering mechanism and technical appreciation.

As the jet age was celebrated in California, the age of reasoning and mathematics demystification was celebrated in the area of the Oriental apostolate. This is why every age has always had a dynamic history that creates the sanctum of vision. The use and application of this sanctum has always encouraged the demystification of sacred institutions, which invariably become the tedium of knowledge that has helped humanity foster a new vision with mission. The classical visions that gave rise to the demystification of mathematics as a sacred knowledge include:

- The use and application of orphic philosophy as a means of leading the populace.

- The way and means the Delphic science was protected as a holy grail.

- The sacredness of mathematics was practically and purposefully mysterious; this is why it was considered by that era as a rational science.

- Students were barred from its knowledge and study. This is why that era encouraged the unification of arts as a means and way of evangelizing honest thoughts.

- Society could not be allowed to study philosophy, mathematics, engineering, and technology until the era of great foundation, which brought a lot of ingenious emengards.

- Honest ability was recognized as terror to the sacredness of mathematics. This is why the challenging vision worked acidulously in order to demystify the sacredness of mathematics with its allied production sciences.

- The use of dialogue was rejected. The issue of arguments was treated with a high level of logical concepts. It was in this tedium that the era of science, technology, and mathematical appreciation (with its philosophical rhythm) started gaining a high level of ground, which later laid a supreme foundation for the era of knowledge.

It is necessary to underscore that the demystification of the sacredness of mathematics was a visionary challenge. The contribution of the Oriental wizards must always be seen and understood as a new beginning that later featured and furthered the course of human evolution and scientific empowerment. As science is working to give the world a globalized community, the demystification of mathematics underscored a universal greatness, because the relationship of science and mathematics can be defined as an electrification of the intertwined relativity of electronic ideas. Our modern age, which is best driven within Microsoft ingenuity, must fall back to the glories and challenges of the demystification of the sacredness of mathematics.

Mathematics is moderate and balanced with the use and application of ascending and descending principles, and a lot needs to be known, observed, expressed, and highlighted on how and why it has remained an intertwined partner to the rhythm of engineering and technology. All aspects of the transmutation of metals into creative economy was naturally seen and appreciated as a sequential mathematical order. That order is natural in philosophy, ironical in logic, futuristic in diplomacy, and argumentative in dialogue. It is here that mathematics was utilised by the Oriental apostolate as a course that has the power of translating human ingenuity into a creative esteem.

Our present era is afraid of knowing the wisdom and knowledge that occult science and technology warehouses, and the era of the Oriental apostolate was afraid to access and appreciate what was contained in the secret and sacred anatomy of mathematics. Unfortunately, a period of punch came in the annals of human history with its scientific evolution. The vision of this punch was to demystify mathematics from its sacred institution. Philosophy was considered a rational science that was meant for rational souls, and all that was contained in mathematics, further mathematics, statistics, research methodology, and engineering technicalities (including all that were anatomized in the warehouse of technology) was equally considered by the unholy beings who were defined as academic hoodlums. This is why this chapter is defined as a careful and calculative investigation into what gave rise to the challenging visions that originally demystified the province of mathematics.

One must appreciate that human evolution has always experienced ups and downs, changes and challenges, revolution with evolution, dynamics with un-dynamic concepts. This is why every effort of man, at any point in time, must be geared towards uprooting standards that are not in philosophical actualization of the dictates of normal standards. This was the vision that metamorphosed into the era of the great foundation.

Nature as the anatomy of all knowledge has always initiated humans with ingenious dynamism to challenge, to change, and to revolutionise what can be said to be the working esteem of a logical idea. This is why the evolution of man has always had the vision to grow, including the desire to drive life concepts with the use and application of science, philosophy, mathematics, and technology. Above all, it has renamed the challenging visions that gave rise to demystification of mathematics during the Oriental era, which was driven and governed by humans who secretly and sacredly mystified mathematics as an oracle.

We must appreciate that growth is the rhythm of nature; evolution has directed the affairs of Mother Nature. This is why there is need to access and appreciate what gave rise to this vision, including knowing what this vision has achieved in order to draw a philosophical benchmark on what could be esteemed as the honourable way forward.

CHAPTER 3

WHY WERE THESE VISIONS IMPORTANT MATHEMATICAL ACHIEVEMENTS TO THAT ERA?

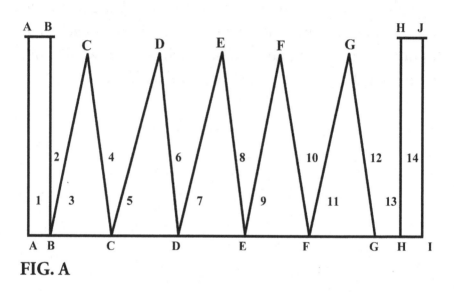

FIG. A

The vision of demystifying mathematics cannot be defined as a mere achievement, because the way and manner in which these philosophical sages and geniuses appreciated the validity and visionary technicalities of mathematics was seen and recognized as a meta-mathematical Oscar, which gave rise to the life of reasoning, knowledge, science, and technology. It equally fostered the dignity of unity in Plato's metaphysics. The achievement of a milestone in any civilization has remained a Herculean task that can only be achieved by humans who are spiritually determined, philosophically fit, economically buoyant, and philosophically and mentally mandated to prove that the universe is not a stunted or stagnated province. This was a challenge that served as a philosophical mandate, which was a stylistic and mathematical achievement to the era that was dwarfed and stagnated with a lot of illogical characteristics.

We cannot have a good education in our present tumultuous society, which is driven neither with philosophy nor with metaphysics, if the efforts of these gentlemen who were committed apostolate are dashed to the wall. That is why the philosophical mountain of fire that ignites the anatomy of choosing souls developed a titanic and tangent vision, which was later utilised as a mathematical incubator, a mathematical versant, and a calculative broadband. This vision became an important mathematical achievement not only to that era but also beyond the existence of that obscure and colloquial period.

The following must be documented, stored, and preferably kept in the museum of philosophical mathematics.

- This vision gave rise to the growth, expansion, and utilisation of visionary education.

- It gained the horizon for the development of science as a purpose-driven partner with Mother Nature.

- The anatomy of science, with its structural philosophy, further developed the natural sector of technology, which has become a benevolent art to all ages.

- Technology, with the combination of science and technology, further opened another sector: the jet age.

- The jet age equally developed and gave a scientific and tactful impact, which the Microsoft community is enjoying which is known as the computer age, with all its atrocities and humiliations.

- In other words, the computer age can be defined as a period that borrowed all the cunning aspect of science, technology, engineering (particularly electrical electronics), and metallurgy.

For the record, the mathematical vision of that era has not stopped. This is why there is a lot of dynamical esteem, which is propelling another age known as the occult engineering age to emerge.

The technological age is defined as the creative and productive age, which will be followed by the Aquarian age. This is why metaphysics and mystical philosophy have always insisted that the human race is developing in material and conscious appreciation. After the Aquarian Age, another luminous age will come into the life of evolution; it will be known as the mystical age, or the cosmic age of human and spiritual development. At any point in time, philosophy does not rest in educating all civilization that evolution is both evolutionary and revolutionary with the institutions of growth and degeneration.

A look at what is contained in the works *Man, the Unknown* and *The Man Who Tapped the Secrets of the Universe* by Glenn Clark will prompt one to appreciate that at the moment man is at a dangerous crossroads. He needs to be rescued, without which the agony of waiting and being at risk of oblivion will become another factor that will challenge the coherence of philosophy and mathematics as dynamic sciences.

A look at the growth, achievements, appreciation, and expansion of mathematics to the present era will certainly give scientific kudos to these noble souls who decide to take up the challenges simply because man is the mission of philosophy. In one of the ornamented and highly authenticated letters from the papacy, it reads,

> Man is at a dangerous crossroad and he is equally enmitied by his leaders simply because he has not understood the rhythm of philosophy as the way back to Godhead.

The wording of this letter from the papacy, which can be described as an exilia of wisdom from the seashell of sacramental theology, reveals that every era has always had great emengards, great heroes, and amortized theodolites. That is why the ingenious thinking of this work is to assure all the ignited facts that have worked against philosophical mathematics and mathematical philosophy.

We can conclude this arena of philosophical segmentation without looking backwards to the contributions of these illumined actors who on their own decided to revise and revolutionize their province because of the evolution of ideas. Revolution has always remained the practical and objective way by which humans have been futuristically fostered for a great advancement; this is why the development of that vision was synchronous with the development and achievement of a mathematical province to that era, which was a colossal threat.

A look at the achievements of the demystification of mathematics will let one appreciate the high level of honour these achievements bequeathed to humanity. We must appreciate that this holistic honour tactfully and technically destroyed all the illogical institutions of the following.

- The Barbados.

- The Delphic and orphic undoing to humanity.

- The militants with their illogical military.

- The Watzinger brothers, who were rude particularly to the knowledge of philosophy and mathematics.

This is why this chapter cannot be sidelined nor omitted. Evolution has stated that history is receptive, and philosophy has a national mandate of educating its province with the use and application of mathematical policies, principles, and practises. At this point, who should be blessed, who should be honoured, and who should be asked to share the monumental contributions of philosophy, particularly in the demystification of mathematics as a great oracle?

The philosophical jubilee in this rhythm must honour the soul of mathematics and the search for philosophy. But the best to be honoured are the Oriental wizards, who out of their ingenuity with practical vision decided to demystify mathematics as an oracle. This is why posterity decided to build a hall of fame that respected and galvanized them as philosophical emengards and great souls who will ever be remembered for their inputs in ensuring that evolution does not lead to a permanent and perpetual suffering. At this point, the human race is thereby admonished and challenged to appreciate that the sacredness of mathematics and the causes of this demystification have paved a way for a proper understanding of the knowledge of philosophy. This is why this book, and particularly this chapter, can be defined and understood as what lies ahead of mathematics, and as why philosophy is made a monument as the balanced occupation of the Oriental wizards.

CHAPTER 4

What Are the Achievements of the Demystification of Mathematics as a Sacred Oracle?

We'll take a look at the universal and cardinal input of mathematics, which is perfectly intertwined in all arts, sciences, and technology. But a few will certainly appreciate that the demystification can be called the glory of life, an honour to the world of knowledge, and an impetus in the best objectivity of what mathematics is all about. This is why one of the best advancements of the demystification can be defined as the scientific rationale that constituted and computed the rational.

It was not possible for mathematics to achieve my aim before now. The struggle to demystify the project was impossible to venture on, and this is why the sacredness of mathematics, even up to this moment, has remained protected in the use and application of the following mathematical formats and concepts.

- The addition principle.

- The subtraction method.

- The division system.

- The multiplication method.

- The compound ratios, which includes simple compound, multiple compound, diversified compound, business-related compound systems, and science and engineering compound mechanics.

The demystification equally gave rise to the appreciation and application of the following.

- The use of factors.

- The use of figures.

- The ability to determine fractions.

- Simple and multiple simultaneous equations.

- The introduction and utilisation of the log table with the following formats: the sine formula, the cosine formula, the tangent system, and the co-tangent (including what I have discovered to be the mathematical mechanism of the tangent method, which is a logical appreciation with a simultaneous approach that will help mathematics to work synchronously with data, research methods, and the use and application of the prudential guideline mechanism, which is the accounting system).

Other mathematical logjams and their log tables were equally invigorated:

- The use and application of ratios.

- The remainder theorem.

- The simultaneous equation.

- The psychic equation.

The demystification can best be defined with a mathematical notation as a springboard that further gave rise to the harmonious relativity of mathematics as a science, engineering mathematics as a noble science, and technical mathematics with its statistical determination. The frequency in this mathematical formulation gave rise to the following: geometry, algebra, statistics, calculus, quantum mathematics, appreciative mathematics, and database mathematics. Other unknown areas were equally discovered, such as economic mathematics, econometrics, analytical mathematics and business mathematics (which comprises the arithmetic of commerce, which further developed an ingenious industrial mathematics), commercial mathematics, financial mathematics, accounting modulus, and mechanics. The demystification opened a domestic horizon in the use and application of mathematics.

Currently, a lot has been discovered in our present interlined computer age and jet age, which includes computer mathematics with its Internet and versant-driven ratios, aeronautic and technological mathematics, and astro-dynamics with its mathematical quantifications. It is here that such noble mathematical laureates must be mentioned: Albert Einstein, Pythagoras, Euclid, Michael Faraday, Galileo, Edward Halliburton, Walter Russell, Aries Felix, Bill Gates, and Chike Obi. They deserve consolidated chairs in this hall of fame that exists with the tedium and rubric of electronic dynamism.

In this respect, the demystification achieved beyond words, above letters, and beyond notes. This is why I must point out the following.

- It gave rise to the assessment of mathematics by all and sundry.

- It empowered the use of mathematics as a commercial and industrial gate.

- It lifted the standard of technology beyond the informal level.

- It gave science protection with technical and technological impetus.

- It intertwined the relativity of mathematics and philosophy as a common friend with a calculative goal.

- It gave birth to a new era of knowledge, designated as the great foundation.

- It made commerce an honourable inventor.

- It uplifted the standard of economics and added the study of environmental economics, developmental economics, and economical policies, principles, and practises as ways and manners of enabling commerce and trade. Cost accounting, financial accounting, managerial accounting, and management accounting, to name a few, were expanded to take care of business-related transactions.

- It equally restructured and fine-tuned the standard of understanding of physics, chemistry, geography, biology, engineering, and technical dynamics.

A look at the achievement of the demystification must award a great honour to the contributions of the Oriental wizards.

At this point, our present era, which exists in a combination of ages that include the jet age, the computer age, and the scientific age, is hereby challenged and at the same time called upon to delve into the best portion of appreciating the functions of mathematics, including knowing how philosophy was honoured as the mother of knowledge and the father of science and technology, and the wisdom of all ages. It is important to ennoble and appreciate this chapter as an educational, compact print that is anatomized in the best achievement of a Britannica, by which all that is de-composited in the encyclopaedia of balanced knowledge can be harnessed, appreciated, and perfectly utilised in bringing the world together as a village, which must be driven with the Internet versant of philosophical mathematics.

This chapter is best defined as an incandescent mathematical alloy that is donated as a philosophical Aloysius. It must remain and exist as a guardian teacher to the world of knowledge, which is still in search of the realities of the perfect forms that cumulate and cultivate philosophical mathematics as a practical and scientific endeavour. That reality holds the key in empowering humanity to a compact and horizontal dimension.

CHAPTER 5

WHAT IS THE HONOUR IN THIS ACHIEVEMENT?

A look at the pros and cons of the demystification of mathematics will let one appreciate that a consummate and immortal honour is embedded in this achievement. We must acknowledge that man is in search of truth and in need of knowledge. One of his greatest challenges lies in discovering his real self, by which the wisdom of the true God can be discovered and utilised.

From various ages, the human race has always been involved in one catastrophe or another: monumental homicide, involvement in atomic suicide, and the inability to appreciate God and the unwillingness to know his roots. This is why this chapter can be defined as a wakeup call particularly to the province of science and technology, and to the sector of mathematics and philosophy, which is understood to maintain an inclined relativity and practicability with all human endeavours.

What we are materially enjoying today is a global village. Economic and Internet principles can be defined as what lie in the advents of the demystification of mathematics. This is why our present era must work hard to partner, protect, and utilise all that is contained in this honour, in order to advance the sentiments of human ability with their dynamic calls.

It is important to understand and experience that what lies in honour of this advent has not been well utilised by our chaotic civilization because of the following reasons.

- Lack of honest ability.

- Inability of the human race to practise truth.

- The use and application of un-objective and un-dynamic facts in determining the welfare of the human race.

- Inability of research methodology to partner with the overall segments of education, industry, commerce, science, technology, and economics.

- High level of poverty, which has created a degenerated melancholy in the affairs of the human race.

- The presence of chronic and multiple sicknesses with their ingenious injections, which are presently responsible for the high level of material and mental incubi.

- Inability of science to perform its honoured and moderate role as a partner with technology and an investigative chancery that honours other human endeavours.

- The problems of dermatology, migration, and brain drain must be taken into serious consideration.

- The use of objective knowledge must be taken as a serious challenge.

- The leadership system must be prepared to structure and restructure the world economy in a way and style that it will be socially and intercontinental productive.

- There must be enhancement in the areas of structural and infrastructural development, which in return will empower the economic system of the one-world purpose.

- Civilization must not be seen as a material achievement but as a natural phenomenon that humans can use in creating a permanent economical feature, which will in turn empower all and sundry to be creative.

- The issue of separation, particularly to the feminine sector, must be abolished.

- Efforts must be made to address all the multiple problems created by religious sects. The gospel of the brotherhood of man's principle must be ornamented and universalized.

It is unfortunate that the efforts of the wizards have not greatly touched the cartel of the academia and the institution of research. A look at what lies in these achievements cannot be appreciated if the human race as a globalized asset is not sensitive in knowing that every institution living as an oracle portends a danger to the growth of human and economic development. It is important for our educational planners to appreciate that original life is structured in the use and application of philosophical mathematics. This is why when it is stated that philosophy is the mother of knowledge, it only reveals that man is a fulcrum of philosophical mathematics and the body of philosophical Oscar. All efforts must be made by all and sundry in assessing, appreciating, and understanding how and why the world has remained practically stagnate despite efforts in the areas of material science and technology.

A look at the prowess of this chapter will certainly reveal that we are at the threshold, which has placed us in a tighter corner and in oblivion. This is why to awaken to the challenges of knowledge can be defined as the consummate honour that lies in this achievement, which can be known and understood as advancement into the endeavour of human ingenuity. At this point, the immortal words of Abel Boulevard and Fidelis Brown must be mentioned.

Every created mind must lift up to the challenges of philosophy and every lifted soul must be purpose-driven in assessing and renewing other souls. This is the honest way by which Mother Nature has ornamented philosophical mathematics as living water by which our lean liquids can be expanded to the honour and rank of a monumental living worker.

In this respect, Abel and Felix were only inviting the human race to the orchard where philosophical trees are planted. I feel greatly challenged by the immortality of this maxim. This is why I hereby donate the following for the upliftment of the human race, which is in need of philosophy as a harbinger of universal thought and as a mathematical agape that is honoured with scientific and technological love. The achievements that are contained in this honour are hereby summed up.

- Creation is creative, sensitive, and principled with the use and application of policies and practises.

- Alone with philosophy is always a mathematical attraction to a greater achievement. This is why all over the world, philosophy is seen as a mortal action that speaks louder than a voice with its vehement victories.

This work is tutored, tailored, trained, and trimmed in order to arrive at what lies in honour of these achievements. It is designated as a scientific and calculative introduction to the demystification of the sacredness of mathematics, and it is seen and utilised as a mathematical principle with its objective ratios that will help our present era in proposing principles and policies. The policies will engineer the new era to practise philosophical mathematics as an enabling and perfect way of honest life that lies in the achievement of this implacental vanguard.

CHAPTER 6

How Can We Reconcile the Problems of the Present Era with Those of the Obscure Era That Mystified Mathematics as a Sacred Oracle?

Philosophy is greatly loaded with perfect and scientific achievements. The past hindrances that mystified it with mathematics as a sacred oracle can be defined as being in the becoming in the rays of the dawn. This is why mathematics in our present era—which is chaotic with scientific and jittery incubi—is still suffering in the following ways.

- Lack of knowledge, which can be utilised in explaining, understanding, appreciating, and assessing its united diversification.

- Inability of the tutors to see mathematics as a scientific esteem, which must be utilised in determining facts and figures.

- Lack of apparatus and what can be defined in this context as a standard laboratory.

- The problems created by un-dynamic culture and social norms that were originally contained in the teachings of religion.

- Inability of the environment to be friendly with the needs of mathematics.

- Inability of the educational sector to protect and empower the teaching of mathematics not as a difficult subject but as a friend of students.

- Inability of students to see mathematics as part and parcel of their empowerment.

- The destruction of the mathematical institution, particularly by the Amazons and barbarians.

- Inability of the educational class to separate mathematics from philosophy.

- Lack of mathematical ontology and appreciating its mystification with a philosophical prologue.

- Lack of a universal standard, which is important for the regulation of mathematical ingenuity.

- Inability of the class system to liberalize the thinking of mathematics as a universal mind.

- Mathematics has equally suffered the problem of trainers and trainees; this is why it has remained an obscure endeavour particularly within the circuit of the uninformed.

- Mathematical materials are often difficult to assess.

- Courses that are being driven by the most deep thinking of mathematics are usually difficult to understand by the teachers, the students, and the colleges of mathematics. This is why it is practically and dynamically asking for a reengineering.

- Both the present and the past have always seen mathematics as an oracle and as a rational science that must be demystified for it to achieve a scientific and technical goal, which is the basis for why its origination and invention has always attacked the presence of inferiority and inequality.

A look at the deep thinking of philosophical mathematics with its dynamism, content, and branches will certainly lead one to appreciate that mathematics is still in the developmental stage. This is why it is facing a critical and primary concern regarding a lack of development. This chapter explains that the demystification of mathematics can best be defined as a scientific synergy that has always presented a titanic goal to help humanity appreciate that it will help evolution to exist in the new age, which is going to be technically driven with occult mechanism.

The universe is practically driven and materially engulfed with the titanic challenges of the Microsoft system, and this mechanism is actually calling for a lot of innovations, renovations, evolution, and revolution. It is clearly seen that scientific cosmogony is lean and critically difficult, simply because the role mathematical moderation should play in achieving the needed globalization appears to be hitting the rocks. If the world as a global village can understand that all that is contained in the science and arts of economics and econometrics is mathematically driven, then it will then help humanity to understand that the mystery that was detached from the sacredness of mathematics can be seen and defined as the golden dawn. This is why the obscure era of mathematical existence was greatly celebrated when it was detached from networking and supervising the province of mathematical mystification.

As Plato stated, what propelled the demystification of the sacredness of mathematics was to assure humanity that mathematics is a freedom fighter and to expose to the evolving man that there is nothing like sacred cow, sacred teaching, sacred knowledge, sacred science, and sacred understanding, in as much as the province of the teachings of mathematics is concerned.

A look at what is happening at different faculties of mathematics all over the colleges of mathematics reveals that even in our jet age, mathematics is still meeting and attacking a lot of human and evolutionary challenges. It is on this notion that Pythagoras decided to demystify the sacredness of mathematics and how humans access the lintious content as an indomitable oracle.

We can appreciate that such mathematical branches like statistics, simultaneous equations, fractions, and compound and simple interests are still being taught at the obscure and un-notable system that most of the time neglects and rejects the institution of mathematical modernization. A look at the way and manner technology is fostering the course of the world economy will certainly reveal that there is high level of mathematical ingenuity. This is why mathematics, engineering, and science are strongly being driven with the ethical principles, policies, and practises of Mother Nature.

It is at this point that our present era must appreciate and understand the following.

- Since the dawn of consciousness, mathematics is an ingenious and practical thinker.

- It is the symbol of science and other creative arts.

- It is the anatomy of technology. This is why other related engineering branches are practically driven by the thoughts of material ingenuity.

- As philosophy is the maker of great thinkers, mathematics is the maker of great arts, great thoughts, and great engineering creations. That is why all the creations of technology are mathematically configured.

- The reconciliation of the problems of the past with the present must always help both our jet age and the expected occult age in ensuring that humanity does not degenerate into another oblivion, which shows clearly the mark and face of poverty in the psyche and annals of human biology.

- Every occupation must understand that creation exists practically on the logos of mathematics. This is why its thinking is harmonized and simultaneous with the expansion of technology.

- It is in the ability of Mother Nature—with its principles, policies, and practises—to intertwine the relativity of mathematics with science, engineering, and technology. Mathematics is the only subject that is practically driven with the lintious thinking of objective dynamism.

It is important to explain these objectivities purposefully and tactfully. Growth is part of human evolution, and human expansion, upliftment, social and universal tradition, and the ingenious thinking of mathematics must be accepted as a working-class tool. For example, if it is stated that A + A = 2A and that A + A + A = 3A, then there is a mathematical reconciliation that 1 + 1 + 1 = 3 and that 2 + 1 = 3. It is here that the coefficient mechanism, or what can be designated as a mathematical principle, will appreciate that A + B + C = 1 + 2 + 3, which will equally exist at the simultaneous thinking of three letters.

The mathematical use of these letters means that mathematics exists with a simultaneous standard, with a unified policy, and with the use and application of subtraction, addition, multiplication, and division. This is why the urgent challenges of our modern technological problems lie in reconciling the problems of the present with that of the obscure era, which mystified mathematics as a sacred oracle.

It is at this rubric that I am empowered by this work, donating with all the inputs and outputs of engineering (which is interlocked and intertwined with technology, with science, and with other objective sciences) to say that no mathematical reconciliation is possible if man does not appreciate or understand the urgent necessity of reconciling his arts, creations, transactions, and reactions that are constantly driven with the ingenuity of mathematics.

This chapter exists as a mathematical moderator and a philosophical ignition through which the global family can start to see the origin and great works of mathematics. Those works are hereby designated as an indomitable laurel that, when properly understood and utilised, will uplift the global village, which is known as a community of humans beyond the present existence. That village has not made any efforts to reconcile the problems of the past with the present, which is naturally asking for philosophical mathematical guidance.

CHAPTER 7

WHY IS THE DEMYSTIFICATION OF MATHEMATICS A UNIQUE ESTEEM THAT ACHIEVED PLATO'S METAPHYSICAL COMMUNITY?

It is important to know that a community cannot achieve unity without employing the dynamic concept of philosophy, whose objective is to create a province of absolute unity, total purity, and ingenious perfection. This is why a lot needs to be derived from the esteemed and purpose-driven community, which was created in Plato's era. This creation would not be possible if the philosophical guru and his aborigines did not unite in attacking and demystifying mathematics as a sacred institution and as an oracle. This is why the demystification was able to strengthen and purge the kingdom of Plato's community.

A fundamental look at the achievements of this feat will reveal that philosophy, psychology, anthropology, and all that is conceptualised in international history are practised under the tedium of community and international propagation. By international propagation, I mean the overall achievement of an evolutionary concept through the use and application of social structures, which include political, economic, administrative, and administrative mechanisms.

This is why posterity boosted Plato as an amalgamated achiever in the overall conquest and demystification of mathematics. A person looking at the practical composition of mathematics in such luminous and production areas as engineering, technology, chemistry, information technology, geosciences, and geophysics will certainly appreciate that the world could have been at sea, and the universe could have existed as a moribund phenomenon, if the concept of the demystification of mathematics was not achieved. Against the backdrop, it could now be seen in the universal psyche of Plato's metaphysics, which is defined by the province of philosophy and logic as an ingenious era for being, bearing, and becoming in the Orpheus and orphic institutions that were sequential to the encouragement of learning with higher understanding, education, and practicalities. The achievements of industry and commerce, and the notation of economics and econometrics, must be seen as a welcome concept that came into existence because of the achievements of Plato's metaphysical community. In this respect, the demystification brought the following into the arena of Plato's metaphysical community.

- Social security was encouraged.

- Political orderliness became a constitutional framework.

- A social feeling for the welfare of all and sundry was cultivated and incorporated as the dynamics for honest existence.

- The community was detached from the inglorious infections of the religious dogmas, creeds, norms, and other un-mystified religious worship.

- Growth with balanced expansion achieved a desired and clandestine order, which later was responsible for the creation of the Order of Ingenious Minds.

- The demystification favoured Plato's metaphysical community and encouraged the use and application of democracy, wisdom, and honest thoughts, which became a serious threat particularly to the province that was designated in that era as the manorial system.

- It was the seed of this demystification that gave rise to the introduction and practise of the brotherhood of man's principles.

- The demystification weighed heavily on the short-sighted attitude of leaders, commerce, industries, politics, and administration. This is why the social order that it encouraged and reengineered was responsible for prompting Edom Markham to state, "Against making cities, let us make man."

The encouragement and building of man was the focal point of Plato's metaphysical community. This is why such other committed ingenious thinkers and creators—like Isaac Palm, Mohammed Paul, and Anselm Peters—started utilising the unlimited powers in that community in energizing a perfect and purpose-driven cord for the welfare of their age and the incoming posterity. All that is structured and contained in the studies of Platonism and Marxism were bequeathed to the new era of ingenious minds, when the oracle of mathematics was demystified as a simplified and purpose-driven knowledge. Because of the long trend and struggle that was involved, the province of philosophy decided that philosophy, as a perfect and purpose-driven science, must not have any attachment with sacred practises. However, philosophy must do its best to ensure that the human mind was energized with aesthetic and logical ignition in achieving a purpose-driven policy with a high level of import practises, which will make the practise of philosophical mathematics portray the logo of Plato's metaphysical community, which existed in a fundamental mathematical segment. This is why philosophical laureates and nobles have always blessed the efforts of these ingenious emengards, who worked tirelessly with assiduous technicality in ensuring that the obscure and moribund opacity was ingeniously removed and destroyed, with a positive result of amplifying and creating a united community that the kingdom and kindred of balanced knowledge has appreciated as Plato's community of metaphysical harbingers.

Based on this, I note with philosophical and scientific admonition that this is the beginning and creation of our heritage, of which the dynamic functions of philosophical mathematics is hereby designated as the beginning, the thrust, the expansion, and the universal consolidation of the era of great foundation.

CHAPTER 8

How Can We Believe the Fact That the Demystification of Mathematics Has Remained the Greatest Challenge of the Past, Including Being the Best Encouragement of the Present?

The above subject matter must be seen and appreciated as a philosophical mandate, a wakeup call, and a logical and luminary antidote. This is why it is not gainsaying that every belief cannot be accepted as reality, without the foremost wisdom of philosophical faith. What is accepted as a great challenge must have been an act that is not synchronous with the social order, the social norms, and the civil liberty. All that is contained in the reminiscence ability of this great challenge must be analysed with a view to ensuring and appreciating why and how such an inglorious act must be rejected.

Unfortunately, the sacredness of mathematics before this demystification created and donated a room for fears and hatred, and this situation attacked knowledge at its face. This period and the pre-Aristotelian period cannot be defined as enigmas in the understanding of mathematics. Going by the ontology of the previous chapters, the crusade and the jihad to demystify mathematics has purely and perfectly benefited the present more than the past.

An honest quest can ask this question: how and for what reason has it benefited the present more than the past? I hereby submit the answers as follows, which are in the best paradigm and opinion of philosophical reality.

- It lifted the understanding of knowledge beyond the former existence.

- It gave birth to liberalism and liberalization.

- It is the beginning of what our modern civilization is enjoying as freedom of thoughts, freedom of worship, freedom of education, freedom of action and reactions, and freedom of national and international relations—which in turn gives rise to freedom of trade, diplomacy, and international treaties with a high level of national and international organization known as the ministry of foreign affairs.

- It opened up a new horizon for the learning of science, new arts, engineering, and technology.

- It conceptualised and compacted the luminous ideas of mathematics in order to synchronize with the figures and features that constitute and compose the science of astrology.

- It brought and encouraged a blending technicality that gave rise to the study of the anatomy of alchemy, which later empowered humans to appreciate the science of transmission, transmutation, expansion, and reduction. Our modern science denoted it as the study of pure and applied chemistry.

In earnest, the alchemy of the transmutation of matter, ions, compounds, and elements can be defined as a scientific chimney that opened the cataract in the best knowledge of the reaction and action of such minimal elements like atoms. The demystification can be defined as an empowerment that encouraged the likes of Dalton, Faraday, Einstein, Galileo, and Bishops Crowther and Crawnwell to come out openly with their scientific theories and practises. Those results originated at the foremost foundation of their empathic research method, which is helping our modern scientists through the use and creation of what is defined here as further developmental theories.

I must not fail to mention that our modern era is not well positioned and exposed to appreciate the joy of the demystification of mathematics. A work to appreciate this must blend the facts of this greatest challenge to the past and its ennoblement to the present, and it must understand that the multiple mathematical segmentations were practically and technically derived because of the conquest of the demystification of the oracle.

Scholars who are only denoted to the levels of scholastic and academic understanding may be in a difficult corner in understanding, with practical assessment, the logical views and contents of this chapter. They are of the opinion of arriving at a comprehensive and coherent conclusion on how, why, where, and for what purpose I feel so dynamic and encouraged to bring this reality into a reasonable bearing.

The answer to this question is simple, but one must appreciate that both philosophy and mathematics are dynamic. This is why the present era must see the truth of the mirage from the workable placement of the plane mirror, which can be defined as the connective part with solar currents that connects the rays of the sun in order to perform this natural and dynamic functions, by which the creation and scientific result has remained the appearance of the mirage, which is linked to the efforts of the Oriental wizards.

A look at the practicality of this will certainly reveal that the human race has come a long way, and it equally shows that creation has suffered a lot of setbacks. This is why our civilizations, acts, and creations must be on the vanguard of ensuring that the negative sacred creations are not imported or transported into the acts and practicality of our daily lives, which are ornamented to be philosophically driven with the use and application of ingenious esteems. The tedium of this comparative analysis, with its metaphysical explanation, is to appreciate that empowerment and nobility of the conquest—which is being enjoyed particularly by the province of the Microsoft world—are gradually and unknowingly regrinding and remodelling a lot of technicalities in order to exist and become another infamous oracle.

The question here is: can this be accepted? The answer can be intertwined into a lot of rubrics with practical segmentations.

- Globalization is tactfully driven in the openness of ideas and thoughts, which means the rhythm of Microsoft technicality is already demystified.

- The economic village with its concept is all about creation with the use of inputs and outputs, which means that a lot of scientific modules and goggles must always be tapped by humans who will drive the Microsoft technicality of this engineering. This means that all mathematical ratios must exist at the tedium of the fountain, and all philosophical rationalities must equally work with the mathematical ratios.

This is why the fears about the dangerous and negative portraits of the Microsoft system must be demystified in the same way. The Orients reduced and demystified mathematics as being the Halliburton of oracle performance, which exists for the choice of the selected few. This comparative analysis is real and important to be gained simply because evolution and lifelines exist practically and purposefully at the tedium and compendium of philosophical and mathematical, comparative analyses.

CHAPTER 9

HAS THE PRESENT EMBRACED THE REWARD OF THIS CONQUEST, WHICH WAS A PHILOSOPHICAL JUBILEE?

Our present civilization has not embraced the wisdom of this philosophical conquest. This is why the universe, particularly the sectors of industry and education, appears to be stunted. If the philosophical jubilee of this conquest has been embraced in the honest dynamism of philosophy, the following could have been a thing of the past.

- Man's inhumanity to man.

- Un-dynamical creations.

- Lack of investment in the areas of philosophical mathematics.

- Uncertainties that affect civilization and social security.

- Crime, which is affecting growth, evolution, economic activities, and universal industrialization.

- High rate of falsehood, which is becoming the tradition of the universal civilization.

- The use and application of unguided norms and creeds, which do not drive the thinking of technological advancement.

- Brain drain, which has given rise to human slavery and the impoverishment that in turn affects human capital development with economic activities.

- Lack of investment in the lintious and challenging areas of research methodology, engineering technology, and production technology. The situation is worst in the third-world countries, which were colonized by the European province.

- The practise of indecent acts, which does not characterize the true nature of philosophical mathematics, is still dominant and predominant in our present age.

- A lack of honest circulation of knowledge has affected evolution, growth, expansion, and industrialization with a high surfacing of poverty, even in our mock celebration of the jet age.

In this respect, the bottom line must be addressed, re-endowed, and revisited. It is unfortunate to see that the present era has not embraced the wisdom of this philosophical conquest, which was a great feat that was achieved as a golden jubilee from the wider province of the Oriental apostolate that tasked itself unanimously in establishing the purpose-driven era of great foundation. A look at the scenario of human activities will certainly object practically, scientifically, technically, mathematically, and even philosophically that the human race is still at sea in the permanent embrace of this conquest.

Earnest Holmes' work, entitled *High Philosophy,* states that:

The universe will remain dominant and dormant with the use and application of obscure concepts until humans decide to revive, revisit, and revolutionize all that is contained in the ethical propagation of philosophical mathematics.

Holmes strongly believes that the task of the future generation is a critical and mystifying target that requires a philosophical reengineering with mathematical dynamism. This is why the concept of high philosophy can best be designated as the greatest assignment befalling the evolution of man of today. In this respect, it is clear that our material civilization still looks upon the demystification of mathematics by the college of the great foundation, by the unity of Plato's metaphysical community, and by the supreme philosophical aborigines as another convert of electrifying ubiquity. This is why the human race is still in search of the following.

- The way and means to discover the unknown man.

- The way and manner to appreciate the immortality of the blessed angels.

- The philosophy that determines mathematics as an ingenious thinker of models and modulations.

- The reason, if any, why occult science and occult philosophy, including the enabled soul of occult engineering technology, can be appreciated as noble creativity.

- The astrological keywords that will help man to gain the inner meaning of why he is a transcendental immortality of Godhead.

Man is still in search of the way to himself, to peace, to love, to light, to knowledge, to understanding, and to procreation with balanced productivity. These lapses prompted some philosophical emengards to make the following statements.

- Abdrushin: "Man, how do you stand before thy creator?"

- Edwin Markham: "In vain we build the city without building the man."

- Ralph Waldo Emerson (with his associated philosophical inheritors, who are defined as the supreme heritage of mathematical thinking with its universal and incandescent knowledge): "Man, know thyself."

- The likes of Walter Russell, Lao Russell, and Herbert Spencer equally focused on the ornaments of this maxim when they said, "The greatest knowledge in life lies in man knowing himself." They equally stated that the greatest thing that can happen to a human while in an earth life is to connect and interconnect his conscience with the illumined and lintious wisdom of the creator.

It must be stated that what has been enlisted in this chapter has remained the errors of creation—particularly man, who has refused to embrace the wisdom of the great foundation, which was the point of philosophical conquest and featured a living and monumental community of balanced thinkers. Those thinkers were formally and purely designated as Plato's metaphysical community, which existed in the honest knowledge of unity in propagation, in diversity, in relaxation, and in understanding. This is why the present era is hereby referred to one of my inspired work *The Electrifying Power of Unity Scientifically Explained*, including the masked mirror of a quantious work, which was at the designated sponge of my philosophical illumination and was named with scientific, mathematic, and evolutionary truth.

With these words, our present era is hereby warned to desist and not accept to exist as a material recharge card.

CHAPTER 10

WHAT IS THE MATHEMATICAL MANDATE OF THE DEMYSTIFICATION, WHICH IS HEREBY EXPLAINED AS THE VISION OF THE ORACLE?

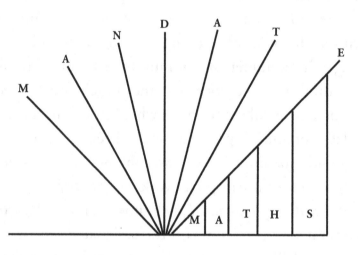

M	1	2	3	4	5	6	7
A	A	B	C	D	E	F	G
N	M	A	T	H	E	M	A
D	T	I	C	X	Y	Z	I
A	P	H	I	L	O	S	O
T	P	H	A	B	C	D	E
E	I	2	3	4	5	6	7
XYZ = M =	A =	N	= D	= A	= T	=E	
XYZ =XY=	XZ =	YZ	=M	= A	= N	=E	

420

Before the urge of a social and spiritual guest to demystify mathematics as a sacred oracle, there was a united and agreeable mandate that formed the beacon of this conquest. Every effort that is geared towards bringing a social emancipation is always conceived through an agreeable and practical mandate, which will certainly bring a change in tradition, in principles, in policies, and in practises. In this respect, the mandate started with a social crusade that originated a forewarning about the preparedness of this highly evolved principle in changing the direction and activities of the social order, particularly that which is ascertained to have kept civilization at the balcony of honest living.

That is why the urgent necessity of appreciating the validity of this mandate must certainly be seen as a wakeup call whose effort is to liberate humans from a lot of obnoxious laws with myopic melancholy. That melancholy later resulted in the order of a great foundation.

In this respect, the following can be understood and appreciated as the order of the mandate.

- Reducing mathematics to become the pioneer of scientific engineering.

- Ensuring that mathematics is used in ascertaining the functionality of facts and figures.

- Teaching mathematics in all fields of human endeavour without class distinction. This is why this mandate attacked the sacredness of its originality directly from the roots of human wickedness.

- The mandate equally expunged its functionality as a universal and practical art, which can be used in appreciating the functions of science, the meaning and definition of technology, the principles, and the practises and policies of engineering. This is why, since the mathematics demystification, it has been taught to a wider level, from the kindergarten stage of human knowledge.

- The mandate was equally a sign by which all and sundry can see the reality of the functions based on what is on ground—including how it affects the lives of humans positively to the era of the great foundation, which was equally instrumental to the development of such highly gifted mathematical wizards like Pythagoras, H. A. Clement, M. Durell, and others. The mandate will always remain a futuristic and achievable endeavour that will help to improve the sectors of science and technology as follows.

- It helped in the creation of a respected social order, which later developed a system of unity. That system was a practise that is in synchronous rhythm with the dictates of Platonism.

- A work to examine the importance of this mandate must appreciate that mathematics is really detached, from being a non-functioning course to a dynamic and functional course. This is why it became the prince of inventions and the anatomy of balanced and objective creations.

A synchronous look at the stylistics in which the appearance of the above drawing is made will certainly help one understand that mathematics is the royal balance of thoughts and of dynamic achievements. This is why the benefits of the demystification is hereby defined as a spontaneous, futuristic mandate that has invariably created a purpose-driven province, a missionary crown, which is hereby originated as the beacon of the great foundation.

In the illumed words of Billy Graham Douglas, every culture and tradition cannot achieve a target if the mission and the mandate of mathematics is not involved in providing a lead for the functionality of such a culture and tradition. In my opinion as a mathematical and philosophical mandate, the benefits and all that man has created within this period of great foundation can best be defined as a scientific call to awaken. The purpose of the awakening is none other than achieving the mission of the mandate of mathematics, which is naturally and philosophically man driven.

Efforts must be made in ensuring that the efforts of our laureates are not destroyed or relegated to that un-scholastic period, which fought and ornamented mathematics as an oracle. Philosophically stated, this is not the glory of our modern age; neither is it the vision of the micro-driven, technological province. Philosophy is purpose-driven with the rhythm of purity and perfection, whereas mathematics is dynamic in achieving results with the use and application of figures, facts, and features which, when further observed, are essential in the objective quantification of mathematical ideas. For the objective understanding and thought, this idea was the vision of mathematics, which later translated into the philosophical transcendence of the demystification of the oracle.

Unfortunately, humans were not bold enough in that era to see reason alongside the ornamented wizards. The wizards were naturally mandated by the forces of Mother Nature with the urge of the social oracle to carefully and technically truncate what can be seen as a mathematical oracle, which was purely in the hands of the wicked leaders. Those leaders were using the astrology of mathematics through some religious hypnotism to weaken and reduce the evolution of their era, which was not driven with the honest functionality of mathematics. This was the beginning of the journey that later resulted in the demystification of the sacredness of mathematics, which was purely at the philosophical proximity of the wizards. The challenges have always been to appreciate that mathematics must not be seen or understood as a sacred oracle.

It is here that the thought for making it a must-know by all and sundry must be welcomed and appreciated as a dignified mandate, which was used to create an era of great foundation and great freedom, and which later revised the standard of learning and the tedium of knowledge. It must be stated for the record that the ordination of this mandate was responsible for tactfully creating a philosophical mathematics mandate that uses wisdom as the wealth for all. Every era must live to understand with social appreciation that the meaning, the purpose, and the tedium for which this mandate was assigned can simply be seen as being in the principle of a new beacon, in the purpose of the great foundation, and in the polices of a new beginning. This is why this chapter is best denoted as appreciating mathematics as bankable mandate, which was at the heroic proximity of the ornamented philosophical wizards.

CHAPTER 11

AN OVERVIEW TO EXPLAIN HOW THE DEMYSTIFICATION OPENED A NEW HORIZON FOR HUMAN ELEVATION

It is in the golden plate of physical rest, practically customized by the universal will of balanced wisdom, that the demystification of mathematics as a constant oracle and with monumental legends opened a practical impetus that is today enjoyed as a new horizon for the human-celebrated elevation. This philosophic jubilee was the origin of science, technology, alchemy, and transformation with transmutation. That is why this part, conceived as a technical summative of the overall parts, is best defined as the dawn of a new day in the tedium of philosophic and mathematical genealogy.

It must be appreciated that this part started with an ingenious definition of mathematics and the philosophical explanation of the mystification and demystification (including what gave rise to the philosophic catechism), which later became accepted as a technical mystification responsible for the flourishing province of the celebrated mystical community. As noted previously, this jubilee was celebrated as an important era that welcomed the period of the great foundation. The foundation practically and perfectly collaborated with the mathematical ingenuity of such practical and celebrated mathematical apostolate as Euclid, H. A. Clement, and more.

This part equally delved greatly into the challenges of demystification, including the import of the vision and its meritorious and mathematical achievements. The achievements later helped man to understand the reconciliation of the problems with a perfect and practical judgment of the seal of that era. With our sealed un-engineering era of Microsoft technologies, the concentration is now on how the demystification is a celebration of the achievements of Plato's metaphysical community, which further downloaded all that was contained as the greatest challenges of the era of science, which is greatly metamorphosing as impact technology. This is why it delved into explaining how the present embraced the wonders of this philosophical conquest, which is hereby understood as the beginning of the new age that could ascertain, buttress, and (if possible) condemn what is not in allegiance with the main thrust of physical, mathematical wisdom.

At this point the demystification with its monumental achievement spared a new court of physical mathematical horizon with its occult and monumental achievements. Those achievements range from A to Z of the life trust with its practical and perfect ascendances, which include:

• Bringing a new knowledge that opened a vast key in the areas of overall educational propagation.

- Liberating humans from the doubts and fears of what are contained in the sacredness of mathematics.

- It brought a new ideology into its unity with science and the creation of technology, which is today understood as protecting and dynamically bringing humans within a new era of being purpose-driven with the creator of a balanced, idealistic universe.

- It brought a stoppage to ensuring that women are given a scientific and social leverage in participating in all acts of human development.

- The Institute of Crotona could not have been established if the choice and task of demystifying mathematics were not achieved.

- The upliftment of humans who propagated the engineering of philosophy as the science of purity and perfection would not have been possible if the demystification was not achieved.

- The fundamentals of mathematics, with its numerous brands, could not have been welcomed or appreciated if these feats were not achieved by the holistic aborigines of physical methodology.

- Other compartmental horizons were achieved as a result of this demystification, including the study of astrology, astronomy, space engineering, and informative and translated psychology. It delved into aesthetic mystery, and this is why our modern age must appreciate the contributions of this demystification purposefully and practically from the ingenious part of physical empathy.

It must be explained that this conquest has actually encouraged the elevation of man's dynamism. It has also opened an intellectual and special horizon that is important for the fundamental empowerment of the mental system, which technically and practically characterizes the working of the human mandate.

At this point, the demystification being enjoyed today is from the point of honesty, physically driven with its mathematical dynamism as a conquest that has greatly empowered the philosophy of universe, including the ingenious world of Microsoft technicality in appreciating that we cannot exist with obscure sacredness. Neither can human dynamism or creation be allowed to become a terror to the world of knowledge. That is why the celebration of this achievement is concurrently being compared to what happened in Edison's room in San Francisco, California, during the famous and thunderous celebration of the jet age.

It is my opinion that the world will continue to celebrate great achievements without apologies to those that had greatly masked the face of human development. This is why all the monumental and political oblivion which the south Africans were subjected to are being celebrated today, because an ingenious statesman named Nelson Mandela, with like-minded men, was at hand to liberate his people from the sacrileges of man's inhumanity to man, under the auspices and pretences of social and political governance. In this respect, the achievement of science and technology, and the growth of religion and spiritualism, must always bless the contributions of the demonstration of mathematics as a social and untouchable oracle.

It is here that this overview brings the likes of Plato, Pythagoras, and their students as philosophic mathematical banks that will always be legends in the analysis of human history. This is why it is in the beauty and blessedness of this part to appreciate and eulogize the empathic contributions that these philosophic monarchs left on the sand of times. In turn they are dynamically engineering and seeing humans through in knowing and appreciating that whatever we have done in uplifting the standard of humanity, it is an honour done to Mother Nature, the Mother Universe, the Mother World family, the Mother World of knowledge, the Mother World of philosophy, and the ingenious Mother World of mysticism, with all the matrices of mathematics.

It is on this tedium that this overview blesses, with practical and philosophical salutation, that the living world at all times and ages cannot be controlled or governed by the limited wisdom of man. Instead, it will always be governed, directed, and willed in that power that is the author, the finisher, the igniter designated as the super-generous power of all the philosophical and mathematical ages, which is hereby designated as the omnipotent active waves.

PART 8

PHILOSOPHICAL MATHEMATICS AS THE LOGO AND SYMBOL OF THE CREATIVE UNIVERSE

"The mathematical sciences particularly exhibit order, symmetry, and limitation; and these are the greatest forms of the beautiful."

—Aristotle

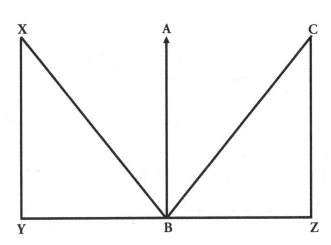

FIG. A

$XY = BY = BA = BC = CZ$

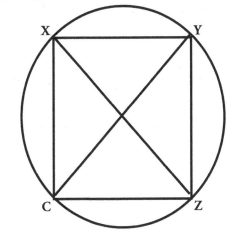

$= YB = BZ = ZB$

$<sYXB = <s\ YBX =$
$<s\ XBA\cdot$

$= <s\ ABC = <s\ BCZ$

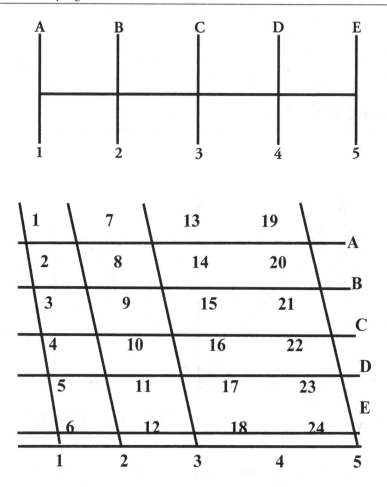

Numerology of the logo format using an acceptable mathematical method. This diagram shows how mathematics (with symbolic numerology) can be used in creating a philosophic alphabet.

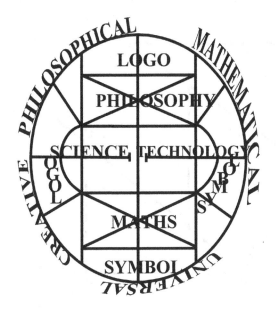

This drawing shows an ingenious and symbolic universe at the proximity of a philosophical mathematical dynamism.

It must be stated that this is the scientific, technological anatomy of the creative universe, which has not been appreciated by the material province.

Mathematical Logo with Symbols

1	2	3	4	5	6	7	8	9	10
AX	BX	CX	DX	EX	FX	GX	HX	IX	JX
1Y	2Y	3Y	4Y	5Y	6Y	7Y	8Y	9Y	10Y
1AXY	2BXY	3CXY	4DXY	5EXY	6FXY	7JXY	8HXY	9IXY	10JXY
1Y	2Y	3Y	4Y	5Y	6Y	7Y	8Y	9Y	10Y

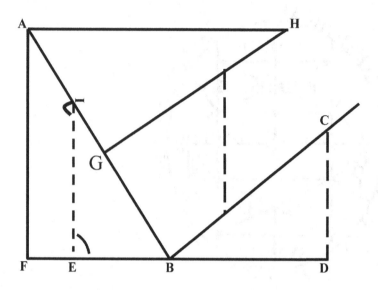

Creation is usually a dynamic, philosophic, mathematical, scientific, engineering, and technological hectare and nectar of forceful activities. These activities are sparely divided into positive and negative, mathematical and super-mathematical, philosophical and ingenious, titans. The logo and symbol of the ingenious, creative universe have remained the greatest estates, which humans are practically fragile in exploiting with scientific discovering.

Neutrally and naturally, philosophy and mathematics are best defined as the synthetic and ingenious logo of the creative universe. This is why the aspect of philosophical mathematics can neither be neglected nor omitted in this super-generous encyclopaedia of philosophical mathematics.

Just as material doctors who are trained in physical medicine prescribe orthodox and materially fermented drugs for patients, ingenuous science with the creative dynamism—which is hereby defined as the symbol and logo of the universe—is freely and practically in the tedium of organizing, recognizing, and sensitizing mathematical arguments that are contained in the best ability of philosophical mathematics. In this context, it is an established mathematical phenomenon that the advancement of a universal cosmogony cannot be possible if the titanic dynamism of philosophy is sidelined with mathematical relegation.

At this point, some questions need to be asked.

- For whom is our universe meant?

- What are the mathematical logos and the creative symbols of the objective subsonic and supersonic?

It is important to explain that subsonic means all the challenging problems that most of the time serve as a catalyst in achieving the creative desire of the symbolic universe. It can equally be defined as the meteoroids and super-genic actions of Mother Nature, which ignites humans to be positive-driven in order to create arts that are in conformity with the symbolic details of philosophical mathematics. Subsonic, which forms the realism of illumined knowledge, is always the mathematical logo and the philosophical symbol of the ingenious creativity. Most of the time it lures humans to be directly connected to the wisdom of the open heavens.

The need to open this philosophical, mathematical Mendel is borne out of the fact that the universal logo is the connective and connected infinitude matrix of balanced creativity. That is why such glorious titans like the Twelve World Teachers were adored, honoured, and appreciated as the main thrust of the universal logo.

As technology is the logo and power of the people, and as science is the symbol and honour of the ingenious universe, one should appreciate that nothing in creation exists without origin and the blending factors that are galvanized in the context of every origin, which has remained the practical symbolism of every anatomy with practical symbolism of its logo.

In the fourteenth century, an enlightened soul named Albert Mianu was sent on a religious asylum, and the consequences of this asylum finally became a deluge to the world of ingenious knowledge. To God be the glory that this rare knowledge was not completely lost to the world. This is why the compacted compendium, which naturally appeared in the meritorious atlantics at the proximity of no other than Bacon, is forever respected and recognized as a philosophical helmet and as a mathematical material, which he left for immortality.

To introduce philosophical mathematics as the logo and symbol of the ingenious creative universe, it is best defined as another factory of scientific investigation that must be research driven with mathematical impetus. Blessing and thanks to the monad of monads, who has retained the occupation of the well informed and is the sergeant of the monumental Sagittarius. Every effort must be made with natural technicality in order to make evolution not suffer in the opinion of the uninformed. As the world is evolving with scientific and technological changes, it is important for the human race to appreciate that we must not be jittery because of the chaotic enmities of material science.

At this point, every human being is hereby enjoined to be a beacon to the world of knowledge. This is the only hope by which the world of philosophical mathematics will creatively drive its natured and natural logo, which is a titanic symbol for the practical utilisation of the ingenious creative universe. The universe is hereby noted and donated as the convex, conclave, convert, and natural hector and nectar of the creative logo, whose wisdom is embedded with the wider knowledge of symbols that make the universe remain a logo for scientific habitation. It is also an acceptably and practical lintious ecosystem that houses the body and soul of creative ingenious, of which man is the icon creative genius and is ornamented as the symbol and logo of philosophical mathematics.

CHAPTER 1

WHAT IS LOGO AND SYMBOL?

There are a lot of meanings for the words "logo" and "symbol" when they are scientifically and simultaneously used. This book is not going to depend, borrow, or determine their meanings from dictionaries, because philosophy is an evolutionary and evolving art and science. It is logically driven in ensuring that standards are maintained with aesthetic and liturgical confirmation. It is the objectivity of this chapter to explain to the world material academy what can be dynamically accepted as the ornamented definition of logo and symbol.

- Logo is a dignified barrage of the people, for the people, and with the people. A symbol is the transcendental beauty of the people.

- Philosophically, logo and symbol in this mathematical transcendence are defined as the queued knowledge and electrifying power of the people, which are natural gifts from Mother Nature. This is why it is usually believed, discovered, and accepted that philosophical technology, with its occult engineering mechanism, is the logo of the people.

- Logo can equally be defined as the superior gift of the people, to the people and by the people. This is why democracy, which is a philosophical incubator that uses the legislature as a symbol, is usually acclaimed and designated to be the upkeep of governance; it came to be at the proximity of the Oriental apostolate.

- In the province of philosophical mathematics, logo and symbol are defined as the ingenious acts and objectives with balanced creativities that extol the beauty and power of the act, the person that is involved in the act, the meaning and mission of the act, and the rhythm and tedium of the act. This is why I am inspired to define and structure the philosophical mathematics logo as the illumined and ingenious architect of its workability, scientifically with some input from engineering technology.

- Logo can equally be defined as the objective cartel by which the functionality of science and technology can be known, using the mechanism of engineering as its practical and tactful symbol.

- In pure and applied mathematics, logo can be defined as the dynamics by which the objective technicalities of mathematics can be understood. Symbol can be defined as those mathematical frontiers by which the fractionate of mathematics can be gained. That is why every nation is naturally gifted with a great logo, whereas every human being is equally created for the sake of using the millicential power of logo and symbol in extolling immensely as a divine and composite personality from, the Godhead.

With these facts, which are definitive in purpose and nature, we can appreciate that every human is a present and perfect logo of Mother Nature. It can equally be said to possess the powers that Mother Nature uses in creating dependable and enduring civilizations, because he is the Orion and origin of the use and application of philosophical logo and mathematical symbol.

Among the apostolic and philosophical foundation, the word "logo" was such a pure talent, a perfect and powerful ornament, that must be cherished simply because it was seen by the illumined world as the snow and consummate breeze from the apostolic foundation of all ingenious creativity. In this respect, the following can be ornamented with millicentia conformation to be the logo and symbol of philosophical mathematics.

- The Ten Commandments.

- Herbert Spencer's code of ethics.

- Lao Russell's code of ethics for a living philosophy.

- My own living code of ethics for the modern generation.

- Other monumental codes with their illumined symbols are the principles, the policies, the techniques, and the sermons that came to man at the mortal and illumined proximity of the Twelve World Teachers.

From that information, we can conclude with objective comprehension that nothing in creation exists without logo and symbol. That is why at all spheres and realms of existence, we see and appreciate ample and divergent logos that show that the confirmation of creation as a practical and consummate logo cannot be disputed. This is why man is made a monument as an extension of logo, which is hereby defined as the symbol of the creator that forms the basis for the introduction of philosophical mathematics as the logo and symbol of the ingenious, creative universe.

CHAPTER 2

A LOOK AT THEIR ORIGIN WITH PHILOSOPHICAL EXPLANATION

We cannot delve deeply into the position of this part if efforts are not made to philosophically explain the origin of logo and symbol. This explanation will actually mandate the whole creation, particularly the province of philosophy and biology, to ask the spirit of Charles Darwin to apologize for misguiding man with the concept of his origin, which was unscientific and un-philosophical. Darwin and other misguided souls have apologized to the creator, who ornamented them as men with monumental gifts—which they rebated to suit the errors of evolution. Those errors invariably mimicked and brandished their ingenuity not to appear in the golden plate of dynamic, philosophical evolution.

If Darwin was not materially driven in thought and objectivity in this ingenious concept and teaching, then the world of knowledge could have designated him among the Twelve World Teachers because in this context he can be defined as a powerful possessor of ingenious intelligence. I also offer more kudos to the Creator, who is always forgiving mankind, particularly when humans have lost track of the eternal and

444

perfect truth that symbolizes them as a cardinal and peaceful logo of the Creator.

The understanding from this context is to philosophically explain that the origin of logo and symbol is a natural and balanced creation to which no human being can lay claim. It is the act of the ingenious Creator who is defined, named, and called in the following humble but respectful ways.

- The Creator.

- God the Father, the Son, and the Holy Spirit.

- The universal One.

- The sacred Cherubim and Seraphim.

- The Holy Trinity.

- The alpha and omega.

- The omniscient, omnipresent, and omnipotent.

- The omni-active one in all, who is equally all in one.

- The united unity of all ages.

- The mirror of all ages.

- The power of the latter-day saints.

- The missionette mind.

- The ornamented sanctious.

- The supreme universal One.

- The sacred sanctum.

- Jehovah Elohim.

- Jehovah Shamah.

- Jehovah Jireh.

- The ornament of the faithful.

- The beginning and the end of all. (In this context, I define him as the most powerful, purest, united one in all perfect and unified diversities.)

He is known as the Holy Grail; this is why all that is contained in the monumental works of the Twelve World Teachers are always showing, with practical and mystical explanation, that God is the supreme logos. His symbols are consummately written in all his acts, creations, and transactions. It is here that philosophy defines God as the threshold of all things. A look at the philosophical explanation of the origin of logo and symbols will certainly make one appreciate the definitions of God by the provinces of metaphysics, meta-science, meta-logic, meta-mystics, meta-occult, and the contents of hermetic science and Kabala which honour and respect the Creator as the infinite matrix. This is why the pilgrim fathers called him the light. They also named him "I am that I am" and represented Him as the sun and moon of their souls.

One must appreciate that the origin of the logo is the origin of life, the origin of man, the origin of truth, and the origin of the world and all knowledge. That is why this chapter practically and spiritually exists as a philosophical reminder, a philosophical admonition, a mathematical think-tank, and a scientific and practical guide. It is here that the introduction to philosophical mathematics as the logo and symbol of the universe is defined and denoted as an ingenious, creative occupation. This aspect tactfully exists as a crusade from the world of facts and truth. It is understood as a philosophical living gospel that will broaden the understanding of man in knowing that all knowledge comes from the universal One, because in him knowledge is driven. He exists in eternity as a permanent bank of philosophical wisdom that material science finds too critical and difficult to appreciate and observe in a manmade laboratory.

For the record, the origin is eternal, perpetual, constant, and immortal by the super-ingenious, universal One who is defined here as "with, more; without, less".

CHAPTER 3

WHAT IS THE IMPORTANCE OF LOGO AND SYMBOL?

As already stated, logo and its rhythmic balanced symbols can be defined as the aesthetic and quantious eye camp, whose vision is focused in ensuring that the point of honest knowledge is given and appreciated without pollution. This is why at any sphere of life, philosophy exists in the objectivity of mathematical purity and perfection. Mathematics is practically ennobled to exist at the illumined proximity of the heartbeat of philosophical aspects. It is at this sequence that the two are compacted, intertwined, and united in practical segmentations. This is a study that the academy should approach, while the philosophical research institutions must utilise its dynamism in assessing what can be known and appreciated as a philosophical think-tank.

From the aesthetic point of originality, mathematics—whether in statistics, algebra, geometry, equations, or trigonometry, with their purpose-driven multiplications—is always symbolic with the use and application of philosophical logo. Engineering mathematics appreciates that the symbol and logo of mathematics cannot be separated from the constituted production of mathematical creativity.

Philosophy argues symbolically that the logo of mathematics can be represented with numerals, angles, triangles, circles, squares, and rectangles; it can involve the use and application of such other mathematical symbols and radians, which include sine, cosine, and tangent formulas. Others include the subtraction sign, the multiplication sign, the addition sign, and the division sign.

Mathematics, with the involvement of a driven philosophy, always looks at creation as an impact and input of actions with objective fractions. It is here that $X + Y + Z$ will finally give XYZ and $2 - 1 = 1$, while $4 - 2 = 2$ and $4/2 = 2$. On this note, the ratio of $2:1 = 2$ and in philosophical mathematics this is always defined and discussed as the lintious figments of letters and fractions.

Let's go back to the focus of the chapter. Mathematical philosophy's importance must be attached and detached as a contingent philosophy of honest appreciation, whose importance includes all that is known in philosophy, mathematics, science, engineering, and technology. This is why philosophical mathematics is best defined as the symbol of production technology and engineering technology. The importance includes the following.

- It is the luminous gateway that devises the course of knowledge.

- It is known and utilise as the overall target that is used in all the comprehensive and cohesive fractional ties of mathematics.

- The importance is purpose-driven in the best objectivity of data and research methodology, and this is why it harmonizes with information technology and information communication technology.

- It is the eye centre that empowers all illumined knowledge in appreciating that philosophy is indeed the anatomy of consummate mathematical knowledge.

- It is known and defined as the supreme heartbeat of universal creativity.

- Great minds utilise it in charting the course of anatomy, astrology, and the mysteries that are contained in space dynamism with space meteorology.

- It is known as the diction of facts and figures, which are scientifically ennobled in assessing the wisdom of engineering technology.

- It is known and assessed as the tutor, the rotor, the mantle, and the engineering goggles of scientific balance.

- It is a teacher who is involved in correcting the lapses of sciences that work to pollute the province of philosophical mathematics. This refusal to pollution has mystified philosophy and mathematics as rational courses and endeavours that are dedicated to the minds and souls of those who are rational.

- The logo with its symbol was utilised by the wise, Eastern emengards who were propelled and directed to visit the light with lintious rays of the stars. This is why a lot is contained in the history and ontology of Bethlehem, which is edified in the infinite-dimensioned matrix of scientific mysticism.

- The symbol is the Mendel of all souls, while the logo is the blended grease that is always energizing practical actions with tedium and balance.

- It can equally be defined as a collection sector and a connecting fold that links the effective production and relativity of all sciences, engineering, and technology as being at the omni-obvious heartbeat of the philosophical mathematical pyramid.

- It is always known by the Orientals as the sacramented and ornamented living basis of illumined philosophical mathematics, which is purely the anatomy of all millicentia creativity and which must be understood, appreciated, and utilised by all and sundry in order to attend the exilia of perfection and purity, which is the mandate and challenge of philosophical mathematics.

The need to explain these concepts is borne out the fact that the human knowledge of philosophical mathematics is minimal, moribund, and myopic with astigmatic understanding. This is why the chapter is inspired to reveal that the importance of the logo and the wonders of the symbol are the cardinal, constitutional rights of man to understand that only in light do all forms dwell, while only in this light is creation driven to see more light. This is why the logo is the will of Mother Nature, and why the symbol is the heartbeat of philosophical mathematics, which has remained an acknowledged, dynamic knowledge that will serve humans as dynamic and practical trailblazer.

The conduction of this chapter is a wakeup call for all and sundry to appreciate that man is a sacramental and ornamental philosophical mathematics logo who is designed to achieve immense, divine purity and perfection. He is evolved dynamically in appreciating that his roots are practically blended with the call of perfection and the use and application of purity, which is the call to know, awake, and appreciate that the logo of philosophical mathematics and its symbol are the ornamented and sacramented forte that he can never reject or neglect in order to exist and live as a millicentia philosophical symbol with mathematical logo.

CHAPTER 4

WHY IS IT KNOWN AS THE SUPREME HEARTBEAT OF THE UNIVERSAL CREATIVITY?

$$= LP = \frac{U}{C} = SM.$$

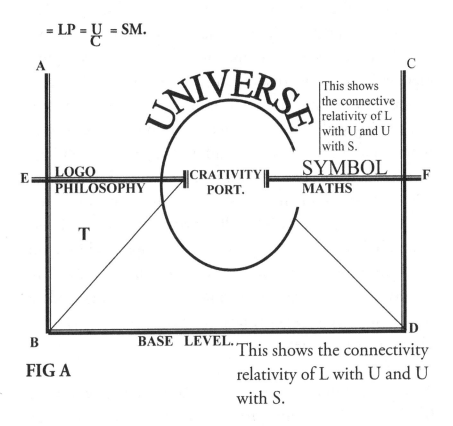

This shows the connective relativity of L with U and U with S.

FIG A

This shows the connectivity relativity of L with U and U with S.

In the illumined words of Ralph Waldo Emerson, "The happiest man is he who learns from nature the lesson of worship."

Man is the supreme heartbeat of Mother Nature. This is why he is gifted with engineering dynamism: to practically create that which will sustain and ennoble him as a millicentia image of Mother Nature. Philosophy is the logo of the universe, while man is its creative symbol. Philosophical mathematics is the authority that directs and moderates the supreme heartbeat of universal creativity.

A look at the philosophical Mendel of these great facts reveals that humanity has not known the mathematical logic that drives and determines the functionality of philosophical mathematics. It is worrisome and unfortunate that humanity is not even making an honest effort with a practical desire in knowing with objective determination how and why the philosophical logo, with its symbolic ecstasy, is of objective creativity. At this point it must be understood that what is known as the supreme heartbeat of any form of any logo and symbol is philosophically fathomed, nurtured, and quantified as a driving mechanism of that form. This driving mechanism propels the functionality of objective creativity. This is why the universe is fathomed, featured, and mandated to function as the supreme heartbeat of the functionalities of philosophical mathematics.

A look at the drawing symbolizes that the universe is hiding between the rudimental anatomy of philosophical logo and the scientific incandescence of mathematical florescent. That is why there is a permanent panel that can be defined as the mechanism and framework of the anatomy. It is here that the immortal words of Ralph Waldo Emerson can be welcomed and invited to remain the supreme heartbeat of the universal creativity, because nature adheres obediently to the tedium of worship.

Apart from this worship, this book is a wakeup call to the province of scholars and research fellows, to stand tall in order to appreciate the scientific and philosophical mechanism of the contents of philosophical logos and all that is serving as practical posters to the figurative and fragmented symbol of mathematics, which is hereby designed as a philosophical heritage.

At this point, I am inviting scholars, research fellows, teachers, and technocrats to start amortizing in natural practicality the coefficient manufacturing of arts, realism, realities, being, and becoming. These are great challenges, particularly in this era where the loading system is stringent and challenged, but downloading is porous with a lot of un-purposeful mechanism.

Let us look at why this is appreciated as the supreme heartbeat of universal creativity.

- Philosophy is the creative desire of knowledge, and this is why it uses mathematics to determine the punctuality and functionality of that creativity.

- Mathematics is an organized, loaded park, and it is here that it involves philosophy as a power that works to aim the desire of Mother Nature in the scientific and practical analysis of achieving practical perfection and purity.

- As both are tremendously obedient in the areas of design, construction, and mechanism, with a view to achieving the purpose of creative engineering, they utilise all the forms and factors of universal creativity in order to arrive at a coherent and cohesive concept.

- Mathematics is always moderate in practical Mendelism, and philosophy is naturally original in the best appreciation and utilisation of logical thinking. This is why a lot is determined to desire and understand all functionalities of philosophical branches.

- The universe does not make friction or create without involving its creations, which are hereby defined as the logos and symbols.

- Creation is the mechanism of universal worship, and so it is always looking at philosophy and philosophical mathematics in order to achieve this organized and rarefied goal.

- Every engine is the creation of organized concept, and the logo and symbol of philosophical mathematics can equally be defined as a universal creative engine.

- As creation is a loaded and downloaded concept; it cannot make friction or feature productively without inviting the luminous and dynamic province of philosophical mathematics.

It is important to explain all these practical, ingenious inspirations. A look at the immortal concepts of the books: *The Secret Teachings of All Ages, The Philosophy of the Oriental Wizards, Philosophy: The Maker of Great Thinkers, Fundamentals of Core Logic,* and *The Secrets of All Light,* and the contents of all certified logos and symbols reveals and represents a great age that is un-quenching. This is why this chapter is at the proximity of downloading the greatest truth that lies at the heart and beacon of supreme ecstasy, by which the creative universe is practically driven with mathematical ingenuity. This is why apart from designating this book as the dynamic form of philosophical mathematics, it can equally be named the immortal ledger of the creative universe, or the philosophical heartbeat of Mother Nature that exists at the Orion-driving mechanism of mathematics. It is coming at an age that the human race requires a reengineering, reassurance, and rebranding, because the universe is facing a lot of challenges resulting from the wrong incursion of the Microsoft province which, when properly viewed from the point of honest knowledge, does not feature, function, or synchronize with the supreme heartbeat of the universal creativity. That creativity honoured and adopted the logo and symbol of philosophical mathematics as the rising point and the horizontal heritage of all ages.

CHAPTER 5

HAVE THESE FACTS BEEN KNOWN BY THE WORLD OF MATERIAL SCIENCE?

It is important to explain that material science has made a lot of cardinal errors that have greatly stunted growth and success. These errors have contributed in ensuring that there is no permanency in scientific and technological creations. It is the goal of this work to explain that the desire of material science to give the world a new cosmogony is not possible, because material science decided to lock the creator out of its creations. This is why philosophical mathematics designated these creations as the in-glories of the universe.

Material science—which is at all spheres being driven with myopic concepts, un-illumined and dangerous creations—faces a lot of challenges, particularly in our contemporary era. It appears not to be making efforts in order to effect corrections that will help the sector to move further, beyond the present state.

It is important to appreciate that most of the errors of material science include:

- Lack of objective philosophy.

- Lack of purpose-driven and desired concepts.

- Inability to involve long-range planning.

- Lack of the understanding of nature's creative and dynamic processes.

- Inability to appreciate the perpetual permanency of the creator.

- Change in principles, policies, and practises.

- Lack of constant research methodology, including how data and statistics can be configured as workable tools.

- The storage mechanism of material science is not realistic in the best opinion of the working of philosophical mathematics.

- Material science has not known that mathematics and philosophy are nurtured, natural, and titanic; this is why their anatomy is the logo and symbol of constructive creativity.

The ways and manner that material science develops facts and figures are not in consonance with the human understanding of philosophical mathematics. This is why what we are experiencing today as the sanctious of chaoticism is borne out of the mistakes of material science. It is in the objective view of this chapter to proffer the following points, which can be translated as a rudimental solution to these problems that have affected and annihilated growth and success.

- Science must be driven with a practical and configured constitution.

- Ingenuity at all levels must be recognized and utilised.

- Material science must be purpose-driven to learn from the dominant province of nature's rhythm, which is the greatest apex point from where the creations of material mathematics are taped and derived.

- Material science must be taught the original lesson of working with nature. This is important particularly at this era, where the enmities of material science have subjected the world to an untold torture.

- It must appreciate the philosophy of open-door policies and practises with its mathematical derivations, which will open doors of scientific derivations and research methodology.

- The conversion and transmutation process of material science must be watched and guided with the use and application of the natural coherence, which directs and motivates the balanced workability of philosophical mathematics.

- Philosophy is pure and perfect in action, and material science must bend to learn the lessons that are contained in this practical functionality of philosophical mathematics, knowing how their intertwined relativity can be galvanized.

These lessons are important—they are the primary and secondary issues that do not require any contestation, and their full knowledge, when gained and appreciated, will achieve the following for human growth and sustenance.

- Neutral and natural cosmogony.

- Permanency in the reality and realism of using facts and figures.

- It will amortize mathematics to function as the eye centre of all human endeavours, which includes economics, commerce, industry, engineering, technology, banking, and finance, and all that is contained in the educational progressive system.

A lot is contained in the phrase "educational progressive system", which includes manpower development, infrastructural development, capacity building, and human resources with dynamic engineering. This is why it is in the focus of philosophical mathematics to energize the thinking and practise of EPS.

It is not important that the human race is not in the convex of this honest thought. The illumined prayers of Manly P. Hall, which is important to be recapped in this chapter, reads thus:

Make philosophy your journey. Respect it as your pathways. Understanding its position as the apparatus of all scientific creativity and with all these man and the entire humanity will be ennobled to appreciate that philosophy is the mother of science.

I conclude that it is the ingenious maker of great thinkers.

Philosophical mathematics by nature is the permanent root of all things, the wisdom of past eras awakening, including being the only supreme knowledge. That fact will make posterity appreciate that philosophy is the anatomy of all things, the chancellor of all creations, and the ingenious VSAT, which will help the world to be configured as a globalized cosmogony.

A look at these immortal words will certainly show that material science requires a philosophical and mathematical blueprint. This blueprint, when developed, will help all and sundry to appreciate that all that material science has achieved is done through the perfect grace of the thinking of philosophical mathematics. It is here that the world and all the scientific and technological practises have been designated as what is contained in the hub and anatomy of philosophical mathematics.

The immortal words of Dr. Walter Russell, particularly as contained in *The Secrets of Light*, is an eloquent testimony that science cannot achieve the desire of the universal cosmogony without practicing and romancing with the wisdom of philosophical mathematics, which is titanic-driven with the ebonite and rudimental forces of Mother Nature. This is why every effort needs to be made in order to detach the errors of scientific balance, which are greatly mimicking and reducing man's consummate philosophical mathematics heritage.

At this point, we must appreciate that the world is still at the teenage level of primary evolution, and it requires a lot of careful handling, systematic and empathic directives, planning, research, and more. This chapter can be defined as another envelope that will help readers know and appreciate that the management of the universe rests on the purpose-driven creations and transactions of man. The chapter is inspired and is propelled to exist as what material science needs to reduce its monumental errors, particularly those that have to do with atomic suicide, nuclear weapons, and the manufacturing of biological and other armaments that do not strengthen the health of the global universe. The invitation of philosophical mathematics to rescue this province with the use and application of its logos and symbols is hereby edified and designated as the greatest act of an ornamented physician, which is the greatest need of material science with its infamous and deadly technology.

CHAPTER 6

WHY IS THIS LOGO AND SYMBOL THE LIVING ANATOMY OF PHILOSOPHICAL MATHEMATICS?

Anatomy is a philosophical science that can be appreciated when there is a perfect purity, which is the sequential mathematical laws that drive the basis of living things. Anatomy can be defined as the structure, frame, founder, and fulcrum of unity of a symbol that can only be obtained and accelerated through the application and assessment of the logo.

It is in this format that "logo" is defined as the living basis of illumined philosophical mathematics. Through illumination we can understand the high level of purity, the transcendental and occult engineering that are involved in the composite constitution of philosophical mathematics.

It is important to observe that the human race and its creations are still in limbo regarding utilising the anatomy of logo and symbol of philosophical mathematics. This is why man's creations are strict and stagnated, and most of times his illicit character and attitude does not represent him as the cardinal image of the consummate, universal creator. The structure and anatomy of philosophical mathematics have remained the driving forces that detect the rhythm and rudiment of its diversified facts, which include metaphysics, logic, epistemology, ethics and morals, and aesthetics.

Illumined philosophy, with its mathematical heritage, is always in agreement with metaphysical transcendental arts and creations. It is here that anatomy is designated as the supreme basis of the infinite dimensioned matrix. In the seventeenth and eighteenth centuries, Reverends Canon and Crawnwell were greatly uncomfortable at the way and manner their civilization defaced philosophy. That era witnessed a high level of massacre of mathematics, policies, and principles. This is why the tedium of canon laws is usually accepted as the beginning of a new era in man's quest for knowledge, truth, philosophy, and understanding. The Oriental churches decided to canonize the study of both reverends' laws as major courses at all level of philosophical schools and seminaries.

For example, Reverend Apostle, who was at proximity of finding the Apostolic Church as a fundamental religion, declared that:

> The church cannot appreciate the high level of purity which is contained in philosophy if mathematics is not used in blending and postulating the dynamic functions of philosophy.

It must be stated that the stigma of that era—which was un-philosophic, un-dynamic, and uncreative—was practically and purposefully wiped off by the apostolic of pure and transcendental metaphysics. For the record, illumined philosophy, with its mathematical dynamism, is defined and understood as an encroachment into the province of metaphysics, meta-scientific, and meta—occult. It also affects a lot of illumined souls because man's consciousness has always defined and designated philosophical mathematics as the anatomy of all, simply because it is the only unpolluted science that is dynamic in enabling appreciating who he is, what he is, and why he is created. This knowledge includes directing his thinking in knowing and appreciating that if his creations are not balanced, he is only explaining and exposing himself to fall at the gate and chancery of the thinking of the monarchs.

Unfortunately man is finding it difficult in appreciating the illumined facts and figures, the data and realism, that are contained in the anatomy of philosophical mathematics. The difficulty in appreciating and understanding these facts has created the following issues.

- Lack of understanding.

- Lack of permanent and dependable knowledge.

- Inability to think and act universally.

- Locking and disturbing the great personality who dwells in him.

- Creating concepts and sometimes developing permanent ideas to help him to be transformed as an extended philosophical heritage.

A look at all these aspects reveals that man contravenes and sometimes disrespects the laws that make him function as a living and an illumined mathematical philosophical thinker. This negative act has ingloriously impacted into him un-lintious arts that cannot help him to progress, to asses, and to appreciate that all in him, with him, and for him is practically driven with the tedium of philosophical mathematics, for which he is purely created in the original anatomy of this basis. A scientific look at this chapter has to come to reveal that creation is still in search of the illumined philosophical talents, mathematical geniuses, and technological icons whose ideas are pure and perfect and sometimes synchronize with philosophical mathematics.

At this point, the logo is the basis of illumined philosophical mathematics, which is purely the anatomy of all living because of the following reasons.

- Man and creation are fathomed as a millicential, philosophical mathematical anatomy.

- All the cells in him are spiritual, and this is why they are all symbolized in the perfect anatomy of the living monarch.

- He is designated as a mathematical extruder, including being a mathematical ingenious thinker.

- His anatomy at all levels is the originality of the philosophical, original alchemist.

- His biology is tenderly spiritual and not only material or physical.

- His biorhythm represents the illumined anatomy of the Creator, and this is why he is designated as the foremost millicential monarch.

- He lives for transition, transmission, and transmutation; this is why advanced spiritual biology designated man as the philosophical horizon of T^2.

- He is created as a co-creator, and it is here that all his arts, reactions, and transactions must be mathematically and scientifically creative.

- As an ornamented porter from the ingenious incubator of the foremost monad, he is divinely designated to be in simultaneous with his purpose.

It is important to explain all these points because a lot of humans are narrow-minded about the constitution and constituent of what the anatomy of a living thing is, particularly when it has to do with man. This drawback is why the philosophical logo and the symbols of anything are hereby defined as the porters wheel, the fulcrum fort, the ingenious and purpose-driven context. This chapter can best be appreciated through the wisdom of transcendental philosophical mathematics.

CHAPTER 7

HOW DID THE WIZARDS USE THE LOGO AND SYMBOL AS THE PERFECT ORNAMENTS OF MORALS AND ETHICS?

Both philosophy and mathematics are practical and intertwined with strict adherence to their ingenious morals and ethics. A work to explain why and how the wizards utilised the logo and symbol as perfect ornaments must ensure that the science of morals and ethics is simultaneous and harmonized with the practical blending of philosophical codes. Professor Patrick of the University of Birmingham has researched the dynamism of utilising mathematical codes and ethics in blending with the anatomy of philosophical morals. The outcome of this research has shown that the wizards were trained, gifted, and disciplined humans who were able to impart their knowledge to the willing souls who appreciate that nothing in life can exist without the transcendental contributions of philosophical mathematics, philosophical engineering, and mathematical directives. This is why the contributions of Professor Patrick were able to open a new horizon that was dynamically used to perfect studies of the sequence of ethics and morals.

Ethics progresses with the use and application of universal culture, whereas a moral can be defined as an immortal attorney that the psychology of conscience utilises in directing the affairs of man. It is here that the works of a philosophical logo with a mathematical symbol are not realistic or feasible without the perfect directives of ethics and morals. At all times they can be defined as the centre point of philosophical and mathematical engineering. This is why the Oriental wizards purposefully used moral ethics as a perfect ornament that was capable of driving creation, the worth and works of philosophical mathematics. It did so to show that the intertwined relativity is the compact pivot that drives the activities in order to show that there is nothing in life without a basis, anatomy, morals and ethics, and symbols and logos. It is here that every work is blended with moral codes that are hereby defined as a psychological, sociological, and philosophical instrument. This is best understood in the original work, which exists in the philosophic ingenuity of Ralph Waldo Emerson.

History has always proclaimed Emerson as a lintious wizard whose gospel was perfect and synchronous with the scientific and ingenious thinking of transcendental moralism with practical ethics. This is why his works were donated to the book of life as the sanctuary of philosophical mathematical anatomy, which exists at the tedium of ingenious moralism. It must be stated that the creation of material science, engineering, and technology does not consider the input of moral ethics as a blending standard. This is why in its construction, material creation is always epileptic and myopic, and in most cases it has recognized and positioned the limitations of the mental intelligence. The same incubus is still attacking the faces of religion, culture, and tradition simply because humans have not been able to access and appreciate what is contained in the unlimited philosophical morals and ethics of the anatomy of the divine, which transcendental philosophy designated as the biology of the spirit or the occult anatomy of the universe.

I wish to borrow a reference from one of our purpose-driven works, known as *The Philosophy of Oriental Wizards,* in order to access and ascertain how these great souls were naturally and spiritually cultured, trained, and pacified in the use and application of moral codes with scientific and ingenious ethics. This is why their footprints have remained glorious and challenging to the province of material world. In this respect, the wizards utilised these logo and symbols as the perfect ornament because of the content, context, power, knowledge, wisdom, and overall energy that were contained in this practical and purpose-driven asset. That asset, when practically utilised, can always ornament a human as a glorious extension of the consummate, universal, moral apostolate whose ethics are symbolized in the powers of universal creation, mathematics, philosophy, and dynamic science. This is why this chapter can be known as an advanced reference to the introduction of the dynamic functions of philosophical mathematics.

$$\frac{\text{MATHEMATICS}}{\text{CODES}} = \frac{\text{PHILOSOPHY}}{\text{ETHICS AND MORALS}}$$

$1 \times 1 = 1; 1 + 1 = 2; 2 - 1 = 1; 1 \div 1 = 1$, etc.

A look at the equation formats and symbols, with their arithmetical explanations, will understand that the wizards were thoughtful people. This is why it is in' the annals of philosophical history, with scientific propagation, to appreciate with all honest understanding that philosophical mathematics is a wakeup call and is the only answer to man's spiritual, physical, and material problems. Most of the time man has succeeded in creating more problems in the course of trying to proffer solutions and answers.

It is at this point that the human race, and particularly the province of material cosmogony, is being challenged to think positively and act dynamically. This chapter is inspired to reveal that nothing can be achieved if the world family does not adhere strictly to principles, policies, and practises, including the discipline that the science and art of philosophical morals and ethics is trying to reveal with the use and application of mathematical codes. The wizards were perfect actors and geniuses who ornamented their creations through strict obedience and adherence to the rules and regulations of philosophical mathematics. They practically protected philosophical mathematics from being polluted, and they created an immortal bank in order to ensure that mathematics, as an intertwined partner of philosophy, did not suffer particularly in the hands of the uninformed.

Institutions, universities, colleges, and research emengards hereby introduced and inculcated the discipline of morals and ethics as a transcendental supporter to all human endeavours, particularly those that have to do with human development and with universal growth. Morals and ethics always work simultaneously with the use of philosophical logo, applying all the mathematical symbols in a way and manner that portrays that the great illumination and growth, at any point in time, can only be achieved when man thinks philosophically and acts mathematically. That action is in synchronous agreement with transcendental mechanism of the universal world.

FIG A.

Shows the T square
of Philo/Maths, etc.

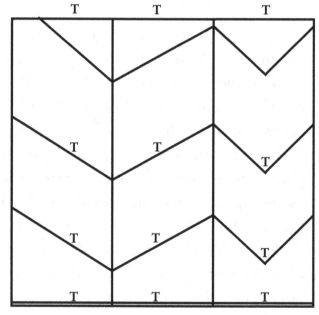

CHAPTER 8

The Urgent Call for the Modern Era to Appreciate the Use and Application of the Symbol as an Ingenious Creator

The modern era has a Herculean task that must be accomplished if the desire of globalization and scientific cosmogony—including living and existing as a common mind—is to be achieved. Our contemporary civilization can neither appreciate nor galvanize the instrumentation, which is institutionalized in the scientific and purpose-driven course of philosophical mathematics. To this end, the urgent call becomes a great necessity, a challenge, an admonition, and a mandate with philosophical assignment. That is why the philosophy of the logo and symbols must be accepted as an ingenious creator.

The creative universe is practically illumined with lots of philosophic and mathematical ideas. These ideas are practically simultaneous with the ingenious concept of the creator, which in earnest directs and moderates the workability of these logos and symbols. This is why Manly P. Hall defined philosophy as the wise counsel to the wise. The ingenious creation of these logos and symbols must be utilised as the first principle of philosophy, the ornamented principle of science, the basic constitution of technology, and the mathematical heartbeat of the ingenious and universal fort.

The explanation of this call, which is a thought-driven concept, will make modern man know himself, the basis of life, the philosophy of balanced reasoning, and the rhythm of mathematics. This is why this urgent call is manifesting the following.

- Creation at all level must appreciate the use and application of philosophical logo and mathematical symbols.

- The universe is a practical and technical structure of living philosophical mathematics anatomy.

- Every act must be symbolic with the fullness of the representation of a natural and united logo.

- The calculation that anatomized the use and application of logos and symbols must be accepted as a primary and secondary concept.

- Institutions at all levels must utilise and mystify the mathematical mystics, because these logo and symbols represent mathematical revelation.

- The Orion, including the mathematical tutors of the logo, must be utilised as permanent and constant trailblazers—particularly in engineering creations, which represent the anatomy of technological configuration.

- Science must relate the art, and art must be utilised in moderating the postulation of technology. The postulation of technology at any point in time must be utilised in a constructive and dynamic engineering fort.

- All scientific mathematical enabling must be lintious and sacramented in the best ordination of the dynamism of philosophical mathematics.

- Enabling laurels must be configured, confined, and utilised in ensuring that the rudiment of the logos and symbols are purpose-driven in objective creativity.

- The genealogy of the symbols must be ornamented in the deep most thinking of ethical rudiments with scientific codes.

- All scientific codes must be utilised in accessing and appreciating the ingenuity of engineering and technical models and modules.

It is important to explain the factuality of this urgent call and the dignity of its involvement as an ingenious creator. This is why I have made references to my other works, which exist at the balanced ingenuity of the periodic chart of information technology and information communication technology. That technology was designated and amortized by Dr Walter

Russell as the cosmic clock, which he opinionated as the supreme sanctuary of his cosmogony.

In this enabling, the universe must be confronted with the following questions.

- What is the safety of our present ecosystem which is Microsoft-driven with a lot of chaotic and enmitied transactions?

- If science can globalize the universe, can it globalize the sophisticated and segmented universe?

- If scientific cosmogony is feasible, is the same cosmogony reasonable and realistic?

- Can philosophy produce a mathematical, purpose-driven objectivity and mathematical balance sheet that will hold the key to appreciating the rhythm of life?

- If the world is tentatively globalized, can the same world be pragmatically philosophized?

- Is there any difference between my information technology and information communication technology periodic chart, with a scientific conclave and engineering parameter, and Dr. Walter Russell's opinion of a scientific cosmogony (which is deeply rooted in the best objectivity of Russell's cosmic clock)?

These and other metamorphosing questions will spur our era to be deep in thinking, to be practical in practise. It is important that the modern era heeds the following philosophical mathematics adjustments.

- Adjustments in thinking and action.

- Adjustments in transaction, reaction, relation, human dignity, administration, and government.

- Adjustments in trade, commerce, industry, economics, and measurement.

- Adjustments in the educational sector, family circle, community, and environmental leadership.

These adjustments, which are hereby defined and designated as philosophic Omongo with mathematical Omomdia, are greatly needed in the areas of science, technology, engineering creativity, econometrics, leadership, and public and diplomatic relations with their bilateral transactions. This is why the synergy of these symbols and logos can best be defined as an ingenious, interactive, and coordinating vision.

In our modern era, it is obvious that the logic and metaphysics of the philosophical mathematics logo, which is the symbol of honest transaction and creation, has not been known by the limited wisdom of the modern man. This is what ignited this call to make a balanced appearance. That appearance will help posterity in knowing that the wisdom of the logo with its symbol is deeply rooted in creating an ornamented museum of knowledge with a monumental bank of wisdom. In the opinion and concept of the well informed that bank is dated and determined to live as the ingenious creator, which is the urgent need, call, and necessity of the modern man, whose call and postulations are eroded with un-philosophical postulations that have forced him to still operate at the obscure level.

CHAPTER 9

WHY ARE THE LOGOS AND SYMBOLS, THE BALANCED KEYS, THAT CAN BE USED IN KNOWING AND APPRECIATING THE PHILOSOPHICAL RHYTHMS AND CHALLENGES OF LIFE?

A lot has been said and explained about the wisdom that is contained in the logo and symbol of philosophical mathematics. Much more needs to be said, however, because the created universe is a practical and purpose-driven logo with its dynamic and titanic symbols. The rhythm can best be defined as the consummate music of the spheres; the challenges are spanned in the best changes of life with evolution. This is why the rhythms and the challenge of life are galvanized as the tumultuous challenges of philosophy, which make it a noble adventure and the way of honest realization. A look at the introduction of philosophical mathematics as the logo and symbol of the universe, including being the ingenious creativity of great thinkers, will show that it is always driven with the tedium and rhythm of life practises, life gates, and life policy. In this context, Philosophy can be defined as the honest challenge of a dynamic activity.

From the age of consciousness to our modern civilization, the logo of philosophy and its practical symbols have remained the power of honest achievements and the purpose of honest creativity. This is why the simultaneous and titanic universe of life is always radiating and falling back on the electrifying philosophy, which is designated as the wisdom of science, the rhythm of religion and other arts. This idea inspired Jerome Pijionee to tell the Italian and German provinces, "Life is a designated and desired MasterCard which is at the proximity of logo and symbols of philosophical ingenuity." The products of this ingenious creativity include the following.

- Mathematical and calculative concepts (MCC).

- Symbolic and sibilate logic (SSL).

- Meta-science with meta-rhythm (MSR).

- Meta-philosophy and meta-logic (MPL).

- Meta-occult and meta-astral (MOA).

These ingenious creations are known to be rarefied, sanctified, and mathematically sacred in ensuring that wisdom governs and moderates the affairs of philosophical mathematics. The logos and symbols are balanced keys that one can use in appreciating the rhythms of philosophy with its numerous challenges. These challenges are always known and appreciated as the gateway to growth, evolution, expansion, concept development, and idealistic reasoning.

That is why the above are contained in the metaphysics of symbolic community.

Charles Rigs and his paternal friend Owen Richard existed in the eighteenth and nineteenth philosophical community. They were practically transcendental with meta-scientific ingenuity when they explained, Life holds its success when the keys of philosophy and mathematics are being utilised in unlocking the rhythm and tedium of the universal bank." These gentlemen opinionated strongly that the future of philosophical development with its mathematical chancery, which is an apostolic cardinal, will explain the logo and symbol of philosophy as a universal databank; it is the supreme genealogy of symbolic logic and symbolic thinking. This is why the logic of symbolism exists as a fraction of the original logic, which is natural, core and crystal, cardinal and critical.

Unfortunately, the human race is slowly advancing in the best objectivity of this course, which since the dawn of consciousness appreciates the following.

- The achievements gave rise to man's awaking as a super-genius, creative image of the super-genic, philosophical creator.

- It is the practical anatomy that gave rise to the discovery of science, which was narrowed to the province of matter by uninformed and material-driven humans.

- What the world is enjoying today as the province of technology and engineering originated from the deep most engineering of philosophical ingenuity, which utilised the postulations, policies, and practises of mathematics with the furtherance concepts.

- Philosophy achieved a blended and love-driven universe that humanity is enjoying as an ecosystem.

- It gave the world of knowledge forceful thought forms and formats. This is why all adventures in realism and reality have always immortalized and honoured philosophy as the ingenious Millicent of practical creativity.

- It is philosophy that named man a composite personality and a composite monad who originated from the infinite monastery of the perpetual universal philosophical bank, with its ingenious and unequalled anatomy.

- The highly illumined were honoured and gazetted as wizards and geniuses, and ornamented as super-geniuses by philosophy and philosophical mathematics.

- All practical creations and inventions of man were developed at the hunches of inspiration, which later transferred its ingenuity into man through a meta-mystical method defined as a monumental transfusion, transmission, transfiguration, and transmutation. This transfer is why philosophical mathematics is defined as the symbolic logo of a permanent pyramid or a crystal square.

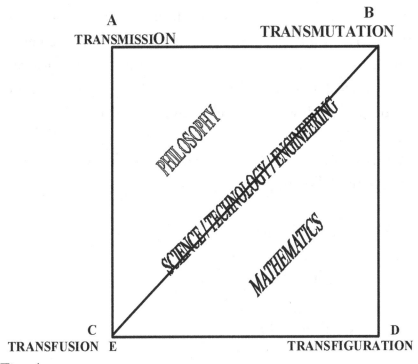

Fig. A

A look at the component structure and frame of this philosophical mathematics diagram reveals that

$$T + T = T + T$$

and that

$$\frac{T}{E} = \frac{T}{F}$$

It must be stated that ABC is angled at E; BDC is angled at E and F, which shows that the challenges of life can only be tackled with the use and application of philosophical rhythms with the ingenious concept of mathematical tedium. This can be defined as the sequential and practical appreciation of philosophical mathematics ideas, which humanity finds impossible to assess because the human race is not driven by inspired and deep thinking of philosophical mathematics, which was responsible for ennobling such idealistic metaphysical emengards like Plato, Pythagoras, Aristotle, Motti, and Professor Michael Nicholas Hudack.

At this point, a look at the challenges of life will certainly narrate the following issues.

- Social challenges.

- Political challenges.

- Industrial challenges.

- Religious challenges.

- Educational challenges.

- Family and community challenges.

- Personal challenges.

- Industrial and commercial challenges.

- National and transnational challenges.

The human race is also facing the challenges of thought, action, reaction, evolution, growth, industry, desire, vision, accommodation, and conception and actualization of facts and figures. This is why this chapter can be defined as honest facts that can make civilization move forward. In our present Microsoft ecosystem, economic and financial challenges in defining the menace of poverty with its cancerous stigma have remained the teething and tedious problems that are demystifying the rhythm of philosophical logos with mathematical symbols.

A look at the ingenuity, which is contained in the above drawing, will certainly appreciate that the purpose of this symbol with the constructive logo is to illuminate the actions of the universe in order to present and portend a better philosophical actor.

It is on this note that I donate the following as a physical step that can be transformed and translated as a mathematical maxim, with reasonable symbols and logos of a centennial eulogy.

If man does not use his right, the future of his posterity is at stake. If the world does not rise to elect illumined leaders as actors and leaders, all actions of the universe will always be subjected to enmities and regrets. Every human being must develop a philosophical keyboard with a mathematical searchlight by which all his actions and transactions can be moderated, directed, and performed with the use and application of ingenuity and philosophical creativity.

The addendum that we leave for posterity in this context is a galvanized logo with an embodied symbol, which is the will and wisdom of the balanced key and can be used in appreciating the rhythm and the challenge of life. It is hereby defined and consummated as the proposed philosophy of the grant, of which all adventure into the world of knowledge must remain and be beatified as honest commitment. This is why this chapter is donated as realism, real, reality, resource, and the reasonability of philosophical mathematics' commitment.

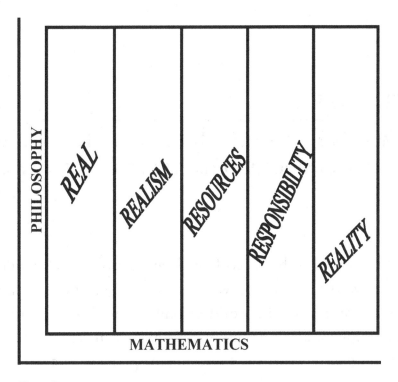

Fig. B

CHAPTER 10

WHY IS THIS SYMBOL KNOWN AND DEFINED AS THE INGENIOUS AND ILLUMINED ACTOR TO ALL ACTIONS?

To the world of balanced knowledge, actions speak louder than words. To the province of the uninformed, action has no reality. To the world of philosophical mathematics, philosophy is an actor and mathematics is an action. This is why philosophy exists as a logo and mathematics as a symbol. The compact unity between the two is best obtained as a balanced actor with rhythmic, balanced actions. A selector cannot work without selection. That is why in automobile philosophy with its mathematical techniques, a selector is an accelerator, and acceleration considers the functions of motion, emotion, and technical mechanics. Most of the time those functions are being driven by the philosophical mathematical appearance, which symbolically is designated as an actor acting on the unity and electrification of acting.

It is here that philosophy is always utilised to dissolve any difference that arises between ability and probability. Again, deep mathematical thought with its ingenious wizardry is always applied to understand the mechanics and engineering philosophy of emotion, motion, and frequency. This is why the purpose of this chapter is to explain what is known as the symbol of philosophical and mathematical illumination, whose actions and results are capable of creating actors to all centennial and sentimental actions.

A thorough look at this explanation (and those of subsequent chapters) will confirm that this symbol is the genealogy of the power of the actor, whereas the logo is structured in ensuring that all actions are best protected as a philosophical asset.

Scientifically and aesthetically stated, it is contained in the honest heartbeat of Mother Nature with universal actions. This symbol is known and defined as the philosophical illuminator, as well as being the mathematical actor to all actions, because of the following reasons.

- Philosophy, as a perfect science and as an honest act, is an illumined actor whose actions are reality in realism.

- Mathematics, which understands the heartbeat of philosophy, is equally a super-genius calculus. This is why the intertwined reality that exists between the two is naturally on par with electrifying balance or parlance.

- Philosophy always exists as an action-driven actor who is moderated or directed with the forces and wisdom of empathy. This is why every action of philosophy produces a monumental result that energizes mathematics.

- Just as philosophy is the power of motion, so is mathematics the engine that drives the thoughts and concepts of motion and emotion.

- Creation originated from the honest heartbeat of the incandescent universal force, and philosophy is always designed, defined, and designated as the supreme pyramid of all ages.

- Mathematics understands the philosophy of quantum and quantious principles; it always involves the wisdom of philosophy in the best explanation of this hybrid philosophical knowledge.

A work to explain the ingenious actor and action of this symbol must appreciate the meanings of the symbols, logos, and drawings. This chapter is an inspired illuminator that exists at the balanced tedium of the philosophical eye centre.

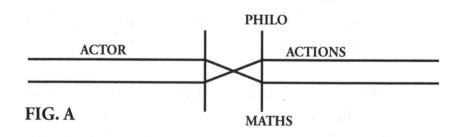

FIG. A

A look with careful study of the blending mechanism, which is contained in the above drawing, shows that an actor is always using the philosophy of symbols in protecting and projecting his actions, which eventually will be understood as a logo. For those who cannot understand the rubrics of advanced logic, or what is best defined as an amalgamation of symbolic and mathematical logic, this work is a practical VSAT that will help them to assess, understand, and appreciate why every actor is an action, whereas philosophy uses the scientific VSAT with the existing T^4, which was explained in the last chapter.

At this point I am reminded of the words of Professor Morris Norway of the Vatican University, who lived in the nineteenth century. He required a recapturing when he told the Augustinian Italian Claudius, "The universe is an actor which plays on the rhythm that motivates the actions of man in order to be creative, philosophic, and mathematical." He also told the Claudius crowd,

The simultaneous result of the actors and actions are responsible for the teething problems that are contained in engineering mathematics and engineering logic with their multiple logos and symbols.

He concluded by saying that the above requires a scientific and synthetic fusion.

Now, is it possible to accept Professor Norway's opinion hook, line, and sinker without some elements of constructive input? The answer to the question can righteously portend and produce a teething and titanic result: Its argument is a cosy attempt and not a loftier concept.

On this note, I believe that no action can be produced without a critical actor and without ascertaining that philosophical mathematics as an honourable, combined actor is action-driven in the blending results of achieving purity and perfection. For the record, these are the centurial and sentimental monuments of the philosophical museum, which are responsible for creating the fortress of illumined wizards who are great actors with monumental actions.

A look at the introduction of philosophical mathematics as the logo and symbol of the universal ingenious creativity explains the creator as a simultaneous actor. Creation is a permanent action that is rhythmically squared with thoughtful expressions and the infinitude concept. That is why it is explained that philosophical mathematics will forever serve the universe as an illumined ingenious actor whose actions are generalized as purpose-driven.

At this point, we must appreciate that the one-world family is a logo, and the creations are symbols. Perfection is equally a logo; purity is a symbol. The comprehensive summative to this chapter is that philosophical mathematics is the honest heartbeat that drives the creation of all actors with their actions, which are best explained in the understanding of Mozart and the music of the philosophical mathematical spheres. The spheres are the comprehensive and illumined anatomy that all philosophical wizards test, and they attest to the anatomy of honest, purpose-driven, universal transactions with a constant and permanent infinitude. The translations give the world the practical meaning of the secrets of light, which are contained in the Orion symbol of philosophy with the logo that propels mathematics to serve philosophy and the whole universe as a bankable databank. It is here that this work acts as an action of a progressive and dynamic actor.

CHAPTER 11

A BALANCED OVERVIEW OF PART 8

·A look at the origination, segmentation, and perfect compartmentalization of part eight reveals that every action is an actor, but every actor is equally a simultaneous action. This is why in this aspect, philosophical mathematics is defined as a simultaneous prodder, sentimental radar, and a concentrated and conjugated framework. Philosophy and mathematics are nurtured and naturally simultaneous in the best expression of facts and figures.

The introduction to philosophical mathematics as the symbol of the ingenious, creative universe started with the definition of what logo and symbol are, their origin, their mathematical philosophy, and their importance—including explaining how and why they are designated as the supreme heartbeat of the universal creativity. It further explained how this fact is not yet known by the world of material science, which further explained how logo and symbol are the living bases of illumined philosophical mathematics (which is the anatomy of all living).

Other segmented chapters that are practically purposeful include why the logo is being used by Oriental wizards as the perfect anatomy of moral ethics, which is the symbol of all mathematical codes. The codes include delving into the urgent call for the modern era to appreciate the use and application of the symbols as ingenious creators, which in earnest is the noble goal and aspiration of philosophical mathematics.

Other chapters that follow are on the explanation of how the logo and symbols are designated as the balanced keys that can be used in appreciating the rhythms and the challenges of life. The concept finally correlated and connected with the last chapter, which explained the philosophic and scientific dynamisms that moderate philosophical mathematics as an action to actor and as an actor to actions, including how the symbolism of these actions and actors are best designated as the ingenious illumination of philosophical anatomy.

A look at the content of my work will let one appreciate that this overview can best be designated as a rendition and philosophical hymn, a mathematical melody, a scientific and technical soul safari that defined philosophy as the way of the illuminates. Mathematics is the humble and absolute occupation of the heart and mind.

In my humble submission, this rendition can best be named the philosophical melodies of orphic and quabalistic thoughts, whose headings can best be appreciated by lintious minds who appreciate that, apart from being the honest mother of consummate knowledge, philosophy is equally the anatomy of mathematics, and that is why it functions in multiple divisions and diversifications. Philosophical mathematics is best designated as the engineering mechanism and model of balanced creativity.

At this point, the human race should have hope and faith that the ingenious Mother Nature, at the proximity of philosophical mechanism, has constructed technical chimneys that are greatly needed by our present era, including the province of posterity. It is believed that posterity will post and create a dynamic philosophical and mathematical intelligence that will help to explain and reveal that philosophical mathematics, with its logos and symbols, is indeed an ingenious creative VSAT whose actions are relative to actors, including utilising the purpose-driven actors with a simultaneous creation of actions.

This overview can best be defined and concluded as a philosophical pillar, with its constructive trees, which will help humanity in understanding that an actor at any point in time is a beacon, whereas an action is a philosophical and aesthetic realism, moderated and inspired to exist as a philosophical and mathematical Leningrad, including maintaining constancy as a blended philosophical mathematical Britannica.

PART 9

PHILOSOPHICAL MATHEMATICS AS THE ENGINEERING MODULE, MODEM, AND MECHANISM OF BALANCED AND CONSTRUCTIVE CREATIVITY

"Empirical sciences prosecuted purely for their own sake, and without philosophic tendency are like a face without eyes."

—*Arthur Schopenhauer*

INTRODUCTION

Philosophy and mathematics have remained honest endeavours that can be utilised in appreciating how, why, and for what purpose natural resources are transformed and transmitted with the use and application of engineering modules and mechanics. The creation of engineering modules and mechanics cannot be possible as a material functionality without the intertwined involvement of philosophical mathematics. This is why philosophical mathematics has remained a practical approach and a compartmentalized module with scientific mechanics. The mechanics understand the dictates of ensuring particularly the aspect that deals with production engineering, production creativity, and construction.

The purpose of this introduction is to scientifically explain the intertwined and natural relativity that exists between philosophical mathematics and engineering as a process of transforming natural and material resources into economic and commercial assets. The material schools of engineering technology, particularly electronics and production technology, do not understand the deep most functionality and how philosophical mathematics is the scientific anatomy of engineering mathematics. Again, the creation of modules, mechanics, and other balanced engineering formats cannot be gained or appreciated without the following knowledge, which is natural and lintious.

- Philosophical mathematics.

- Philosophical engineering.

- Philosophical economics.

- Environmental economics.

- Engineering mathematics.

- Engineering economics.

- Mathematical mechanics.

- Mathematical modules.

The simultaneous relativity between the anatomies of engineering modules and mechanics has remained one of the practical functionalities of philosophical mathematics. If one takes a practical look at the composition and segmental disposition of engineering mathematics that exists at the proximity of Stroud and his colleagues, one will appreciate and attest that these mathematical calculi materially explained the dynamic modules and mechanics that drive the balanced functionality of philosophical mathematics, philosophical engineering, production, and technical mathematics. It is here that philosophical mathematics has remained the dynamic approach by which the content and context of engineering mechanics can be harnessed with scientific objectivity.

Professor Singh of the University of Kalakuta and Professor Vicky of the University of Bombay, both in India, gave engineering and logical advice:

The tedium of modules and mechanics as engineering assets cannot be understood without involving the detailed explanation of philosophical mathematics.

They structured and defined this as the anatomy of engineering modules and mechanics, but the colleges of material engineering production did not strictly understand the opinions of these gentlemen. This is why the practical and natural science of mechanics and modules, including their convention and conventional systems with practical modules and moderation, cannot be understood by material science.

A work to explain the practical relativity of philosophical mathematics to engineering dynamism, as it relates to how modules and mechanics are balanced and intertwined, must certainly know that engineering creativity is not possible without the input of philosophical mathematics, philosophical econometrics, mathematical modules, and mechanics. The noble province of science defines Sir Isaac Newton in this context as a chancery to the workings of modules and mechanics, because he strongly opinionated that engineering is a practical data and the creation of physics and the moderation of chemistry, with some inputs of technical biology. It is important to explain that his roots and aborigines could not understand this great scientific emengard. The province of physics and electro-mechanics have always renamed and appreciated him with a dominant chair in the hall of engineering mechanics and modules.

To appreciate the relativity of philosophical mathematics, this part explains what constitutes the frequency concept of engineering modules and mechanics, and how the logic and anatomy are driven with the balanced wisdom of philosophical mathematics. Modules and mechanics are productive alchemy and technological and scientific creativity. A work to explain the dynamic engineering system—which can best be defined as that which determines the powers of engineering modules and mechanisms—will certainly appreciate that philosophical mathematics is the scientific moderator and the ingenious concept that technically drives the creativity of engineering modules and mechanics. That concept contains the anatomy of frequency modulation, motion and emotion, tangents, and the powers and formats of engineering functions (particularly that which deals with polymer technology). Architectural constructive techniques with their methodology adhere to the directives of philosophical mathematics.

At this point, we must appreciate with objective technicality that philosophical mathematics is the balanced key that can be used in knowing and understanding the rhythm of engineering modules, techniques, moderations, and frequencies. This is why the objective view of this introduction is to explain that philosophy is greatly intertwined with the functions of engineering mechanics, modules, and techniques. Thus it depends on an acceptable result that upholds the tedium of its balanced creativity as a system and concept that is naturally allowed the rudiments of engineering functions, which include directing the concept of modulations and mechanics, and hence the use and application of philosophical mathematical mechanism.

This introduction exists alone as an engineering concept that will help the world of science and technology to appreciate that philosophical mathematics is a naturally gifted engineering calculus that works to achieve the design and desire of purity and perfection. That perfection can be defined as the hallmark of engineering measurements, modules, and mechanics with scientific and practical dynamics that galvanized engineering as a scientific and technological endeavour existing naturally and practically on the natural exigencies of the conversional and transmutational techniques.

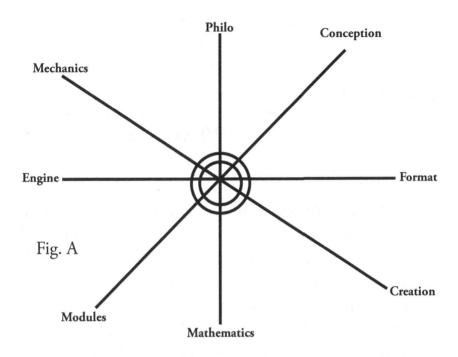

Fig. A

The above diagram is a true explanation of the engineering works that involves the lintious powers and forces of philosophical mathematics in order to achieve its desires with practical nobility.

CHAPTER 1

WHAT IS ENGINEERING?

There are a lot of definitions attached to engineering, to convey what can be defined as the construction or management of machineries, roads, bridges, and more. *Webster's Dictionary* defines engineering as "planning, designing, construction, or management of machine, roads, bridges, etc.". That definition is practically narrow and myopic on what the concept and definition of engineering could be, to the best of my knowledge.

To point out the true definition of engineering from existing dictionaries, we have to consider the age of the dictionaries, as well as the environment, including the terminologies used for the sake of these definitions. To obtain an acceptable and futuristic definition of engineering—including its monumental contributions to the province of technology—the world economy must be considered as a serious factor. The economy moderates the definition and dynamism of engineering creativity. In this respect, I give the following definitions of engineering as a science, an art, and an anatomy of technology, including holding an intertwined connectivity with the directives and detects of philosophical mathematics.

- It is a science and an endeavour that deals with constructive creativity, with a view to creating and transforming what Mother Nature has given in order to be utilised as a technical and economic asset.

- It can equally be defined as the logo, the symbol, and the mathematical dynamism of creative arts, which must first and foremost be conceptualised with plans and designs in order to obtain an acceptable form of the engineering creativity.

- Engineering can equally be defined as an occupation, that utilises the processes of designs, concepts, and planning in order to achieve economic and technological assets (that can be translated into resources) and wealth (including physical and monetary assets).

- From the heartbeat of Mother Nature, engineering can be defined as the process of transformation, transmutation, transmission, and translation of existing resources into other forms that can be econometrically useful in uplifting the human standard.

- It can equally be defined as the management of material and natural resources in a diversified, creative concept in order to ascertain its technical, scientific, and management usefulness.

- Engineering as a science, as an art, and as a philosophical mathematics is the process of creating wealth through procedural concepts.

- In our contemporary society, which is charged with globalization, millennium development goal (MDG), information technology, and information communication technology, engineering can be defined as a mechanism of moderating and managing resources for effective goal appreciation and actualization.

A look at these ingenious definitions of engineering as a science, an art, and the goal of human endeavour with objective creativity will certainly appreciate that the definitions of *Webster's, Learners,* and *Oxford,* including what is contained in Americana and Britannica encyclopaedia, will appreciate that the age for which these dictionaries defined engineering can be assessed and defined now as the machine age, which worked to grind humans into the gears of machine. This constant grinding was against human development and engineering, and it motivated Edwin Markham and his engineering apostolate to state, "Against making cities, we should make man who will in constant and efficient mechanism make cities simply because man is more superior and more dynamic than material cities."

My challenge to bring this as part of the appearance of philosophical mathematics is borne out of the fact that the teachers and professors of engineering, in its diversified and multifarious areas, are not naturally ingenious in appreciating how, why, and for what purpose the permanent avatars of engineering as a human occupation were motivated to discover this important life occupational concept, which everybody practises every day because engineering is a part of nature. Everyone practises it either ignorantly or intentionally.

The next chapter will delve into knowing with adequate explanation who an engineer is, the intentions of engineering, and engineering's intertwined relativity with technology. We must appreciate that the class of humans designated as industrialists are real philosophical engineers whose large minds are naturally occupying the tedium of engineering as a science, as a philosophy, as a human endeavour, and as a craft and act. This is why every human being is defined as a super-genius of dynamic engineering who retains and maintains his cordial and coordinate relativity with the use and application of philosophical mathematics, which directs and motivates his crafts in order to use his ingenuity in creating resources and wealth that are needed in sustaining the ecological development of our existing contemporary ecosystem, which is driven and directed by the forces and functionality of engineering mechanism.

CHAPTER 2

WHO IS AN ENGINEER IN THE LOGO AND SYMBOL OF ENGINEERING?

The universe originated at the careful and dynamic construction of a super-genius engineer who naturally extended his consummate ingenuity as a way and means of maintaining the continuity principles. Mother Nature does not function or conceive without the propelling concept of the super-genius engineer, who is honest, dynamic, pragmatic, and titanic with his highly conceptualised engineering methods.

The necessity and challenge to define who an engineer is, is borne out of the fact that the dynamic system in which our ecosystem is operating is sequentially being managed and governed by the super-genius engineer who, in his great and honest science, is defined as the consummate universal One.

In this respect, the following definitions from the existing Britannica and dictionaries will serve posterity as dependable definitions of who an engineer is.

- He is a trained actor who drives and designs all the actions of engineering.

- An engineer is one who is trained to train, to design, to conceptualise, and to conceive all the modulations and mechanisms that drive the structure and anatomy of engineering as a perfect product of ingenious philosophical mathematics.

- He can equally be defined as a transformer, a constructor, an organizer, and a researcher, including being a titanic human who is skilful enough to handle all the challenges of engineering as a human and commercial endeavour.

- An engineer can be defined as a fellow who has knowledge of the designs and postulations of engineering.

- He can equally be defined as one who is trained to be an engineering guru in order to appreciate and understand the concept of natural resources and material environment. This is why the logo and symbol of an engineer is designed to appreciate with technical understanding the challenges of engineering.

These definitions are borne out of an inspired vision, because if an engineer is defined as a craft maker, it naturally and originally demystifies him as a mere technician who is not trained for the purpose of engineering. This is why engineering as an industrial and technical endeavour is always an expert in producing technologies that drive the creations of technocrats.

At any point in time, scientific and material technocrats are faced with the challenges of management, human endeavour, industry, economic changes, and commercial actualization. This is why the present global meltdown, with its multifarious challenge, is always revealing that our contemporary system is not driven with naturally ingenious engineers. As the universe is dynamically driven with engineering and material challenges, so it is here that our Microsoft technology, including the province of information technology and information communication technology, must be charged with a lot of engineering modulations and mechanisms. This is important because the challenges of globalization and scientific cosmogony cannot be gained or actualized if the task of driving the natural engineering concept is not at the proximity of highly ingenious engineers who are broadened and enriched with the VSAT postulations of philosophical mathematics. Those postulations include using the logos and symbols in ascertaining the functions of engineering and the functionality of an engineer.

Since the dawn of consciousness, the human race has known few prominent and ingenious engineers. These gifted geniuses include Leonardo da Vinci of Italy, Oscar Dares of America, and the famous and structured engineering emengards named Thomas Edison, Austin, and Fredrick Ferdinand. They are always remembered in the annals of human evolution with engineering dynamism as noble laurels that are titanically embedded in the book of engineering.

Our present era has not produced such lintious, luminous, and titanic persons. This is why the universities and research institutions are churning what is known as purported and glorified engineers. In our contemporary civilization we must respect such highly anointed engineering wizards like Damian Anyanwu of Mbaise, Ahmed Zubairu of Kano, and this humble personality, A. U. Aliche, whom the world has designated as an illumined great thinker whose engineering prowess is yielding results. This is attested by all and sundry as what a noble and millicential wizard should be. To be an engineer does not require a lot of academic ascension, because engineering is an honest appreciation with perfect blending and usage of natural resources in order to create products, assets, wealth, and lots of resources.

In countries like China, Taiwan, Thailand, India, and Japan, these countries have naturally grown edified and highly illumined engineers who have succeeded in turning their economy from a poverty-driven state to a luxurious and boisterous standard that is succeeding in restructuring the economy of these countries to fit and compete with economies of the world powers.

At this point, the provinces of science, technology, and engineering are naturally looking for highly gifted talents. What is happening in America's aeronautic yard, the space garden, are events and concepts that were not developed within the university walls. Neither were they obtained from most of the unpurposeful research institutes. Instead they are from an in-depth search and research into the nature's bounteous workshop.

The task to explain for whom and for what an engineer stands will always appreciate that Chancellor Mitsubishi and Lord Peugeot were not products of any renowned institution. For this reason, Shakespeare opinionated that the best education is that which is acquired at the proximity of Mother Nature. He further stated that this is the education that uplifted the likes of Mozart, Paracelsus, Pythagoras, and all that are enlisted as the Twelve World Teachers. It is this education that is empowering my ingenuity to give the world a detailed and consummate definition in the enduring philosophical concept of philosophical mathematics, which exists at the fontious and fountain of every mathematical and philosophical rhythm.

CHAPTER 3

WHAT IS THE ORIGIN OF ENGINEERING?

Engineering, as a perfect and constructive endeavour, did not exist without an origin. It is its origin that mandated philosophy as an enduring logo to explain with practical symbolism that nothing created is an accident. Both the science of philosophy and philosophy as a science appreciate that engineering, in its parallel and pluralized structure, is an applied human and economic endeavour that follows through a constant practise. This is why the structure indicating the nature of engineering makes it an endeavour that has monumental challenges. It is here that engineering is always utilising human, material, and natural resources as symbolic catalysts in its consummated creations. By consummate creation, I mean a vestry and a vast, consolidated chancery that is enabled and ennobled in order to satisfy the quest and zeal of humans.

Mother Nature, which is hereby defined and designated as the porter's wheel, acted on the super-genius intelligence of man in order to conceive how things could be designated, modelled and structured. This follows a natural conception of engineering technology, which is only possible and feasible at the infinite laboratory of the consummate Creator. This is why the science and technology of engineering understand the following languages.

- Transformation.

- Transmutation.

- Translation.

- Trans-modulation.

- Transition.

- Trans-analysis.

- Trans-mathematical appreciation.

Engineering is focus-driven with the following natures and rhythms.

- Anatomy

- Structure

- Quality

- Quantity

- Qualitative analysis

- Mechanics

- Quantum

- Modifications

- Design with aesthetic structures

- Research

- Data with statistics

This is why the advancement of Michael Faraday was designated as a creation of a super-genius thinker, which that era called the creative engineering of all ages.

As a philosophical mathematical analyst, I have come to appreciate that nothing created or invented by man is really an invention of its own. This is so because creation moderated all engineering concepts through the utilisation of its perfect incubator, which material science has not been able to prove, define, or understand. The engineering technology of Mother Nature—which does not make any mistakes and does not require the utilisation of material resources—is abundantly anchored with monumental wealth. Its understanding has greatly ruled that in Elohim is all that made the holy alpha an engineering omega. which designated the holiness of his omnipotence, the actions of his omnipresence, and the consummation of his omnipotence. All these must be presented in this analysis as being responsible for the creation and constructive origination of what the science and technology of engineering represents as something that has come to be.

Professor Abel Bruce of the University of Germany, who is known as one of the best experts in mathematical engineering with its scientific philosophy, stated that his creations (particularly with the use and application of metals) are only possible because he is naturally possessed by an engineering, intelligent spirit whom he appreciates as being the foremost Creator in his creations. Abel's position cannot be accepted to be colloquial statement, but it is practically ingenious because he appreciates that all he is has already been created by a super-genius engineer, whom he defined and designated as the ingenious, intelligent Creator. Going by Abel's analysis and understanding, the Creator is simplified with engineering ornamentation as the most intelligent spirit.

For the record, the philosophy and wisdom of this recognition is good, but it is not enough. This is so because reducing the Creator to a mere spirit only makes a tactful and un-engineering analysis of him.

In this respect, I say that the origin is of the oneness of all holiness, the oneness of all bodies, and the oneness of all principles, policies, practises, and methods with permanent research. This is why he is forever known and designated as a living structure and constancy that permeates, permits, structures, and creates with the ingenious engineering of his wholeness engineer. Hence the Creator is ornamented in this marble as the ingenious One who is the perfect and consummate engineer of human souls.

It is appreciated that the origin of engineering is the consummate engineer who is perfectly known as the choice of immortalized creation and a living monad who must live in appreciating the dynamism of his greatness, which is practically diversified in great and monumental areas. That is why the origin of honest engineering is the origin of the Creator, the honest purpose of life, and the symbolism that lies in the intertwined relativity of translation, transmission, and transition.

Material science is already at a difficult deluge simply because nature's process of engineering and reengineering its creations is not fondly recognized or understood by it. This is a great task that is daily facing the limited science called material engineering, whose principles, policies, practises, and methodologies usually experience constant changes.

Philosophical mathematics as a science and as an engineering calculus appreciates that God is the engineering mechanism as well as the engineer of all great creations, great inventions, and immortal arts. This is symbolic in the way and manner he reengineered humans to create and establish the wonders of the immortal pyramids, which have remained the marbles and mysteries of architectural engineering concepts.

The structure, frame, anatomy, and composition of the engineering marble defined and designated as man is a symbolic architectural ornament that is performed as an illumined craft of the immortal engineer. In his divine cartel, he originated the science and technology of engineering for the sake and purpose of man's ennoblement, creation, and construction in order to develop the universe as an ecosystem that will be greatly driven with the engineering activities through the use and application of human and natural resources. The resources will help man to establish a technological and purpose-driven life whose transactions will be greatly established on the engineering law of constancy and balance, which will be a yardstick in measuring the dynamism of the universal economy.

For the record, engineering is God's procedure of industrialization, instrumentation, and technological development, including how alchemy and all that is involved in chemistry, chemicals, and metrology, can be enhanced in creating a dependable, world-class economy.

CHAPTER 4

Has Creation Known and Appreciated Its Roles and Functions in Directing the Universal Economy?

As a comprehensive and engineering entity, creation has greatly drifted from the honest understanding and appreciation of philosophical engineering, philosophical mathematics, and the dynamism that models and directs the mechanics and mechanisms of balanced engineering creativity. This is regrettable and has led to the non-functionality of engineering policies, practises, and principles, which has further created a vacuum. The vacuum has then led to the problems of engineering and scientific principles, policies, practises, and standards that are not constant.

Engineer Eric Rose, who lived in the eighteenth and nineteenth centuries, regretted this situation when he told California and Chicago that the institute of Microsoft dynamism would certainly pilot the universe into a regrettable oblivion. This is understood from my own universal point of view as the beginning of human creative anarchy, which has further developed a chaotic system that science and engineering find too difficult to resolve, balance, direct, and moderate. This chaos is why humans have not been able to appreciate the roles and functionality of engineering dynamism.

The first principle of philosophy and the dynamic principles of mathematics do not appreciate that material science will give the world a productive cosmogony. The cosmogony will be greatly utilised by the ingenious provinces of engineering and technology because a creation that refuses to acknowledge its creator cannot effectively recreate. The philosophy of engineering mathematics is always at a dangerous crossroad because of these lapses. That is why the present globalization is best defined by the well informed as a mockery of human ingenuity. It is a mockery because it is short-sighted in the views of balanced philosophy, balanced metaphysics, balanced logic, electrifying understanding, ingenious aesthetics, and theurgy. Again, globalization in a material aspect, with the anatomy of material scientific system, is founded without living ethics and moral codes. That is why it is being recognized by the well informed as the engineering jamboree of unproductive mediocre.

It is important to explain that engineering has a universal call with a global temperament, and it uses this fathomed model in its designs and creations. The philosophy of these models is to ensure there is a simultaneous and systematic modulation of its mechanics, which can further give rise to the postulated modulation of its mechanism.

When one looks at the workability of the science of mechanism as opinionated by Sir Isaac Newton, Albert

Einstein, and Smith Caroline, one appreciates that motion and emotion are sequential factors that moderate the science and philosophy of mechanism. This is why it is protected and taught at the province of physics with engineering mechanism. I define mechanism as the simultaneous and synchronous amalgamation of motion in emotion. This mechanism can only function where the electronic and electrical emotions are balanced in utilisation, in order to cause a mathematical and philosophical sensation that will be philosophically quantum in relating and electrifying all the agents and reagents. That relation produces mechanism at a moderated frequency, which is honestly defined at the proximity of science as the ingenious postulation of balanced motion and emotion.

Let's get back to the tedium of this chapter as a philosophical mathematical curvature. The lack of appreciating the roles and functions of philosophical mathematics as the dynamic engineering model and mechanics to balanced technical creativity is a result of the following misconceptions and myopic factors.

- Lack of universal standard.

- Non-civilization of nature's principles, policies, and practises.

- Involving mediocrity in the province of energy and mathematics.

- Inability of material science to know what is defined as the philosophy and mathematics of engineering.

- High level of abuse and negligence from the academia and research institutes, including a lack of commitment to excellence.

- Lack of research and the system not being able to establish engineering laboratories with scientific incubators.

- The orchestrated errors of the machine age, which were cycled by the jet age and further recycled by the computer age.

- The frequent errors of the Microsoft sector, the industrial sector, the economy, and the lack of an accounting standard, which makes auditing engineering a difficult task.

- The neglect of philosophy by the province of engineering and science, without knowing that it is the anatomy of knowledge and science.

Creations at all levels are hereby advised with philosophical admonition to know and appreciate the:

- The true knowledge of engineering either as a natural endeavour or a manmade endeavour.

- To know and utilise the sequential factors and features of Mother Nature that are responsible for nature's creations to be error-free.

- To know how to download all that nature has created with the use and application of inspiration, which is hereby defined as the most intelligent spirit.

- To advance in order to learn the occult coding system that nature uses in creating and directing the affairs of ingenious engineering.

- To advance in understanding the intertwined relationship that exists between philosophy, mathematics, engineering, science, technology, and the nature and structure of the methodology of chemistry (including the science of alchemy, which must be well-known by the province of future engineering).

This knowledge will greatly solve a lot of problems that science and technology are finding difficult to contend, to put into context, and to withstand. What is defined in the science and engineering of metrology as the problems of containing the ocean cloud, the aquatic province, and the ecosystem must be known to the fingertip. Creation must be able to know the structure and anatomy of the following natural resources.

- The kaolin

- The iron ore

- Ionized salt

- The pebbles

- The stones

- The marshy sands

- The ocean and its whales

- The calcium family

- Limestone and other rocks

Agricultural science must advance to appreciate the natural content of carbon dioxide, carbon monoxide, hydrogen, and all that is contained in the family of sulphate and phosphorus compounds. This knowledge is important because there is an electrically intertwined relativity that naturally exists among this family, but during extraction and transmutation, science and technology do not take this into a serious consideration. That lack of consideration is why there is a great challenge in production technology, engineering technology, and engineering mathematics. The science and philosophy of the lithorial forces is greatly being abused simply because the electro-intertwined relativity that exists among these minerals and compounds has not been objectively understood.

Creation has a great task particularly in the areas of technology, engineering design, and creativity. This is why philosophy and mathematics, as co-agents of science, are always driven with the natural polyester and posters of achieving perfection and purity—simply because humanity is not a causality but a monumental engineering creation that must exist and be driven by the simultaneous voice of nature, which speaks to the universe that creation is a fountain thrust of engineering academy and must be protected at all cost. This protection is naturally feasible when man appreciates that life is a scientific and technical adventure, including being the first principles of philosophy and the foremost ingenious philosophical mathematics.

This chapter is best defined as the philosophical mathematical route to balanced engineering knowledge, which exists practically and purposefully at the engineering frequency of Mother Nature, and which philosophy defines as the simultaneous chancery of the universal one that must exist wholly and totally with the use and application of the immortal laws. The laws are technically and scientifically driven in order to achieve a constant and dependable engineering province, which is the view and vision of globalization, millennium development goal, and what technology has named a technical village, an economic village, and an information technology and information communication technology province. Philosophical mathematics in this ecstasy and structure is defined as the first principle of balanced and dependable government.

CHAPTER 5

WHAT ARE THE PHILOSOPHY AND LOGIC OF ENGINEERING MATHEMATICS?

As logic is blended, propelled, inspired, and consummated in realizing the concept of conversion, construction, and more, philosophy is essential in ensuring and achieving purity and perfection. It is here that the Twelve World Teachers configured an intertwined relativity with the objective dynamics of engineering. Material engineering at all levels does not understand what is defined as the philosophy of engineering or the engineering philosophy of logic and technology. Science has remained the extractive and interactive energy that is hereby ornamented as the anatomy of engineering technology, and that is why an objective explanation appreciates that engineering, no matter the vision and concept, cannot create or produce without involving philosophical mathematics, including the science of mathematical philosophy. All that is explained in engineering mathematics is a material furtherance of the intertwined relativity of using calculations, and that constitutes what is learned and taught as mathematics.

At this point the philosophy of engineering is best explained in the following lintious concepts.

- The philosophy of engineering is the science and the technology of its anatomy.

- It is the wisdom that serves as a driving force of engineering technology.

- It is an ennobled goal that examines the position of philosophy to engineering as a constructive and creative endeavour.

- It is best known and explained as the rudimental factor that propels the functionality of engineering as a human, material, and natural endeavour.

The philosophy of engineering can best be defined as the fountain and fortress of this creative force that must utilise all the data contained in science, nature, logic, and abstract knowledge. This is why one of the engineering faculties that utilises philosophy is quantum mechanics.

The question here is, how do mechanics and quantum engineering utilise philosophy? The answer is mathematically simple because both mechanics and quantum engineering are driven with objective and technical logic, with technical and scientific mathematics. The anatomy is always framed and formatted with the deep thinking of philosophical mathematics.

Richard Ernest of the University of Amsterdam advised that engineering as a constructive creativity, and as a material science, cannot achieve a determined and dependable goal if it neglects the dynamism of philosophical mathematics. That dynamism provides all the natural and needed data and all the statistics, including opening an engineering horizon that utilises the logic of science, the concept of philosophy, and the monumental mechanism of quantum appreciation and explanation.

In eighteenth and nineteenth centuries, Professor Rose of the Vatican University, after examining practically and perfectly the networking concept of quantum and engineering techniques, opinionated, "All that is contained in physical and applied physics is a rudimental technology of engineering concept." He further explained that this knowledge can only be understood both as science and technology with a high involvement of philosophical mathematics, including bringing mathematical logic to serve as a lintious lenninger by which the constitution, the anatomy, and what is best known as the logic of science can be understood.

A work to examine what Isaac Newton and Albert Einstein created as a laboratory and library of mechanism with quantum science will certainly appreciate that engineering cannot function as both a course and dynamism, without bringing into balanced reasoning all the contacts of chemistry, physics, and philosophy. It must be stated that the fundamental course of engineering technology is conceptualised in the rhythm and course of philosophical mathematics.

The advice of Professor Abel Collins from the University of Czechoslovakia must be noted as a monumental advice for future engineers.

The strength and measure of mathematical engineering is limited to material science, and the power and creative engineering is equally limited to the purpose and concept of what it is conceptualised to be.

This professor of inspired mathematical engineering also stated, "Philosophical mathematics must be allowed to drive the thinking of engineering as a science and as a creative and constructive human endeavour."

Professor Wagner of the University of Virginia admonished engineering students to be always prepared to correct the lapses of material engineering technology, which works to neglect the high level of structure, wisdom, and knowledge contained in philosophical mathematics, both as a natural technology and as a nurtured engineering. We must appreciate that the global universe is facing a lot of enmitied challenges, simply because engineering technology reframed, restructured, and renamed the thinking of Internet wisdom in the objectivity of Microsoft technology. This is what propelled Abel and Anselm to refuse to be designated as engineers, because their thinking was that nature is the foremost founder of engineering, the fulcrum of technology, and the foundry of science.

A work to examine the philosophy of engineering with its logical enabling must appreciate the following points.

- That philosophy and logic are the power of engineering technology.

- The science of engineering, with its technical configuration, cannot foster or be constructive with a high negligence of philosophical mathematics.

- The Orientals were designated as better monads in the area of ingenious engineering simply because they appreciated the workability of philosophical mathematics as an orchestrated chancery and as the booster station of all facts, figures, and research methodology.

- The philosophy of mathematics must work with reasoning and must format with the logic of technological science, philosophy, and other relative areas.

- The human race must be prepared to face the challenges of engineering, which have created a universe of hybrid erosion, corrosion, and mathematical chaos.

- The fall in engineering standards is a result of the negligence of the involvement of philosophical mathematics, which holds the determining key by which engineering standards can be configured.

- The logic and philosophy of engineering, both as a science and as a creative art, must be appreciated from the point of scientific and constructive empathy.

Engineering as a scientific and a technological prowess does not deserve to be taught in the classrooms, simply because the open field and the laboratory of Mother Nature created a balanced and sensitive parlance through which the enabling and ennobling of engineering science can be harnessed. The achievements of Michael Faraday, Galileo, Einstein, and others were not gained at the platform of academics. This is why most of the concepts that are contained in academic engineering have endangered human growth and evolution. In the immortal words of Spinoza and Bash, Mother Nature is the hallmark of science, whereas Father Nature is the foundry of engineering technology. With this philosophy in the back of one's mind, I opinionate technically and practically that what is constituted and considered as the science of engineering, the philosophy of science, and the logic of mechanism and quantum engineering is only human action, which determines and drives the workability of natural and material resources. For this reason, project management and technology insists that the province of philosophical mathematics is a natural heritage and an engineering pact by which advancement on creative technology can be harnessed.

After critical examination, this chapter can best be defined and designated as the sensitive foundry of philosophy to logic, logic to science, science to engineering, and engineering to technology. It must be comprehensively concluded by the mathematical notation and technology of philosophy, which is practically constant to moderate, direct, and appreciate that the philosophy of engineering is the analytical and practical creativity of its objective empowerment. The chapter will always remind human and material engineers that Mother Nature is the fulcrum of science, the art of philosophy, and the podium of engineering technicality. Nature tells the functionality of the science and the logic of engineering, in a very simple and modified way, which material engineers have constantly abused, mimicked, and rejected. Thus the understanding of what is known and considered as the philosophy of engineering, the philosophy of science, and the ennobled reasoning appears a sophisticated concept to the province of material technology.

For example, natural technology states that the philosophical mathematical format for a balanced understanding can start with $0 \times 0 \times 0 \times 0 = 0^4$. It also explains that $2 \times 2 \times 2 = 8$ and that the reverse of this can be $8 - 4 \times 2$. A look at the practical simultaneous equation—which can best be defined as an engineering enabler—reveals that there is no complication in nature, there is no confusion in philosophy, and what is considered as complication and confusion has remained the problems of man through his enmities in the way and manner he handles and practises engineering and mathematical standards.

It is here that the functionality of the philosophy of engineering, logic, and science, and the standard of the practicality of philosophical mathematics, is to establish a constant standard that will make all and sundry to appreciate that nature is the fulcrum of permanent enabling, whereas engineering and technological standards understand the philosophy of engineering and use this philosophy to approach the province of science.

CHAPTER 6

WHY DOES PHILOSOPHICAL MATHEMATICS CONTROL THE FUNCTIONS OF ENGINEERING TECHNOLOGY?

Engineering technology, as a creative art with scientific concepts, cannot function without the use and application of philosophical mathematics. Stroud Kant explains this, and H. A. Clement conceptualised it; Lawrence Durell also examined this concept. That is why I am bringing philosophical mathematics to the foreground of the academic world, particularly the engineering sector. Engineering as a material endeavour, and with its intertwined unionism to technology, believes in the application of philosophical science and mathematics. The aim of this chapter is to explain how and why philosophical mathematics controls the diversified functions of engineering technology. In this regard we can note the following.

- Philosophy is the mother of knowledge, particularly science and technology, and it empowers the functions and scientific functionalities of engineering technology.

- Mathematics is a calculative function of figures and letters that involve the use and application of simultaneous and linear equations, algebra, geometry, calculus, and statistics. Engineering technologists always employ mathematics as a creative concept in determining the concept of standard, the principles and practises of methodology. This is why engineering research depends greatly on the functions of philosophical mathematics.

- The nature of philosophical mathematics makes engineering technology utilise its diversified knowledge, data, and statistics in performing its mathematical functions.

- It is philosophical mathematics that directs the creative concept of engineering technology. This is why there is a high-tech unionism that exists between the creations of philosophy and engineering.

- It is philosophical mathematics that detached engineering completely as a technological endeavour from the empowered understanding of science.

- Engineering technology is always at the beck and call of philosophical mathematics. A work to explain the detailed concept of engineering standard cannot neglect the ethical principles of philosophical mathematics.

- The ability of Cons and his aborigines to celebrate the jet age in San Francisco was purely and perfectly at the instruction of philosophical mathematics.

- Statistics, as the cheapest engineering mathematics, is determined and controlled by the wider and philosophical wisdom of philosophical mathematics.

- Every aspect of PowerPoint is determined and decided by the ingenious philosophy of dynamic mathematics with electrifying understanding.

- All engineering concepts are determined by the use and application of philosophical modules with mathematical mechanisms and methods.

- Standard performance with engineering reference is guided by the powers of philosophical mathematics

- Engineering technology cannot achieve a dependable standard at any point in time, without involving philosophical mathematics to determine, explain, moderate, and construct a standard framework that will propel the functions of engineering technology, engineering mathematics, engineering philosophy, engineering logic, engineering science, and all that is conceptualised in objective and creative arts.

It is important to explain that philosophical mathematics can best be defined as the motion and emotion of engineering technology and the diversified technique through which this technology can perfectly and practically function. In order words, philosophical mathematics can be accepted as the scientific basis of engineering technology, with its scientific formulations. A lot of scientists and engineers do not understand that engineering technology has an underground physics, including a surface chemistry whose combination determines the activities of its biomass, biorhythm, and biotechnology. Lawrence Anselm of the nineteenth century stated at the University of Arkansas that engineering technology is still in the primary stage of its development.

A work to examine the concept and movement of space shuttles, including the diversified materialism of aeronautic science, reveals that philosophical mathematics is as old as creation and as important and enabling as any enrichment. Unfortunately, due to negligence and myopism the human race is not in the carnival and concept of its illumed creativity. Just as the material pendulum is a determinant factor in assessing and appreciating the movement of weather and other meteorological techniques, so is philosophical mathematics constant and permanently engineered in accessing and determining the creations and creativity of engineering technology.

When it is stated that science is giving the world a new cosmogony, it is a practical statement that reveals that philosophical mathematics is empowering the thinking and objectivity of science. It must be remembered that philosophical mathematics has remained a master sever and a creative asset; this is why it is always moderating the functions of engineering technology, mechanics, and mechanism models and frequencies. A work to decide and determine the ability of its functionality must be prepared to count from 0 to 1, 1 to 50, 50 to 100 and 100 to 1,000. The engineering technology of these number shows that the combination of philosophical mathematics with engineering mechanism will help one to ascend to the infinitude understanding of what scientific engineering define as a technological matrix.

In this respect, every enabling of man must be prepared to satisfy an agenda to protect, propel, empower, and be a determinant factor in the future concept of engineering technology and engineering dynamism. It is always stated that philosophical mathematics controls the functions of engineering technology and ensures that these functions are advanced in giving material technology an objective and acceptable standard. For instance, if $X \times X \times X \times X$ can be reduced to $X + X + X + X$ and converted to $X / X / X / X$, then the same formula can be amalgamated with scientific directives to reason that this can equally mean $X - X - X - X$. This reasoning shows that X maintains an engineering constancy and can be used in determining the functionality of technology as a keyword and a purpose-driven partner that must be utilised by the objective functionality of engineering as science, technology, and database and research institute.

The aim of this mathematical formula is to reveal that philosophical mathematics is practically and scientifically intertwined with the logic of engineering technology.

The creative craft of engineering technology is practically determined by the absolute functionality of philosophical mathematics. That is why this chapter is best termed the monumental Osborne of engineering logic.

CHAPTER 7

WHY ARE ENGINEERING CONCEPTS BEING DETERMINED WITH THE USE OF MECHANICS, MODULES, AND METHODS KNOWN TECHNOLOGICALLY AS THE MMM CONCEPT?

Professor Harrison lived in the eighteenth century and worked at the University of the Czech Republic. He was confused regarding why and how engineering concepts are not universally uniformed. He was equally disturbed by the detrimental features of modules, mechanics, and methods. Thus he worked assiduously to introduce what is universally accepted today as the basics of engineering concepts. The question is, what are the features, the functions, and the scientific enabling of these concepts? Harrison answered this question in a lintious and futuristic manner.

- Engineering concepts are the pivots that determine the mechanisms, the modules, and the methods of engineering and technological standards.

- By the use and application of these standards, the engineering wizards of his time were able to develop material theodolites, which were futuristic in the use and application of standards that work to adhere strictly to the pavement and pedestrian concept of philosophical mathematics.

What bothered Harrison constituted what may be defined as a material melancholy to the twentieth-century engineer who is defined as America's Leonardo da Vinci: Dr. Walter Russell. His confusion was on how scientific principles, policies, and practises do not elongate the principles of engineering standards and the modules, mechanisms, and methods. This situation triggered his obsolete engineering concept, where he told the world of science that what is considered today as modern mechanics is already obsolete, simply because there is nothing like permanent engineering in construction and constructive technology.

The titanic base of these arguments were materially resolved by the alchemist St., Germain, who noted a transmutational mechanism that dealt with the process of transforming metals into arch metals, which include gold, silver, diamond, and other peripheral and non-conformational metals. This is why what is contained in the sector of engineering mechanisms, modules, and methods is hereby defined as a technical and material bane that can be an endemic assault on the face of engineering, which is orchestrated by humans who are not well informed in the use and application of engineering concepts, technological standards, scientific methods, and technical methods with a view to enhancing material mechanism.

In my ebonite and studied thinking, engineering concepts can be determined by the following concepts with their structural and dynamic wisdom.

- The use and application of engineering mathematics

- The use and application of engineering and progressive philosophy.

- Aesthetic logic is important in achieving what can be defined as an engineering concept.

- Substandard concepts must be eliminated in order to pave way for universal and standard concepts, which will further mobilize the functionalities of mechanisms, modules and methods.

- An engineering databank must be created.

- The agitation of general research methodology must not be circumvented.

- Engineering must refuse to accept being a teacher of the classroom, including being a student of the four walls.

- Engineering diversification must be determined with the use and application of scientific classifications.

- Architectural techniques must follow a universal roadmap.

- Both construction and production techniques must be directed with a common engineering concept and with technological standard.

- The science of behavioural engineering must be utilised at all times in assessing and configuring data.

- Engineering enabling must not be seen or accepted as an obsolete Westminster.

- All engineering loads, particularly those that have to do with electrical-electronics, must be handled with practical perfection.

- Titanic utilisation of scientific standards must be utilised in determining the issues that empower and function as mechanisms, modules, and methods.

- Engineering objectivity must not be taken for granted, particularly in methodological engineering, metrology, and astronomy.

- A universal engineering laboratory is overdue for construction and creativity

- A world-class engineering library is urgently needed.

- The achievement of the Oriental engineering apostolate needs to be revisited.

- The content and concept of celestial engineering, particularly in our jet age, is the greatest need of the universe.

- The use and application of occult engineering technology with occult engineering science are important needs of the present age.

A look at the engineering Hansons that are constituted in these points will certainly appreciate that humanity is deadlocked in rebranding and recycling obsolete engineering concepts. The resultant effects of these errors are the visible application of wrong concepts and standards that are not advancing the course of engineering mechanisms, modules, and methods.

If it is believed that science is giving the world a new cosmogony, why is engineering not giving the world a configured and confabulated standard? If the Oriental engineers and scientists were able to determine all the space events in the moons, the stars, the sun, and all the electrifying galaxies, why is it that engineering has not been able to create a dignified galaxy that will comprise the mechanisms, modules, and methods by which we can access the powers of the stars, the moon, the sun, and the super galaxies?

It is in the wisdom of this chapter to explain that engineering, as a human endeavour and with material structure, is lopsided in concepts, standards, and functions. This is why there is no uniformity in the creative enabling of engineering. A look at how we can appreciate the philosophical and mathematical functions that determine and drive the engineering mechanisms, modules, and methods will certainly appreciate that a lot is required from the human angle, from the social sector, and from the environment.

The way and manner engineering as a science, an art, and an endeavour reconstructs the structure of natural resources is not in conformity with the blending philosophy of honest mechanisms, modules, and methods. This is why Professor Aaron Hare of the University of Kalakuta opined, "One of the greatest undoings of man both to himself and Mother Nature is the invention and introduction of soil engineering." Aaron further stated that a look at what is considered the soil reveals that soil engineers make a mockery of their inventions simply because the soil is naturally and monumentally engineering. He stated that its natural engineering is responsible for the growth of the crops, herbs, and leaves, including other micro—and macro-organisms.

A look at the nature of this chapter, the explanations, and all that is conceptualised therein will let one appreciate that the chaotic nature by which material engineering has failed is as a result of the lack of balanced knowledge, which can be derived if man is determined to assess the rudiments of nature's engineering techniques. That is why nature's mechanisms, modules, and methods are not subjected to changes or become obsolete, and why they are not meant to obey the features of material principles, policies, and practises. It is here that engineering at any point in time is naturally looking for a monumental lubricant that will help it to function as a universal galvanizer, which all practical engineers will always contact, consult, and conceive. They need that contact in order to analyse and appreciate the determined functionalities of engineering mechanisms, modules, and methods, which can only be possible when philosophical mathematics is invited naturally to determine the standards, performance, and concepts, including the validity of engineering.

With the use and application of all this information, there is naturally going to be a driving force that will load, offload, and download all the systems of engineering. That action makes this sector a willing and humble servant and an art that humans use in creating features, favours, and wealth, which is important for the objective sustenance of the one-world engineering family.

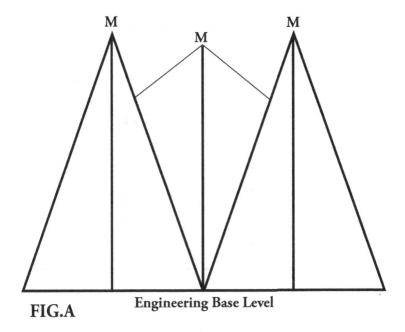

FIG.A   **Engineering Base Level**

A practical study of the drawing above shows $M = M = M$ with a standard base level, which shows that $M^1 = M^2 = M^3$. That result can best known as the structure of material construction. We must note that the function of the solid base level is to achieve balanced concepts with standard structures. Engineering must work in an acceptable, harmonized base level.

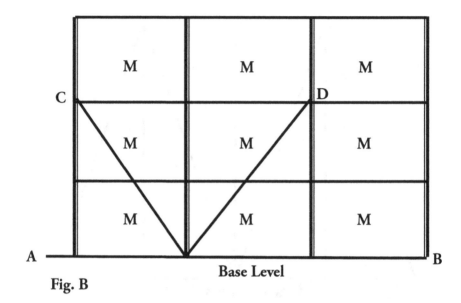

Fig. B

A look at the drawing above shows a great engineering foundation that will serve posterity as a technological foundry. It equally shows the applicability of $M^1 = M^2 = M^3$.

CHAPTER 8

How Does Philosophical Mathematics Determine the Standard Performance of Engineering and Material Endeavours?

Engineering as a science, an art, an occupation, an economic prowess, and a certified human endeavour cannot synchronize with the objectivity of mathematics if both are not intertwined in the use and application of concepts, standards, modules, and mechanics that conform with the dialects of engineering performance. A look at the rhythm and tedium of philosophical mathematics portrays a practical and scientific enabling with its engineering argument and its mathematics—particularly that which confirms mathematical performance as an engineering standard. That is why engineering mathematics is lovingly and highly driven by the standard and anatomy of philosophical mathematics, philosophical logic, aesthetics, and architectural performances and designs. The rudiments of material engineering are always at the proximity of standards and concepts.

At this point, the rubrics in this diversified logic explain that philosophical mathematics determines the standard performance of engineering. The questions that raises are as follows.

- How can this be possible?

- What possibility is this reality aimed to achieve?

- What can be defined as the engineering derivatives of these standards?

The science and technicalities of instrumentation are in a better position to explain this logic, to appreciate the relatively of this data, and to understand the concept of this material standard, which is considered to be enabling engineering. Apart from instrumentation, engineering drawings with designs, and material cycling and recycling (particularly in the areas of polymer and textile) are naturally structured in appreciating the technicalities of this engineering sequence. Sequentially, standards work with the use and application of balanced data, objective statistics, logical performance, and engineering mathematics, particularly the use of metres, millimetres, centimetres, kilograms, tonnage, and other functional mathematical engineering tools that cannot be neglected or rejected.

The compound performance of engineering in the use and application of philosophical mathematics is structured to galvanize the system and synergy of standard performance in actual and non-actual performances. Those performances include modification, modules, mechanisms, and information technology, which are nurtured and natural in the use and application of frequency modification, frequency mechanism, and frequency methods. This is why philosophical mathematics comes in the coordination and cooperation of the concepts of engineering mathematics with a standard performance.

In the illumed words of Ourslo, who is known as the best frequency moderator and modulator, the sector of information and information communication technology has never had standards that do not contravene its concepts and methods. But it works to control the modality of the frequency. Philosophical mathematics, as a roadmap to the actualization of the vision of standard performance, works with engineering oscillation to provide a controlled system with modest, technical rudiments. All these functionalities are achievable by the use and application of philosophical mathematics with engineering standards and performance.

A lot has been said about performance, particularly in respect to engineering as a human endeavour, and it is important to appreciate the relativity of performance with philosophical mathematics in order to determine what can be defined as the intertwined scientific principles, which coordinate them in order to function as that aggregate, standard performance. Just as an engine works on the structure and anatomy of its standard performance, motion, emotion, locomotion, and frequencies (which work with movement) are structured and anatomized on the standard of performance. Without performance, there will be no action, no reaction, no creation, and no construction—including the ability to thrive and retrieve.

For example, an aircraft going from Nigeria to the United States will certainly perform on an aggregate standard, which must be in conformity with the rules and regulations of space science, aeronautic engineering, and frequency regulation and modulation, including ensuring that the rules and regulations of the engineering performances are strictly adhered to. In this respect, all the crewmembers and passenger must cooperate with the rules and regulations, which will help the journey to be performed on the engineering aspects of the aeronautic standard.

This concept is applicable to all aspects of human endeavour, and the issue of using a journey from Nigeria to America is only to establish a case study that will actually examine and explain the engineering coordination of performance and standard.

Let's get back to the question of the chapter: How does philosophical mathematics determine the workability of engineering standard and performance? The following ideas provide the answer.

- Philosophical mathematics is a balanced mystery of science, engineering, and technology.

- Engineering cannot function without the use and application of philosophical mathematics.

- Every road has a map, and every map is created by the use and application of constructive performance.

- Philosophical mathematics is a databank and engineering kit whose parts and instruments do not fail to actualize the vision of this creation.

- Philosophical mathematics is working to create an engineering anatomy and a universal standard laboratory.

- It is examining the data and conceptualising it in science, in order to actualize the vision of a practical globalization with scientific cosmogony.

- Philosophical mathematics is an institution that works to reduce a lot of destruction in the creative concept of engineering.

- It is known and accepted as a sequential and correlated analyst.

With these points in mind, we can understand how philosophical mathematics drive and determine the standards and concepts of engineering mathematics, engineering methods, modules, and mechanics. That includes the practical utilisation of logical mathematics with statistics, which utilises engineering at all points in time in order to arrive at a practical creativity.

The way and manner engineers (particularly in the third world) go about the profession of engineering prompted me to look at the enabled environment and the circumstance that is making our engineers produce substandard materials which do not conform to the world standard and which are responsible for their non-performance.

The other question here is, what is the way forward? The following answers suffice.

- Technology is not transferable; it cannot be stolen. This is why any attempt to mystify or demystify its nature will certainly work against the environment and culture, which are trying to modify the originality of its anatomy.

- It is the symbol and logo of the people, including being the pride of the nation.

- It is the economic forte and foundation that determines the workability of a nation's economy.

- Technology is the Rosé vine tonic that is responsible for either uplifting the standard of the people to be technologically driven or marring their aspiration if they are bubonic in the objective reality of technological prowess.

In asserting the philosophy of this configuration (including its mathematical performance), philosophical mathematics is of the opinion that objective standards must not be mimicked or destroyed. Philosophical mathematics must always be utilised as a natural high-tech PowerPoint that will enable engineering to accept the service of utilising standards to achieve, create, satisfy, and perform honest mechanical and material works, which are important and are needed daily. This chapter can best be defined as a standard performance of engineering rendezvous, which works with the scientific detects and technological directives of philosophical mathematics.

CHAPTER 9

How Can We Explain the Engineering and Philosophical Mathematics of This Work?

It must be stated that the purpose, the philosophy, and the realism of a work of this nature is to develop new concepts that will help humanity in appreciating that life is not stagnant or porous, but is all about being objective, pragmatic, realistic, and dynamic. In the best opinion of dynamic nature, philosophical mathematics does not work to thwart, contradict, or negate existing modules, models, methods, and concepts. Instead, it works to enhance the workability of these intertwined relatives. That is why every human endeavour must appreciate that engineering is all about creating and constructing an economic enabling. By economic enabling, the universe can be opened to access other horizontal concepts that will help humanity in furthering the course of evolutionary concepts.

Unfortunately, humanity appears to be deadlocked in the best understanding, utilisation, and assessment of engineering principles, policies, and practises. The high level of enmity that has become part of human transaction is eroding the occupation of positive agents. This situation must be addressed if humanity is to survive a chaotic sea.

Logically, every human being is created, ornamented, and gifted to be symbolic with the use and application of mathematical prowess. But as stated in an earlier chapter, it is unfortunate that humanity misjudged and misconceived the work of symbolic engineering mathematics. This situation can best be known and accepted as the earlier incubus that worked to reduce intellectualism, morals, and ethics. The conjugal and aesthetic natures by which the intertwined relativity of philosophical mathematics and engineering technology can be determined were practically negated.

At this point, humanity—particularly the Microsoft community—appears to be confused with a lot of wasted and rustic ideas. This is why Microsoft engineering cannot boast having practical and purpose-driven principles, policies, and practises.

If Professor Owen of the Kalakuta University in India were alive to experience what is happening now, this ingenious but universal thinker would have beaten his chest simply because any act or craft that does not exist on an acceptable, universal principle and policy cannot withstand the acid test of Mother Nature.

A lot of people may ask, *Can Mother Nature have an acid test that can determine the workability of engineering mathematics, technology, and logic?* The answer from an objective and calculative point of view is simple, because nature with its motherhood is fathomed and dependent on the symbolism of ethical moderation, ethical principles, ethical policies, ethical practises, ethical relations, and ethical transactions. That is why engineering craft with mathematical moderation must adhere strictly to the rules and regulations of this act. In this context, we can explain with scientific appreciation that the following points can be utilised in addressing most of the confused issues in engineering.

- Appreciation of the imports and importance of research methodology.

- Understanding the importance of standard, which can best be defined as an engineering instrumentation.

- Being able to study the environment in which such crafts or designs are made.

- Developing an acceptable framework that will be utilised in ensuring the objective framework, which will be utilised in ensuring that the objective locus of the standard, is not neglected.

- Ensuring that the use and application of engineering equivalence, particularly in constitution and craftwork, are not neglected.

- Studying the problem of the future through the use and application of engineering forecast, which must be purpose-driven with the use and application of data and statistics. That is why construction and production technology are practically driven with the use and application of equivalence.

- Engineering factors that must have the organic and inorganic matter of mathematical dynamism must not be neglected.

- The use and application of philosophical engineering, which can best be defined as an organized trade to the workability of engineering technology and anatomy, must be taken seriously as a structural instrumentation.

- The use of mechanics, models, and methods must equally conform to the policies and principles that determine the workability of engineering.

- All that is conceptualised in the structure of engineering logic with mathematical prowess must be used in ensuring that errors are minimized.

It is in the wisdom of these points that engineering involves statistics and data both as a recorder and collector. A look at what can be defined in this context as a mathematical ledger, which directs the structural aspect of engineering technicality, must be appreciated and understood as a factory of fathoms into which all concepts and ideas must be rotated and revolved.

This analogy can be represented with a constant engineering mark that reads $X = A + B = X = X + X + X$. That means that X as a letter maintains a constancy and can be used in directing and narrating the functionalities of their fractions and factors.

The urgent necessity of explaining the dynamics of these details are borne out of the fact that engineering technology is greatly looking for a holistic approach towards solving its problems. This is the aim behind the conception of a monumental work of this nature. In this respect, if mathematics is involved in all aspects of human endeavour, then philosophy is working towards achieving the purpose of purity and perfection. It could be argued that the world of knowledge is in dire need of a work of this nature, which has expanded all the lubrics of science, all the monuments of engineering, all the functions of technology, and all the rudiments of philosophy.

Delving to explain the anatomy of philosophical mathematics is a scorecard. The purpose of this chapter is to provide solutions on how most of the engineering problems and their technological lapses can be solved through the use and application of moderate and mathematical philosophical thinking. Mathematical philosophy's enabling functions will remain to sort out all the scientific deluge that has made the province of creative sector, which is known as engineering technology, to remain a coastal terrain that is so difficult and so inaccessible.

It is here that I must state that the dynamic concept of philosophical mathematics is the dynamic concept of engineering technology, engineering science, and engineering philosophy. This brings to mind a maxim: Creation without the use and application of engineering policies and principles will certainly negate the constructive techniques of Mother Nature.

CHAPTER 10

How to Use the Thinking of Philosophical Mathematics in Determining the Concept of Engineering Creativity

Engineering creativity, which extended from the natural anatomy of balanced engineering production, is about the use and application of quantity and quantum arts with actions, postulations, principles, policies, and practises that are in conformity with the philosophical mathematical rubrics of engineering modules, mechanisms, and models. It can equally be defined as a structured and scientific galvanizer. It must be stated that science neglected, rejected, and separated its functions from the illumined functionality of philosophical mathematics. That is why we are suffering today from what can be defined as monumental, scientific errors. These errors are the origin and synopsis of our present chaotic system.

Engineering modules and mechanics are always in synchronous agreement with the laws of constancy and permanence. Since Isaac Newton originated and postulated the gravitational laws, the institutions of science and the colleges of engineering technology have always understood that gravity is nothing but the technical and skilful omnipotence of natural engineering action. In this respect, philosophical mathematics is in agreement with all data of creative and constructive technology. This agreement is responsible for the study of the following.

- The philosophy of engineering.

- The philosophy of science.

- The philosophy of technology.

- The philosophy of mechanism and mechanics.

- The philosophy of techniques of modules and quarts, including quantum engineering dynamism.

- The philosophy of mathematics.

- The philosophy of physics.

- The scientific and mathematical philosophy of astro—science, including what is added in this content as the philosophy and mathematics of ingenious engineering biology (IEB).

Ingenious engineering biology is mathematically treated in most universities of the world as biomass, biochemistry, and bio-polymer, including all that is configured and structured in the engineering website of chemical and production engineering technology. We cannot sideline the titanic functionalities of instrumentation engineering;

electrical-electronics; environmental, mechanical, civil, and structural engineering; and what Leonardo da Vinci designated as the craft creation of man, which fathomed creation as an architectural representation of an ingenious architect. Today da Vinci's concept is being studied in the universities as architectural engineering. Other areas of human creative engineering include:

- Surveying.

- Estate.

- Building technology.

- Oceanography.

- Automobile.

- Petroleum and petrochemical.

- Agricultural engineering with animal husbandry.

Soil engineering, with all the contents of solid mineral technology, always bounces and rebounds on the permanent and constant laws of philosophical mathematics, which enable man to translate, transform, and transmute his original pattern and anatomy in order to give a material form that will further give it a commercial form, which will finally make it accessible and consumable for human sustenance in his gifted ecosystem.

This overview will certainly recapture the simultaneous and ingenious definition that defines engineering and technology as a science, as an art, as an endeavour, and as a human occupation, including producing what is acceptable universally as an illumined definition of who an engineer is. The dynamic concept of philosophical mathematics is actually patterned to serve the academic province as a province of illumined philosophical structure with mathematical enabling.

The ways and manner the creations of man are stunting his growth requires that man should make honest efforts in creating arts and crafts that will make the system be ingenious with a lot of diversified engineering structures. It is here that this work rejects and neglects all arts and facts that reduce and cause him to remain an ephemeral structure of a doctor of science.

The new part—which will deal with introduction to philosophical mathematics as a balanced anatomy of a living knowledge dynamically explained with philosophical explanation—will certainly make one appreciate that philosophy and mathematics are naturally ornamented as perfect oracles to other human endeavours. That is why engineering is always purpose-driven with practical and mathematical enabling, with philosophical ennoblement. For the record, this overview cannot be accepted as a trailblazer, but it could be utilised as a millicent of crystal balance. I hope to create a museum of knowledge for the balanced wisdom of posterity, which is naturally framed and foremost in the wisdom of truth, which is the highest knowledge that engineering as a science, engineering as technology, engineering as a creative, and productive art are monumental, matrixed, and mystified naturally and practically in the views of philosophical mathematics.

The balanced dynamism that originated this overview will help one-world engineering technology to appreciate that man is still at the kindergarten stage of understanding what is already established in the nectar and annals of nature's engineering workshop. That workshop is willing to hand over to all desiring apostolate who are also willing to use these institutions and instruments in creating an era with permanent age, which will extol the unequalled wisdom of the dynamic concepts of philosophical mathematics.

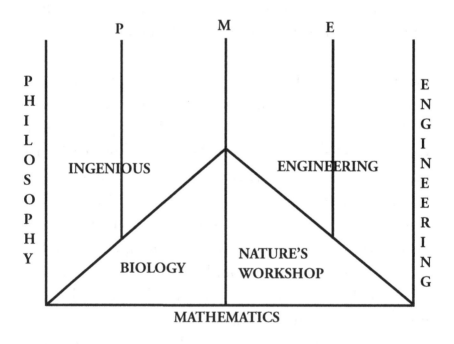

The above drawing shows how Mother Nature, which is the fulcrum of the philosophical mathematical engineering incubator, does her arts and crafts with the use and application of natural enabling, which do not contradict or contravene each other. This is a lesson of truth that humanity must know with great utilisation.

Addendum to Part 9

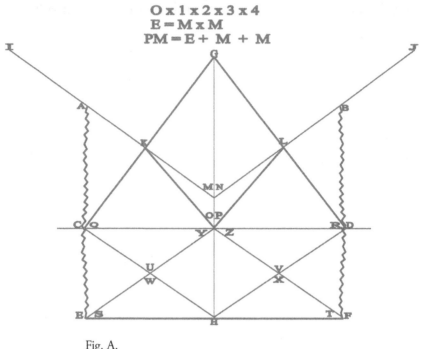

Fig. A.

In part 9, which dwells greatly on the philosophical mathematics as the engineering module, modem, and mechanism of balance and constructive creativity, the drawing (which is galvanized in a structural, mathematical engineering concentration, with cumulative and mathematical construction) reveals the nature, the purpose, and the structure of engineering as being bedded from the original philosophical mathematical anatomy. That anatomy is symbolically exhibited as aggregate mathematical features, facts, findings, and foundries that, when utilised in its objective enabling, is capable of appreciating the technicality of downloading engineering issues from the original creativity of philosophical mathematics.

To further buttress with scientific and engineering technicality the module and modem of this drawing, constructive and engineering logic can be utilised to actualize the vastness of its symbolism. This is why the drawing, with its unending designs and labelling, can be naturally utilised in creating a thematic and diversified engineering foundry. That foundry can be utilised as a nuclear and nucleus of scientific dynamism with technological ingenuity.

I define this drawing as a caster of engineering and mathematical creativity, which is broadened with the ingenious fittings of philosophy. That is why this drawing can be abstractly proved, disproved, qualified, mystified, and demystified—all depending on the objective engineering cradle that the sailing mathematical philosopher is enabling to sail. But with the honest thinking of mathematics, 0 is a factor, 1 is a foundry, 2 is a form, 3 is a fitting, and 4 can be fortified to remain as a mathematical font. This is why the calculative mechanism can start with $1 \times 2 \times 3 \times 4$, which is practically harnessed as a mathematical engineering modem in modules, and modules in modem, which can practically be represented with

$$PM = E + M + M$$

Conversely, the alphabetical connectivity of these letters was sequential in the originality of the mathematical thoughts, which gave forms and foundry to the original thoughts of engineering. That is why the eminent scholars who started this active and meta-materialistic construction must always be remembered for possessing and inspiring galvanized souls in the best objectivity of creative engineering modules, which further gave rise to modems, which globally galvanized the utilisation of the WWW.

Philosophical mathematical proof of the scientific diagram starts with

IGJ = CYD = EEF
AC = CE = EH = EF = FD = DB
BD = DF = FH = HE = EC = CA

Therefore IAKM = JLN, and ΔGKM = ΔGLN with

KMO = LNP and KM = LN
with NP = MO = OP = MN = YZ

Therefore CQ = YZ = RD with
ΔQUY = ΔUSH = ΔHZX = ΔZVD = ΔDTV
= VHT = €s CEFD produced €s CYHE and DFHZ.

Therefore G = MN = OP = YZ = H
With GRC = CYD = DLG
Therefore C = UW = H = D = VX = H

Therefore, ACE = MYH = BDF in scientific mathematical realism. This diagram is capable of creating a lot of philosophical mathematical engineering foundries.

At this point, the concept of philosophical mathematics as the engineering modules, the modem and model of balanced constructive creativity, understands the universal postulation of scientific cosmogony, which has remained a dazzling issue within the rubrics and annals of practical science, foundry mechanism, and engineering and technological concepts. This is why the drawing, which is figuratively and mathematically fontious, is donating some multifarious angles, triangles, squares, and geometrics with algebraic equations. A wider donation to calculus is capable of galvanizing a foundry mechanics, which is the urgent necessity of philosophical mathematics.

In the original understanding of the author of this work, philosophical mathematics can best be defined as a scientific and practical modem of engineering and technology that naturally fathomed the fontious ornamentation of generalized science. That science utilises the compartmental ingenuity of Mother Nature in the sun, the moon, the stars, the oceans, the mountains, the forests, and the aquatic and geophysical structures of our ecosystem in producing its engineering creativities, which are naturally and structurally galvanized in the objective and scientific alphabets of $0 \times 1 \times 2 \times 3 \times 4 = E = M \times M$, which under the mathematical philosophical radius quantifies the ingenuity of practical galvanization with the objective and dynamic postulation and presentation of

PM + E + M + M to un-calculative, infinitude mathematical methods.

At this point, no other thing is preponderantly at work than the energy of Mother Nature, which originally structured the functional and balanced anatomy of the EMM.

PART 10

INTRODUCTION TO PHILOSOPHICAL MATHEMATICS AS THE BALANCED ANATOMY OF LIVING KNOWLEDGE DYNAMICALLY EXAMINED WITH PHILOSOPHICAL EXPLANATION

"When knowledge is electrifyingly balanced, it creates wealth and produces wisdom."

"A man's gift maketh a way for him, and this is only possible when that gift is utilised in presenting man as a true vessel."

"Honest life consists of penetrating and discovering the unknown and fashioning our actions in accordance with the new knowledge that is acquired."

"Only honest knowledge enables us to be ennobled in the absolute wisdom and knowledge of balanced ennoblement."

—All quotes derived from the author's balanced and electrifying bank of inspiration

In introducing philosophical mathematics as a balanced anatomy of acquiring knowledge, it must be stated that knowledge is power—it is creative, dynamically enabling, and ennobling. That is why philosophical mathematics is in agreement with examining what can be defined as the anatomy of a living knowledge. The contributions and achievements of philosophical mathematics, which is relative and intertwined to all areas of human endeavour, has not been well harnessed, utilised, and understood. This issue is one of the foremost reasons that has made scientific principles, policies, and practises impermanent in performing the functions naturally assigned to them.

At this point, we must appreciate that as knowledge is power, its anatomy needs to be examined and explained with the use and application of philosophical mathematical dynamism. The institution of knowledge was instrumental to the discovery of mathematics as an Oriental and occidental achiever, and of philosophy as a scientific, religious, theosophical, Oriental, apostolic, and apostolate foundation of great arts. It is in the tedium of this introduction to balance the thinking of philosophical mathematics with the will and opinion of dynamic knowledge. In this respect, we must define the anatomy of a living knowledge as the bridge and Bond, and the mathematical concept of knowing in the absolute knowingness of all great and ingenious arts. This is the simultaneous work of charity that philosophical mathematics means to achieve by the use and application of aesthetic perfection with mystified purity.

We must appreciate that the Microsoft system, which is driving the Microsoft cell of the world, is not in reality a product of Microsoft anatomy but a sanctified lenninger that was materially restructured by the universe of the uninformed. The uninformed's quickest intention is to amass material wealth at the expense of balanced and dependable knowledge; this can be summed up as another version and factor of human melancholy. That is why their material ingenuity has only pegged them at the hands of WWW which has greatly blocked them from knowing the mighty ocean of knowledge contained in the ominous and omnibus UWW.

There is absolute, scientific excellence and philosophical ingenuity regarding the wishes of Mother Nature, the wisdom of knowledge, and the anatomy and ingenuity of dynamic philosophical mathematics. That is why this part of the book is principled with purpose-driven facts in defining balanced and electrifying knowledge in the rubrics of its origination. If we are living as showcased by material science in a dynamic system, why is this system producing obscured and absurd concepts, machines, equipment, and ideas? Why is this dynamic system always being propelled with the use and application of short-range planning, which is driven with the application of minimized and obnoxious research methods? The answer is that humanity has never followed nature's processes with ingenious production, which perfectly works with the use of long-range planning.

It is naturally believed that the incubation of an embryo in the womb at a permanent period of nine months or beyond (which is at the proximity of nature's protection) is beyond nine million years, when we start to analyse what one second or minute or hour, or even a day or week, is before the almighty, ingenious philosophical mathematical knower.

At this point, a work in this province of knowledge must be seen as a tedious and challenging gardening whose gardens must be lintious in creation. Judging from the pattern and concept we have followed, the driving force and the concept of philosophical mathematics reveal that new ideas are always bona fide gifts to new eras. That is why this ingenious work is not done with the wisdom of the ordinary mental intellectualism.

It is important to explain that dynamic knowledge is a balanced actor in action, and it is equally a tender philosophy. When intertwined with the blending factors of mathematics, it is always an astute assurance that knowledge is wealth, power, and wisdom. It is an electrifying, honest education and a light that lights others. We must appreciate that every creation is fathomed on an existing anatomy, which is already created for creation so that illumined creatures can translate them into realities, actions, and benedictions, and into philosophical utilisation with the power of detailed mathematical engineering. Thus, this introduction can best be defined as the philosophy and mathematics of the anatomy of a living knowledge that is dynamic, objective, and inspiring with monumental engineering.

All and sundry are hereby enjoined to preserve knowledge as a hope of human sustenance with practical continuity, which is in synchronous agreement with the wishes and caprices of nature's balanced rhythm. That rhythm is sequential and synchronous with the illumined laws that govern, protect, transmit, translate, and transmute dynamic knowledge, which is best defined as an organized and ingenious alchemy of all ages.

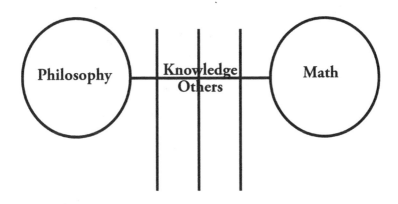

The drawing above shows the anatomy of knowledge and how it warehouses mathematics and philosophy in order for the two to function as one. It equally shows that knowledge is an incubator to the works of mathematics and philosophy. It is here that nature provides the driving force that does its works.

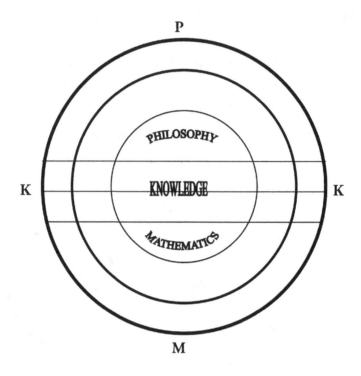

KP   =   PK   =   MK

$$\frac{KP}{KM} = \frac{PK}{MK}$$

∴ K = P = M = KPM

The above drawing shows the intertwined and electrifying relativity between mathematics and philosophy by balanced knowledge with the objective and dynamic forces of Mother Nature, which warehouses the two in its incubator as a futuristic occupation.

CHAPTER 1

WHAT A LIVING KNOWLEDGE IS, DYNAMICALLY EXAMINED

Since the dawn of consciousness, evolution has systematically invested greatly and honourably in material knowledge, which has been part of human life. This high investment in material knowledge has greatly mortified illumined knowledge. That is why human civilization, evolution, science, and technology (including his purported knowledge of philosophy) is stagnated. Introduction to philosophical mathematics, as the balanced anatomy and the sensitive and living knowledge of all hallmarks of wisdom, explains what can best be defined and understood as living knowledge. In this respect, knowledge can be defined as an acquired experience, an acquired power, and a sensitive and sentimental understanding of facts, figures, and factual forces. This is why it is greatly stated that the power of every human being lies in how much honest and dynamic knowledge he has acquired, which in turn will help him to become a teacher, a noble and ingenious thinker. That is why it is clearly stated that noble living knowledge is creative.

In my illumined and inspired understanding, a living, dynamic knowledge is best defined as the quantum objective inspiration, which is translated into absolute reality in order to gain man the best and objective ennoblement of truth. In this respect, knowledge is the manifestation of balanced and electrifying truth. It is the utilisation including the fulcrum of appreciation of eternal factors and facts, which are in harmonious and monumental synchronization with what is considered honest reality.

It is here that the best way by which a living knowledge can be defined is: what is beyond material creation? That which is beyond material creation is living knowledge, which is purposeful, propelling, purpose-driven, and understanding with a high level of monumental impacts in anchoring constructive and balanced creativity. A look at the tedium of living knowledge, which was at the proximity of the wisdom of the Oriental wizards, reveals that our present civilization is still stealing, mimicking, and wondering about how they can derive from what the Orientals acquired with mutual and honest cooperation with Mother Nature, which is an evidence to show that Mother Nature is the compound and fulcrum of living knowledge. That is why the province of philosophical mathematics, its occupation, and the function (including what can be defined and determined as its electrifying anatomy) is practical at the philosophical orchestra of living knowledge. It is important to appreciate that the universe is momentarily controlled by the forces and powers of Microsoft knowledge, which in earnest is not permanent, constant, and dependable. That is why what can be considered as the principles, policies, practises, and intertwined workability of the Microsoft system is not monumental with absolute guarantee.

Dr. Walter Russell, who examined the contributions of material science. His words are incorporated in the symbolism of the following definition. The greatest error of science lies in locking the Creator out of its creations. This error is responsible for the newest machines that are being produced today, which easily become obsolete, obscure, wasteful, and condemned simply because the knowledge that created them is not in the absolute reality with living knowledge. In this respect, what can be understood to be the anatomy of a living knowledge is best designated as the reality of the universal One, the universal existence, the universal operation and cooperation, and the mathematics that form the bedrock, including the sedimentation of transcendental science.

It is unfortunate to observe that humanity is not in the tedium of the utilisation of what is given to creation as a living knowledge. This is clearly seen in the way and manner science underscores the anatomy and philosophy of hydrogen, carbon dioxide, and carbon monoxide, including all that is contained in the carbon family. The way material science treats the knowledge of solid minerals—which include kaolin, sodium sulphate, limestone, and other important gifts of Mother Nature—is an eloquent testimony to show that our knowledge of these gifts are not only myopic but erroneously primitive without learning to appreciate what can be defined as the condensing factors of their original anatomy with sedimentology.

A look at what can be defined as a living knowledge that is free for us to appreciate will certainly delve into asking such questions as how much and to what extent humanity has advanced in the knowledge of the following areas.

- The firmaments with their celestial galaxies.

- The illumined forests with their aesthetic beauty.

- The deserts with their uncountable dominance of sand, dust, and other particles that, when appreciated, can be useful to humanity.

- The mighty oceans with their extensive, oceanographic rules, which expand their volume to form rivers, lagoons, and other constituencies.

- The big mountains, which are known to be part and parcel of creation. Unfortunately, the knowledge of what we can utilise them for is not yet harnessed.

- The knowledge of the solar system, which comprises the sun, the moon, and what is contained in this text as celestial and occult ignited lights, is not yet known.

- Unfortunately, man's understanding of astrology, astronomy, and space science with engineering appears to expose his idiosyncrasies of these wonders beyond explanation and scientific examination.

A work to examine the balanced harmony of philosophical mathematics with a compact, tactful, living knowledge must certainly exist as a pragmatic resource concept that will enhance and understand what can best be defined as the philosophical examination of its dynamism. That is why such topics as the demystification of mathematics were treated with the use and application of a living knowledge and dependable wisdom with the use and application of scientifically and technically intertwined concepts. Therefore the definition—including the ornament that appreciates the concept of living knowledge as the roadmap to honest and dependable understanding of man's evolution and civilization—becomes the urgent necessity, including acting as a practical and philosophical eulogy by which the appreciation of what is defined as a living knowledge can, with its mathematical synchronization, be philosophically gained with the use of honest dynamism, which is the pragmatic and practical reality that makes the science and philosophy of living knowledge the occupation of the well informed.

CHAPTER 2

THE ORIGIN OF LIVING KNOWLEDGE, PHILOSOPHICALLY EXAMINED

When I say "living knowledge", I mean wisdom beyond words and thoughts, knowledge that is transcendentally illumined and quantified in absolute and supreme mysticism. That is why it cannot originate from any other thing than that which is perfected in the anatomy of the holy alpha and omega. The supremacy of his omnipotence, omniscience, and omnipresence is best defined as the knowledge of his ingenuous fluorescence. Material, academic, and intellectual knowledge can best be defined as an iota of what a living knowledge is all about. That is why the occupation of the Twelve World Teachers, which was anatomized in the perfection and reality of living knowledge, has remained the glorious Koran, the monumental living Bible, the wisdom of all sermons, and the sacred wisdom of all ages, including anointing and driving the knowledge of universal culture in that tender gospel and celestial knowledge.

A lot of people are not developed in appreciating that philosophy is the maker of great thinkers, of the Oriental wizards and the apostolic function of ingenuous knowledge with the use and application of celestial mathematics. This is why the concept of donating this work as a Britannica to posterity and our present Microsoft-driven age was conceived as a thoughtful and purposeful project whose activities, productions, and extensions would certainly employ the ingenuity of living knowledge—which originated from the holy sea of all seas and from the holy emengard of all apostolate, which has a lot of names from different cultures and traditions. That is why Buddha, Zeus, and the illumined and unequalled Christ anchored their strength on the power that originated the living knowledge.

When philosophy is showcased and ingeniously keyboarded in maintaining its lintious and locus standard in achieving purity and perfection, then it is explaining the preponderance of philosophical mathematical engineering, which is quantious and quantified in the absolute and pragmatic hall of living knowledge. That is why to make humanity appreciate the living knowledge is God, including being the way of the absolute holy sea. Every alphabet in creation starts with "A", which represents alpha. The concept of omega, which is another great and immortal name of God, equally has its last letter as "a". This expresses a great wisdom that only in that consummate alpha is the omega of living knowledge.

To explain the wisdom of living knowledge is as good as explaining the origin of man, the origin of life, and the origin of spirit, mind, soul, and body. It is this transcendental science that has the power and capacity of explaining other branches of living knowledge, which include:

- Transcendental wisdom.

- Transcendental science.

- Inspirational philosophy.

- Monumental technology.

- Transcendental engineering.

- Compact and inspirational meta-mathematics and mathematics.

- Transcendental physics, which can best be defined as the locus being answerable to all visible controlling concepts, and to all configured and conceptualised in the absolute and dynamic anatomy of living knowledge.

When it is stated that knowledge is power and the mark of a wise man, many people do not understand this maxim, including the eulogy that considered it as a permanent and perpetual epigram. It is here that I designate and immortalize the anatomy of living knowledge in the following inspired donations, which the province of philosophical mathematics will continue to use as a way and means of regulating the concept of Microsoft province of other knowledge whose functions, principles, and policies are contemporary without being conventional.

- Living knowledge is the anatomy of truth.

- It is the immortal power of all things both seen and unseen, particularly those that are permanently created for the works of absolute permanency.

- It is the living occupation of Mother Nature. This is why natural science and natural law (with living philosophy) are always in synchronous and balanced knowledge with the wider and universal wisdom of living knowledge.

- It is a living power that empowers the functions, the acts, and the creations, including the monumental achievement of light.

- It is the mission and the vision, including being the philosophy of the universal One.

- Living knowledge is the only authority that answers that aging and challenging question of what is beyond truth.

- It is the celestial Britannica that answers the glorious question of: What is the grail? Is there an alpha? Are we the product of the Holy Grail?

- Living knowledge is the authority that great and ingenuous scientists use in explaining content and the concepts, including the dynamism of occult engineering technology.

- It is the meta-mathematician of all ages, the meta-philosopher, that empowered the Oriental wizards.

- It is all about "Know ye whom you are", simply because man's knowledge of himself has remained the greatest science that he has not been able to prove, even in his laboratories and in his libraries. That is why living knowledge has remained the immortal occupation of inspiration with transcendental wisdom.

When it is stated that *the universe is still a newborn baby with the heart of a new heaven,* humanity is still at sea in the understanding of the statement. Man, in his evolution with science, will certainly appreciate the wisdom that is contained in living knowledge, right from the day all souls and minds turn honestly and transcendentally to seek, to ask, to knock, to appreciate, to investigate, and to understand the Creator. The Creator is hereby defined as the authority that drives the concept of philosophical mathematics, including being in the becoming and reality of living knowledge that he is the origin, including being the best excel in its excellence.

Consequently, no authority can mortify, disorganize, condemn, or pollute the power of living knowledge, including its authority, which is hereby defined as the pinnacle and podium of all electrification of other ideas. The best definition of the origin of living knowledge is the original in consummate originality, who always appear to the elected in that ingenious rock that he represents as a divine Iliad.

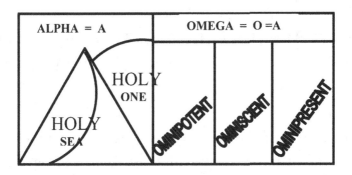

The drawing above shows the origin of knowledge and the perfect source of all things. This is best known as the wisdom of one who is known as the authority of all knowledge with absolute oneness.

CHAPTER 3

What Are the Parts of Living Knowledge and Their Functions?

Having explained the origin of knowledge, it is important to further the inspired investigation on what can be accepted to constitute its diversified and monumental parts with absolute achievements. A lot of people can only stop at what knowledge is, without knowing that knowledge is beyond our present understanding of its composition, its philosophy, and its engineering with practical technology. It must be explained that knowledge comprises all that is configured in material and non-material science, philosophy with its transcendental homogeneity, metaphysics with its consummate extensions, psychic science, astral science, and technology, celestial mathematics, and mathematical inspiration (which we are being urged to adopt and involve as a business necessity). Knowledge also contains all that is in chemistry, which includes organic and inorganic compounds; physics, which involves organic and inorganic; and terrestrial physics. Philosophy, which is omnibus, is beyond our present understanding and is equally inclusive in the anatomy of living knowledge.

In this respect, our concern and consideration is going to focus on transcendental knowledge, which includes transcendental science, transcendental philosophy, transcendental metaphysics, and meta-mathematics. This is why a work to demystify the engineering and technology of the occupation of occult science is all about investigating beyond our present knowledge to what is designated and defined as the unknown hard sciences. When it is stated that the *unseen governs the seen*, material science in particular has permanently locked and lacked the courage and the mathematical engineering in investigating what is contained in the anatomy of the unseen sciences. It is here that metaphysics, which is essential in the investigation of the unseen physics with unseen chemistry, has always considered material science as a beginner.

A look at the compound and universal workings of infinitude mathematics (with its philosophical kit) has logically come to disapprove the existence of QED simply because the existence of QED can best be defined as an inglorious myopism in the province of mathematics that was investigated and orchestrated by mathematical Lilliputians—who never knew that mathematics, philosophy, technology, and engineering (and other blending sciences) exist and function at the cost effectiveness of the infinitude concept with mathematical and philosophical dynamism. In that respect, this chapter is going to reveal how and why living knowledge functions with the illumined wisdom of transcendental philosophy with its illumined science.

It must be stated that when something is transcendental, it becomes pure and exists practically at the celestial province, where creations are made with the authority of perfection and purity. That is why their actions and creations have remained cardinal with crystalline authority. Such creations include man, who is designated as the head of all created. It also includes the sun, the moon, the stars, the seasons, the wind, and the light—all was created from the authority and ingenious VSAT of transcendental knowledge. The authority that is contained in transcendental knowledge is the power of all things, the wisdom of all missions, the philosophy of all gospels, and the cardinal science with its functional technology, which can best be defined as the engineer of all ages.

In the super-genius wisdom of philosophical mathematics, the functions of transcendental knowledge with its philosophy are what man is utilising at the myopic knowledge of material science with its short-sighted technology. That is why scientific errors and glories have permanently lied in locking the Creator out of its creations. What an error! If this error is named a deluge, it is not a mistake. The way and manner that living knowledge (which is all about transcendental activities in the realities of the hallmark of balanced creations that make its activities simple and inspiring) can best be defined as a wakeup call with the greatest strength that humanity could have considered and utilised as a greater power in empowering his own gifted dynamism. To work against this, humanity decided to choose the opposite—mortified and not modified. This can be considered as the beginning of the fall of man, including being placed at his own request to prosper in the ephemeral sector of material science with material technology.

If God is knowledge, and if he is the author of science simply because he is a composite and consummate scientist, then how and why is our knowledge not lintious and luminous in relation and in balancing that which Mother Nature has bequeathed to us? The answer to this question is simple, and it reads thus: Man is not patient with himself. He is not being propelled to be purely patient with his Creator, and his understanding of creation is minimal. This is why he has lost his original concept of what is contained in living knowledge, which is transcendental, monumental, and inspiring. The living knowledge can never be equated or tested in his ephemeral laboratory, which he created out of a lack of knowledge and which includes transcendental wisdom, transcendental science, transcendental philosophy, and transcendental technology.

Thatiswhythemonumentalchallengesthatmeta-mathematics, meta-occult and meta-science have suffered are because man is not patient with the driving and original force of living knowledge, which is the authority that authored him, including being the super-genius esteem that created other things. This esteem is why the work is best defined as the intertwined and electrifying relativity of transcendental knowledge with philosophical mathematics.

CHAPTER 4

WHAT ARE THE FUNCTIONS OF LIVING AND TRANSCENDENTAL KNOWLEDGE?

Living knowledge, which is defined by the world of wisdom as the pinnacle and mother of everything, has not being greatly known, particularly the aspect that deals with the world of purity. The Microsoft province talks about SMS, IT, ICT, and more, but living knowledge is configured and galvanized to explain and propagate the course of purity, perfection, and transcendence. That is why other forms of human endeavour define living knowledge as the anatomy of all millicentia creativity.

If this definition is not true, a noble course like philosophy, in its mathematical driving energy, could have refuted this assignment. Ingenious humans who are hereby defined as the portals and portraits of Oriental wisdom could have equally rejected and neglected this honour. It is here that the woes of material science and material knowledge are not lucidly accepted in the powerful enabling of transcendental knowledge.

At the moment, the only human endeavours or achievements that tries to explain, with logical examination, the functions of living and transcendental knowledge are philosophy and mathematics. That is why the power of the two structured and restructured the enabling understanding of meta-mathematics and meta-philosophy; consequently the furtherance in their achievement was responsible for the lucidious ornamentation of meta-occult transcendence, with its engineering and technological prowess.

When the challenge of writing this book tapped at the ornamented secrets of my thought, which is best defined in the enabling of my ingenuous inspiration, there were a lot of rooted and monumental challenges. It must be stated that the call to awaken and the mission to appreciate knowledge with its compact and diversified wisdom are accepted as the bases for the formulation, the formation, including propagating what Mother Nature, has been incubating at the crystal philosophical anatomy of mathematics, including blending the newness of philosophy as warehousing what can be defined as the functions of this noble course.

At this point, we will examine with balanced explanation what can be considered as the functions of living and transcendental knowledge.

- Ensuring that strong and balanced institutions are created, which work to weaken and disintegrate the negative incursions of strong humans.

- Living and transcendental knowledge is the only garden, or the only ennobled farmer whose responsibility is to produce raw materials for the use of allied industries—which fall under the umbrella of foundry technology, production technology, engineering creativity, philosophical engineering, mathematical production, and statistical and database engineering creativity. This is why the definition as an allied farmer is best appreciated by the world of consummate knowledge.

- Living knowledge performs the function of ennobling humans to be purpose-driven with the use and application of inspiration as a business necessity, and also as a way and means of living an adept and purpose-driven life.

- Transcendental and living knowledge can best be defined as the Christ of the Blue Ridge, the Buddha of the world, the Mahatma Gandhi of the universe, the Rotary of honest thoughts, and the universe made real.

- The world of practical and balanced knowledge utilises it as the monumental perfection of the gospel, which lies in the understanding of the message of the divine Iliad.

- Transcendental knowledge is the super-genius Bill Gates, which comes to ennobled vessels. Those vessels utilise the wisdom therein in creating products that are useful to human life.

In this respect, all that material knowledge has achieved since the awakening of human consciousness is only an iota of what is anatomized in that infinitude province, which warehouses the un-quantifiable and monumental factors, forces, powers, and resources of transcendental knowledge. When Plato told the Grecian and Asian provinces that knowledge is power and the maker of all wise thinkers, he was explaining what is contained in the wisdom of transcendental knowledge. That is why it is ennobled in the best explanation that all powers belong to God, who is designated by the apostolic foundation of the Twelve World Teachers as the consummate universal One who has, without revocation and equivocation, invited us to seek, knock, and ask of him. He also presented the monumental light in the ecstasy of his illumined and ingenuous transcendence to be our guide, our teacher, and our mentor. It is here that the notation of this work is being propelled in presenting to humanity the urgent necessity of using the wisdom of transcendental knowledge with living philosophy.

At this point, I see the transcended nature of philosophical mathematics as a concept of science, an art including being dynamic science, which material science has avoided and neglected. This is why material science—which is an iota of what real, practical, and transcended science is all about—is not worried about investigating what can be defined in this context as the forces and factors of Mother Nature and the factories and fathoms of transcendental mathematics. My work is purposeful in presenting what the future of humanity will be using as enabling resources of engineering creativity.

A look at the intertwined relativity of philosophy and mathematics, including what can be accepted as their practical and universal functions, reveal that because philosophy is copious in achieving success (which is essential in the enabling of purity and perfection), mathematics as a universal and electronic calculator is dynamic in the use and application of philosophical thoughts to ensure that there is a permanent constancy in its principles, policies, and practises, which in conviction and convenience are equally determined to succeed in the best achievement of purity and perfection.

For instance, if $1 + 1 = 2$, $2 \times 1 = 2$, and $3 - 1 = 2$, there is every constant and practical belief that at any point in time, mathematics is always falling in the best understanding of the use and application of philosophical structures, actors, and actions. That is why the academia must readjust in order to appreciate what can best be defined as understanding the existence of transcendental science with the use and application of ingenious living knowledge. A look at the demystification of mathematics by the world of the Oriental apostolate will certainly agree that part of human error practically lies in self-inflicted fears, including accepting falsehood, enmity, and stagnated thoughts as ways and means of advancing his course. The fontious ornament in this context is to reveal that humanity and the whole universe are in dire need of philosophy and transcendental science, with the application of living knowledge.

We must be determined to fall back to our original roots simply because the invitation of Mother Nature, which is the anatomy of dynamic and surrogate living knowledge with its transcendental understanding, reads thus: The one-world family is at sea simply because the sea in which they dwell has sinned and runs short of the glory of the Creator.

Any knowledge that is not ennobled with the millicentia fluorescence of the Creator will certainly create a stunted growth beyond words and thoughts.

This reasoning is why the philosophy and the transcendental colloquium, which is the mathematical and philosophical VSAT that ornamented the epigram of this eulogy, makes all and sundry appreciate, understand, and propagate the course of humanity with the use and application of living knowledge, which is being empowered by the greatest force of transcendental philosophy. The symbolic esteem that will serve as an immortal admonition in comprehending this work is for all creation to think and to create philosophically.

When philosophical mathematics is adopted as an honest, living knowledge, it will certainly appreciate the world as a united family, which is more advanced than a global village. On this note, I make a further case in the honour, protection, and preservation of this wonderful province that living knowledge with transcendental philosophy is the only way and process by which the ennoblement of creation as the Creator's lucid millicent can be granted. The reasoning is because creation has remained a perfect and millicentia asset that is only known and authored by one who is known as a transcendental and living author who is not created but who uses his creative, mathematical, and philosophical processes to advance the course of the multiplication principle, which is in immortal agreement with the functions and features of transcendental knowledge with living philosophy.

CHAPTER 5

WHEN IS KNOWLEDGE DYNAMICALLY CONSIDERED TO BE LIVING?

Knowledge that is already known and accepted as the hallmark of wisdom is segmented into a lot of extensive subsets. That is why a lot of what man has known and accepted as knowledge does not stand the test of time. Philosophical mathematics is a knowledge that is transcendental in action and in reaction, in becoming and in appreciation. This is why its dynamism is utilised in assessing and accepting what is contained and blended in the anatomy of living knowledge. From the dawn of consciousness to the present Microsoft era, civilization has seen a lot of knowledge. But ironically, some of those civilizations could not stand the test of time. In this respect, our concern is not going to be focused on moribund and myopic knowledge, which has become the bedrock of the academia. Our vision in the content and likeness of philosophical mathematics is going to depend on the monumental and magnificent knowledge contained in the anatomy of the One defined as the super-ingenious electrification of all ingenuity. This is the knowledge that is defined and considered as power—dynamic, living, working, expressive, explosive—and such knowledge is only seen in the following nature's architecture.

- Transcendental knowledge, which comprises transcendental education, which warehouses transcendental science and philosophy,

- Other forms of knowledge that are living, immortal, and configured in the following lintious and monumental sciences: metaphysics, meta-logic, meta-mysticism, meta-mathematics, meta-dynamics, meta-occult engineering technology, and the purest transcendental knowledge, defined as super-mystical knowledge.

This is the knowledge that extolled the Twelve World Teachers, particularly Christ, who is known as the son of man, the son of God, and the son of light including being in the becoming of the holiest alpha and omega, which extended his super-mystical functions in actualizing the vision and mission of his omnipotence, omniscience, omnipresence, and omni-oneness wholeness.

A look at the tedium of this reveals that humanity certainly lost track of using the education of living knowledge, which could have extolled evolution and civilization beyond the present stage (which is chaotic). It is the ignorance of this permanent knowledge that could not lift the likes of Charles Darwin and his colleagues into the book of immortal knowledge, which warehouses the authorities of such omni-mystical emengards like Jesus Christ, Guatama the Buddha, Plato, Prophet Mohamed, and Lao Tzu—the permanent becoming of the Twelve World Teachers.

It is this class of omni-mystical beings that honoured the wonders of knowledge by the wholeness of One as being the pinnacle of power, the function of knowledge, and the heartbeat of the electrifying unity, of which living knowledge stands tall to speak and to protect as a way and means of understanding the rhythm and the tedium of absolute knowledge. A lot of people may not understand the ingenious and practical meaning of the omni-mystical emengard, and it is in the vision of this work to explain that another word or phrase that edifies and explains this is humans who are possessed with the wisdom of the grail. They are mystical and supersonic philosophers who are engraved with the knowledge and wisdom of the sermon, the living gospel, and the holy church of the saints; they possess the power of the mystical angelus. This explanation is important because it delves deeper into explaining the introduction to philosophical mathematics as a balanced anatomy and living knowledge. This power, which is from God, does not change or expire, and it is purpose-driven and practical. That is why the songwriter designates God as the "immortal, invisible, and the only wise God".

A lot of people have not known or appreciated that the purity and perfection agenda of philosophy, which has remained the hallmark of its advancement, is titanic in its appreciation of living knowledge. The task of writing this work, which is based on the ingenuity of philosophical mathematics, is hereby denoting God as the living knowledge that creation, particularly material science, has ignored. What a calamity! The task of explaining what is a living knowledge is to reawaken scientific consciousness in a super-mystical dimension by which science will approach philosophy and mathematics, while technology and engineering will approach science in earnest. The orbit of the wholeness of this achievement will be to establish a province of knowledge that is living, that is all in one, and that must depend on the authority of one who is authored for the purpose and function of inspiring vessels for the task of extending the millicential nature and authority of living knowledge.

It must be stated that the aspect of living knowledge—which involves transcendental science, philosophy, metaphysics, meta-mathematics, and meta-occult—is contained in a unionized format known and transcendentally defined as knowledge in the wholeness of a living force that uses pure and applied mathematics, particularly in the tedium of the rudiments of achieving the philosophy of perfection and purity. That is why knowledge is considered living when it is perpetual in action, pure in function, and protected and productive in nature. Such knowledge must confirm and conform to the mystical details of truth, and to the mystical ingenuity of metaphysics. This is why honest knowledge is hereby defined as the super-genius power and portrait of balanced rhythm with balanced interchange.

CHAPTER 6

WHY IS PHILOSOPHICAL MATHEMATICS DESIGNATED AS THE ANATOMY OF A LIVING KNOWLEDGE?

A lot of people, particularly in the ephemeral academic sector, do not understand that philosophy (with its intertwined relationship with mathematics) is immortalized and designated as the anatomy of a living knowledge. Knowledge as a living institution cannot function as winner's pride if it is not structured with the natural rudiments of philosophy, mathematics, technology, production, and all that is contained in sequential enabling. A look at the perfect introduction of *The Sacred Teachings of All Ages,* which came to man through Manly Palmer Hall, as well as *Journey into Knowledge, Pathways to Philosophy,* and *The Twelve World Teachers,* reveals that philosophical mathematics is authoritatively anatomized with the blending functions of philosophy. This blending is why it is a living knowledge with a perfect wisdom that engineers and propels humans to think philosophically and act universally. When we are being challenged to think philosophy and to create philosophically, it is a mandate for us to adventure into knowing what is contained in the anatomy of living knowledge, of which philosophical mathematics stands tall and alone with balanced, electrifying productivity.

The following points are eloquent and purpose-driven in explaining, accessing, examining, and conceptualising why philosophical mathematics, as a noble course and monumental endeavour, is designated as the anatomy of a living knowledge.

- It is a noble world that exists alone for the noble souls.

- It is productive in creating and restructuring what is needed for the betterment and benefit of the one-world philosophical mathematics.

- It helps a living mind to investigate, with honest appreciation, why and how philosophy works to achieve the concept of purity and perfection with monumental authority.

- It answers the *why* and *how* mathematics is splendid in the use of equations, algebra, geometry, data, statistics, and cosine and sine formulas in resolving and dissolving calculative difficulties.

- Again, the use of multiplication formulas, substitution, and division are equally in sequential agreement with the purpose-driven understanding of philosophy.

- The use and application of theorem methods, principles, practises, and policies is a titanic and truncated formula that designates philosophical mathematics as a super-genius anatomy of a living knowledge.

- It is a living knowledge that drives the thinking of other multifarious ideas into compact actualization.

- It holds the vision of creating, strengthening, and establishing the immortal vision of ensuring that humanity is empowered through the honest use and application of living philosophical knowledge.

- It is designated as the anatomy of a living knowledge because the Oriental apostolate appreciated it as the fulcrum of all things, including being the dynamite that makes noble souls exist in the immortality of philosophical mathematics.

- The anatomy contains what material science uses today as an expanded and extended production and engineering production. That is why the designation of philosophical mathematics as the anatomy of a living knowledge is best understood by the famous and dependable world of transcendental science.

- Philosophical mathematics is a mobilizer, a motivator, a task team leader, and an appropriate and ingenious acquaintance. That is why it is best defined as the super-genius foreword to every human endeavour.

- It is a talent that is transcendental in action, reaction, and creation. The anatomy is practical and electrical in ensuring that human evolution is empowered with the use and application of honest, transcendental enabling.

- As a co-creator with nature, it nurtures creation in the best ability of composite intelligence and compact empowerment. This is why philosophical mathematics maintains an intertwined relationship with concepts and compacts of all compositional factors, which are designated in this aspect as the CCC of philosophical mathematics.

- It is an ennobled thinker whose thoughts are pure, perfect, and purposeful.

- It is designated as the anatomy of a living knowledge because of the purest power it possesses, including the high level of forms and factors that it emit when it comes into contact with anything that would pollute its province.

It is here that the academic province is still not in the objective and lucid wisdom of the originality of philosophical mathematics. All over the world, particularly in the transcendental universe of grace and balanced rhythm, philosophical mathematics is designated as the supremacy that motivates the ennobled souls in appreciating what is contained in that meta-mathematical anatomy, which stands tall as words for the wise. That is why it is always defined as a wise counsellor, a teacher, a creative actor, and an ingenuous thinker.

After a thorough and honest investigation of what philosophical mathematics is all about, I designate this noble course as the supreme anatomy of truth and the dynamic philosophy that provides and protects the functions of truth, both as a living counsel and as a living and tender absolute.

A lot of people may ask: Where is the anatomy of philosophical mathematics explained in this context? The drawing below will certainly provide a message with a balanced answer, not only for the use of this purpose but also for the use, application, and understanding of the academic sector, which is not driven and moderated with the functions and ingenuous abilities that are contained in the dynamic anatomy of philosophical mathematics. A work to explain this concept must purely and practically detach itself from orchestrated pollution, which includes the in-glories of enmities and the myopic understanding of ephemeral science. That is why both philosophy and mathematics share a common bond with a living and immortal boundary. The purpose of these bonds and boundaries is to make honest seekers know that there is a sequential relationship that blended the authorities of mathematics and philosophy to exist as a compact and composite concept. It is another conventional formula by which we can appreciate the overall originality of philosophical mathematics, which is always interested in the functions and powers of knowledge, whereas knowledge is always interested in the activities of wisdom. It is here that wisdom at all times is always falling back to living knowledge to perform its perfect, natural, and nurtured acts and functions.

The academic province must be reengineered in appreciating that the wisdom of the past ages, at any point in time, must be used in determining what the modern man and posterity will behold as a frame of noble work that will guide all evolution (particularly the contemporary thinking) in knowing that a system, concept, nation, and person is at sea when the works of honest, living knowledge is being abandoned or polluted. The only institution that gives the universe power is knowledge. This living knowledge gives the one-world family a living philosophical mathematical engineering that, when properly used and harnessed, will make the world think philosophy and create philosophically in order to act universally, which is the honest mandate of philosophical mathematics.

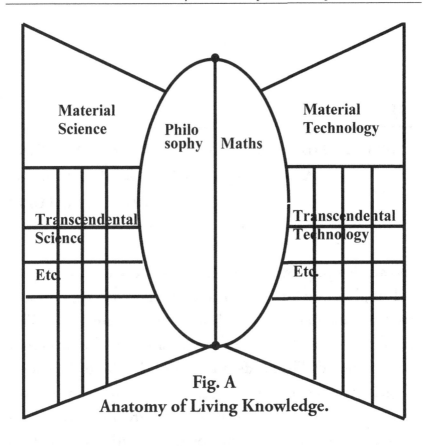

Fig. A
Anatomy of Living Knowledge.

CHAPTER 7

WHY IS THE ACADEMIC PROVINCE NOT USING THIS KNOWLEDGE IN THE WISDOM OF ITS ORIGINAL NATURE?

The academic province, which is possessed with the use and application of mental and intellectual wizardry, is strict and stagnant with the honest use of knowledge. This stagnation is responsible for not moving the academic sector forward. The academic sector exists in the license of material science, which is stagnated in knowledge simply because its actions do not involve the great Creator or his mystic ecstasy and omni-philosophic mathematics. That is why a lot is greatly lacking within the academic sector. The academic sector is against inventing and investing in what Mother Nature has given to us; it is only spending time, resources, energy, and personnel in rebranding, reshaping, and reordering what already exists. In the wisdom of living knowledge, that action is an imposition of melancholy, which has further degenerated the creations of the academia.

The academic province, even as materially electrifying as they are, is not using the living knowledge that fortifies philosophical mathematics because of the following reasons.

- Knowledge at any point in time is only enduring when it is understood as being the Creator's engineering mechanism.

- Living knowledge is always the lubricant that drives the productive engineering of technology. This is why the academia is not being driven in the tedium of this medium.

- *Living knowledge depends on God in order to function; the academic province depends on man and visible materials. That is why it cannot be accepted in the notation of philosophical mathematics as a living knowledge.*

- The academics have mocked and mimicked themselves to ephemeral understanding against depending on the absolute understanding of one who is designated as everything in all things.

- As a living knowledge at every moment is research-driven, both in enhancing and maintaining the vision and mission of creative possibilities, the academic sector is involved in mental and material rigmarole.

- A living knowledge is always being patterned in the objectivity of long range planning with tactful vision and mission, and so the academic sector is propelled with the ephemeral knowledge of myopic and Microsoft technicalities.

- Because loving knowledge is practically strict in adhering to the rule of nature's principles, policies, and practises (of which she is a stock and stakeholder), the academic sector is not strictly adhering to the rules, regulations, and principles that drive the electrifying creativity of living knowledge.

- A living knowledge is always moving to configure the universe as a systematic and practical unity; the academic knowledge works to disintegrate the united beauty of the one-world family.

- Living knowledge is propelled to function in the best dynamism of philosophical mathematical principles through the use and application of transcendental science with transcendental philosophy. The academic province is always detached from the deep most understanding of this knowledge, which could have helped it to appreciate what is known as the anatomy of Mother Nature.

- Living knowledge always dance to the tunes, wishes, and wills of inspiration. Material knowledge, which is the product of the academic sector, finds it difficult to hide its arrogated sins simply because it could not achieve what the sector of living knowledge is designed and designated to achieve.

- Academic and scientific principles are always changing, and living knowledge (in their principles, policies, and practises) are purpose-driven in achieving the desire of wholesome perpetuity, including ensuring that the mandate of philosophical mathematics is ornamented with practical and balanced achievement. This is why transcendental knowledge, which is a branch of living knowledge, holds firm and is fortressed with the illumined feature that the academic sector requires regeneration with a practical and strict restructuring that will certainly reorder and redesign the myopic visions and actions.

The Bible, which has sold for over ten trillion copies since the dawn of consciousness, is still preaching about God, creation, apostles, prophets, seers, and the activities of some selected people who understood that God is the fulcrum of living knowledge. As old as the academic sector is, the academics have not been able to produce works that have stood the test of time; even what these ingenious minds have given to posterity are difficult if not impossible for the academic sector to utilise.

This work is using that information as a case study because ingenious and noble souls like Christ, Buddha, Plato, and the other Twelve World Teachers are practically souled in the reality and realism of configured, living knowledge. It is here that the anatomy of living knowledge designated them as messiah, sages, seers, prophets, mystics, and adepts, including calling them candescent and cardinal extensions of the choice.

That is why their actions are creative in philosophy, and their thinking is productive with the wisdom of the universality principles.

Further in this list is the class of people that are recognized and classified as geniuses. They could not appear in the book of the Twelve World Teachers simply because they were not as cosmically driven, consciously donated, and mystically initiated as their previous counterparts. This list includes the likes of Beethoven, Bach, Mozart, Paracelsus, Socrates, Pythagoras, Aristotle, Emmanuel Kant, William Shakespeare, Isaac Newton, Isaac Pitman, Galileo, Isaac Armstrong, Dr. Walter Russell, and St. Augustine. The likes of Herbert Spencer, Edwin Markham, Walt Whitman, and Ralph Waldo Emerson, who are known as social crusaders and human-rights activists, were greatly ennobled simply because they were touched and challenged by the monumental contributions of the living knowledge. The super-genius Twelve World Teachers donated that living knowledge to creation as a way of life, including being the honest achievement that should inform the bedrock of the academic and educational anatomy.

It is important to explain that living knowledge has greatly eluded the academic sector because of other uncharacteristic reasons.

- High level of poverty.

- Arrogance and pilfering among the academic managers.

- Operating in an unfriendly environment with rudimental and regulated government, and obnoxious policies.

In the class of the esteem of the academics who are designated as professors are some humans who are hereby defined without apology as poor thinkers, poor managers, and lopsided and unpractical humans who are not oriented or structured in the rhythm and ecstasy of the fundamentals of living knowledge. Our professed academia has a case to answer, a mission to achieve, and a vision to protect and prosper—simply because religion and honest philosophy have given us a living maxim that is a mystical admonition: all souls must come to me in due time, but there is the agony of waiting. The scriptures tell us practically and vividly to train up a child in the right direction, and in his life and future, he will never leave that path.

A work to determine what is important, including the woes and the errors of the academic sector, must certainly be able to show the world of the academia the original score sheet, including ensuring that all the errors in the academia balanced sheet are pointed out with dynamic correction. That correction will help the sector to forge ahead and to relate very well with the candescent and practical ingenuity of living knowledge because most of the students who could otherwise be ennobled great thinkers and deep natural researchers are being deceived with most academic concepts, which do not encourage them to develop the ingenuity lying dormant in them. Does it mean there are no more

Pythagoras' in the midst of our students? Are there no more deep thinkers? Why does most of the academia discourage their students to be inventive and innovative? Why quench the electrifying philosophical mathematical powers in your students? Do you not know that in them lies the future of our posterity?

For the record, living knowledge at any point in time must be appreciated as the honest and practical vision of God, including being his wisdom that he utilises in extolling vessels who are philosophically and mathematically driven in appreciating and extolling him as the only authority whose universality and omnipotent wisdom dictate the tune of the Microsoft province;, the academic, the engineering, and the technological sectors; the scientific and technological province. Living knowledge is always looking unto the absolute knowledge of all knowledge whose works are authority unto life. God's actions are living wisdom unto creation; his philosophy and mathematics always configure facts and figures in the best of his thinking.

Therefore it is stated that knowledge with God is perfect and practical, and any knowledge that is not the product of the Creator will certainly make mockery of he that uses it. That is why the urgent necessity of this chapter, which is a dignified challenge, is to ensure with dynamic assurance that humanity appreciates God as the fulcrum of living knowledge.

CHAPTER 8

WHAT ARE THE CONSEQUENCES OF THIS, AND HOW CAN WE EFFECTIVELY RESTRUCTURE THESE MONUMENTAL ERRORS WITH BALANCED CORRECTION?

Effective correction, monumental restructuring, and balanced understanding with the use and application of living knowledge will certainly redress and readjust the infertile errors that are fatal to the growth of human dynamism. From the cradle of human consciousness, man has always found himself making and multiplying grievous mistakes that work to reduce, mimic, destroy, and push him towards oblivion. The following questions should be asked.

- What are the benefits of biological weapons, nuclear armaments, and disunity among nations, which are the products of material ingenuity with its infamous engineering production?

- What are the benefits of slave trade and transnational and national impoverishments, including corruption and cheating in high quarters and at government levels?

These enmities, which are being practised by humans who are possessed by the ephemeral knowledge of material and academic prowess, work against the sustainability of philosophical mathematical principles, policies, and practises. In earnest, they defile the practical principle of living knowledge with its natural methodology. It is not in the rhetoric of this chapter to neglect what is utilised as acceptable philosophical mathematical standards. This is why the following have remained dangerous consequences of the errors of the past.

- Lack of practical actions with objective dynamism.

- Inability of the academic sector to develop concepts with ingenious ideas, which will help the educational province in providing a loftier and sound ground for the overall emancipation of titanic ingenuity.

- Uncertainty in science, technology, and engineering principles, practises, and policies.

- The government of theory, which is not theoretical, is always destroying the esteem and natural tendency of principles.

- Reduction in human growth, including producing what is defined in this context as inglorious academics that make noise on the streets of the world simply because they have passed through the materially driven educational sector without being learned or possessed by the features of living knowledge.

- Introduction and growth of acts that do not promote the longevity of living knowledge in its overall practicality.

- One of the consequences that is stagnating human growth and dynamism is the introduction of bribery and corruption, which introduced and multiplied poverty, cheating, and enmity, and which can best be defined in this context as an indirect human enslavement that neglects primarily and practically nature's law of living knowledge in the brotherhood of man's principle and human relations.

Consequently, the restructuring that must be effective and monumental in creation, correction, building, and empowerment must follow a guided and natural process. In turn, hat process must agree and harmonize with the guiding principles that are contained in the balanced and objective interchange of living knowledge, which works to enhance the objective and pragmatic performance of philosophical mathematics—which is designated as the anatomy of a living wisdom, including being in the becoming of honest life.

- The science of ethics and morals must be introduced at all levels and facets of the human educational system.

- The academic sector must be restructured in order to agree with the vision and mission of inspiration, which must be adopted and utilised beyond the vision of a business necessity.

- Educational managers at all levels must be taught to appreciate why and how philosophical mathematics is practically bent on achieving the serious mission of purity and perfection, which is in agreement with the scriptural eulogy of "Be thou holy, for I am holy".

- The characteristic use of mental wisdom, including the myopic academic knowledge, must not be considered as attending and ascending to the peak of intellectual dynamism.

- The academic system must be energized with rules and regulations that must be conventional and universal, including being purpose-driven with a high level of professionalism.

- All applications by the would-be academic managers must be assessed and scrutinized in order to find out whether such applicants are prepared to take the system to the next level, which is the hope of philosophical mathematics, including being the sacramental unity of living knowledge.

- Governments must be prepared to accept the challenges of educating our youths as the bona fide trustees of posterity, as a mandate which is a task that must be accomplished, and without which the universal world of commerce (including the sector of Microsoft prowess) will certainly be devastated and destroyed, simply because every institution must adhere strictly to the concept of making man against making cities.

- The quality and lifestyle of the Twelve World Teachers must be set as a standard for all.

- The lives and times of the immortal geniuses who were lintious in the use and application of living knowledge must be utilised and involved as a business necessity and a wakeup call, including being the honest examination in which the academic sector must involve themselves.

- Scientific engineering and ingenuity, which work to mimic and mock human progress, must be discouraged.

- Educational management with academic administration must be placed in the hands of those who appreciate that living knowledge is the honest way by which the Creator ordered all his arts, creations, and transactions to produce immortal wisdom, and to which is the supreme mandate every earthly occupant must strictly adhere.

For the record, this restructuring syndrome of philosophy must not be used as magical wand; neither will creation involve a crash programme in its affairs. The procedure must be gradual, systematic, organized, and pragmatic with philosophical dynamism; this is why it must not be used or treated as a material reform agenda. In my opinion, living knowledge impacted into humans is capable of producing the following results.

- Honest reasoning.

- Balanced transactions.

- Appreciating the use, the wish, and the will of wisdom.

- Objectively leading one to obey natural laws simply because these laws are in agreement with the dynamism of living knowledge.

- It produces a platform for human growth, particularly in the lintious objective areas, which include human empowerment in education, agriculture, science, engineering, technology, commerce, and industry.

The empowerment is an anchor point that will help humans appreciate the virtues and values of management, which is a lesson and knowledge that exists practically at the ingenious postulation of living knowledge. It is important for all and sundry to appreciate that as man is from Godhead, created to observe and practise purity and perfection, so is living knowledge a consummate hall that is famous and immortal. That is why it is donated to the wisdom of this work as a museum of knowledge, including being a lintious library with an immortal Britannica that warehouses the perfect workings of philosophical mathematics.

The onus and honest assignment of man in his endeavour as a co-creator with the ingenious universal One is to know that he is a varsity to knowledge, a university of ideas, a profession to wisdom, a Sagittarius who is created to know that only in that beginning, if well nurtured and known, will he appreciate that knowledge, when it is living, is the pinnacle of practicing the light principles. Those principles are the way of the Creator. This is why every human endeavour must agree with the monumental wisdom that the anatomy of living knowledge defined in this context as the fountain thrust, including being the heartbeat of philosophical mathematics.

CHAPTER 9

HOW CAN WE USE PHILOSOPHICAL MATHEMATICS TO APPRECIATE THE STRUCTURE AND ANATOMY OF LIVING KNOWLEDGE?

Every act of Mother Nature, including all that is created—is galvanized with the sensitive structure of philosophical mathematics. That is why human actions are driven by the forces and immortal laws of philosophical mathematics. Life is logical, but it is equally mathematical. Logic and philosophy produce the best of blending for the smooth functions of mathematics. Ordinarily, it is not easy to appreciate the structure and anatomy of a living knowledge when viewed and approached from the point of academics, ephemeral technology, and science. This is why the best university that can authoritatively determine and explain the structure and anatomy of a living knowledge is Mother Nature.

A lot of people do not know that nature is the wholeness of a compact extension of the living universe. This is why a work to appreciate the anatomy of nature as an extension of living knowledge will certainly know that humanity has therefore not been able to appreciate the structure of mathematics and the anatomy of philosophy. This lack of knowledge has actually degenerated evolution and civilization to remain in a dangerous coma. It is a negated threat to knowledge and to understanding, including an outright rejection of the functions of philosophy in its intertwined relativity with mathematics. It is here that all that is considered and configured in the province of the academia, including the hall of technology, is so robotic to the level that ideas failed their purported inventors, including being used to destroy their stakeholders.

We can use honest and practical philosophical mathematics to appreciate the structural anatomy of living knowledge with the use of the following.

- Honest examination of philosophical mathematics.

- Ability to use data in assessing what is defined as living knowledge.

- The use of philosophical mathematics and research methodology must be adopted as consultants, advisers, keyholders, and stakeholders. It must be approached with the grand style that Plato defined as the wisdom of a perfect teacher.

- Mathematical modules must be constructed and defined with philosophical concepts. These modules must be used in assessing what is contained in the structure of knowledge, including the anatomy of its blending force.

- The use and application of transcendental mathematical philosophy must be involved when assessing the structure and anatomy of a living knowledge.

- The use and application of philo-mathematical humanity and rhythm must be consulted as a scientific searcher whose mandate is to create technological assets with its philo-mathematical ingenuity.

- The use and application of mathematical unity, with its gratification of truth, must be assessed as a way by which the pillars and structures of a living knowledge can be determined.

- Honest assessment of the powers of wisdom, including the philosophy of nature, must not be sidelined.

- The contributions of material science must be examined with the use and application of an honest approach.

- All purported invention of technology must be examined in order to determine whether science and material engineering have actually invented anything separate from what Mother Nature has bequeathed to humanity (which is the product of living knowledge).

- Philosophical mathematics is best known as the pathway or adventure into life, and so we must be able to understand whether science can be defined and understood as journeys in truth.

- The use and application of transcendental logic must be allowed in the appreciation and assessment of the structure and anatomy of living knowledge.

- Researchers at all levels must be focus-driven in ensuring that what is considered as the anatomy of living knowledge is greatly protected from being polluted, particularly by the uninformed.

- Living knowledge must be severed and preserved from the dangerous incursion of commerce and industry, along with other figurative and mathematical calculation, that makes statistics and accounting look like dangerous endeavours in the wish and will of living knowledge.

It is not the wish of this text to humiliate accounting, which a lot of people consider a noble profession. But it is a mandate of this work to explain that most of the cheating and enmities that are associated with trade, commerce, industry, and government, and including other areas of human endeavour, originated from the myopic input of accounting profession. A noble chartered accountant of the eighteenth century, Edward Desmond, saw the funky acts of accounting and likened the purported noble profession of accountancy to a sharp knife in the cheater's pocket.

Living knowledge is the monumental structure of mathematics, including being the perfect anatomy of philosophy. That is why the goal of philosophy and mathematics is to attain the brotherhood of purity and perfection. Mathematics is greatly concerned with the adoption of accuracy as its mandated mechanics. The ornamented mathematician Pythagoras, who developed and expanded what is utilised today as the Pythagoras mathematical theorem, opinionated that the mission of mathematics is to create a province of crystal accuracy. H. A. Clement moved a welcomed and supported motion that is fair to Pythagoras' mathematical epigram: that philosophy and mathematics are blended, are naturally structured, and are celestially intertwined. This is why every creation of the Creator is mathematically quantified and dimensioned, including being philosophically directed through the power and honest authority of living knowledge.

It is here that I appreciate the following concepts, which are in correct and balanced consonance with the vision and illumed authority of the past eras.

- Living knowledge is a supreme product of the Creator. This is why honest and practical knowledge is powerful, glorious, and the perpetual mark of wise men.

- It is an eye camp to science and an eye centre to technology. This is why it is nurtured and structured as a titanic canopy by which the future of human endeavour guarantees creation success, particularly when man appreciates that knowledge is from God. That is why the wisest among men still seek wisdom.

A look at the structure and anatomy of living knowledge will certainly appreciate its milicential and transcendental nature. Philosophical mathematics, at any point in time, is always in perfect, ornamented agreement with the mother creator, who is hereby defined as living knowledge or Mother Nature. Philosophical mathematics can be used in appreciating and assessing the structure and anatomy of living knowledge, which man is in agreement with the wisdom of Mother Nature. Every invention and creation that we are enjoying today is simply denoted as an ignited and electrifying art that is from one who is known since the dawn of consciousness as being whole with unity in diversifying. Living knowledge is hereby defined and denoted as the journey in truth, including being the anatomy of millicentia and crystal creativity that all and sundry must accept, desire, understand, and appreciate.

This concept gives us the perfect road to help us know that life at any point in time is adventure into wisdom, including being understood as the golden jubilee that celebrated the successful achievements of philosophical mathematics. The achievements are nurtured and naturally structured with the anatomy and rhythm of natural knowledge, which presents God as the comprehensive and immortal creator who is one at all levels of balanced creation and creativity. This chapter is a lesson for all and sundry to know that creation is the authority of the immortal structure of living knowledge.

CHAPTER 10

How Can We Define This Knowledge as a Journey in Truth, Including Being a Pathway to Philosophical Mathematics?

Rare knowledge, which is borne out of true philosophical wisdom and the act of being transcendental in action and reaction, is always donated to the world of empathic knowledge with balanced reasoning as a journey in truth and as a pilgrimage for truth. It is always known as the monumental curvature, which helps us to appreciate the lintious contents of philosophical mathematics, and which is purpose-driven with the blending factors and forces of nature's purest institution. Knowledge at any point in time is given to creation by the only lone authority, which is hereby designated and dedicated as a super-genius, consummate knower who is perfect and pure in knowledge, truth, procreation, and provision. At this point and hour of evolution, we must understand that philosophical mathematics is essential as a journey in truth, perfection, and protection. It is here that both mathematical and philosophical adepts use and appreciate the relativity of its intertwinedness.

At this point, knowledge known and designated as a journey in truth must possess the following factors with consummate functions.

- Its ability to energize creation in order to be knowledgeable.

- Luminous knowledge is the anatomy of transcendental wisdom, candescent and incandescent science, and practical and purpose-driven technology. That is why it is known as the mark of wise men and the synergy by which we enter into a philosophical journey.

- It must be inspiring and at the same time educative.

- Knowledge in this respect must be luminous to be divine, mathematical, and philosophical. This is why it portrays the light, which is hereby defined as the journey in truth.

- This knowledge is the power and anatomy of all practical and positive enabling.

- It is known at all times as the indomitable force that engineered and originated the wisdom of the Twelve World Teachers.

- This knowledge is practical with ethics, morals, and mathematical psychology. That is why it is best defined as the dozer and dossier of all things.

- It is symbolic; this is why the wisest still seek and ask for it.

- It is the way of life and the balanced way of truth. That is why it is the way of all and sundry.

- It is the living and monumental journey of all and sundry. This is why it explains most of the difficulties that challenge the human journey to truth.

- It is the practical way by which our evolution can be explained with scientific ways.

- It works with the directions and detects of Mother Nature in ensuring that evolution is not being misunderstood, particularly by the wizard and wizardry of intellectual poverty and its melancholy.

- The well-informed adepts know it as the consummate Homer of all Holmes. This is why its functions are practically quantious, quantum, and quadratic with the use and application of functional and applied mathematics.

- It is this knowledge that explains with practical perfection that $X \times X$ is permanently denoted as x^2. It is the same knowledge that explains that $2X = X + X$. The same knowledge let us know that $X + X + X$ will certainly give $3x$. That is why it involves the functionality of the principles, policies, and practises of pure and applied mathematics.

- Philosophically, it is the same knowledge that inspired the Oriental wizards to state that all in creation is simultaneous, balanced, and galvanized with the use and application of the omnipotent realities of the Creator.

The question here is, how can this philosophical tele-max with psychological tele-elements be explained? The phrase "philosophical tele-max" is defined as the purpose and the rhythm by which humanity is assigned to gain the knowledge of what creation is, with the use and application of the philosophies of the astral world. That includes the incandescent sun in its orbital elements, the lintious and orbital stars that exist in their homes, and the sandalwood moonlight, which forms a cool atmosphere for the ethereal planet. This is why philosophy is configured as the science of purity and perfection, which explains how and why it is the maker of great thinkers.

In this respect, the philosophical tele-max is the origin of the knowledge of philosophy. It is the absolute knowledge that is defined as the journey in truth. It is the radar by which the elements of philosophy can be gained with perfect appreciation. Among the philosophical wizards who were fortunate to know the celestial exposition with the use and application of philosophical tele-max was Peter from Crotona, who is known all over the world as Pythagoras. It is the symbolic knowledge of this fact that memorialised him as a philosophical tele-mathematician whose knowledge of astral letters, astral forms, astral knowledge, and transcendental mathematics has remained unequalled since the dawn of consciousness.

A work to define how this knowledge is a perfect journey in truth must appreciate that truth is divine. Truth is philosophical, the enabling anatomy of science and technology, the ingenious practise that empowers engineering protection. That is why the summary and sum total of what truth is has remained the origin and the bell of the omega principle, the alpha practises, and the united fort of alpha and omega in balanced proportion with philosophical mathematical proposition. A work to define truth must always follow through the rhythm and tedium of mathematical ascension with philosophical understanding. How does this stand, and how can it be gained?

0	1	2	3	4	5	6	7	8	9	10
A	B	C	D	E	F	G	H	I	J	K

A scientific look with technological appreciation of this arrangement shows that the following mathematical formulas with philosophical rhythms can be gained, which means that the order of arrangement can follow thus.

0	1	2	3	4	5	6	7	8	9	10
A =	B =	C =	D =	E =	F =	G =	H =	1 =	J =	K.

This is the ascending order, in perfection and protection, that can end up in philosophical principles with philosophical practises.

In this descending order or what we can define as the converse mechanism, 10/K or K/10 can bring us to 0/A. At this point we can now appreciate that philosophical mathematics is defined as a journey in truth because of the following reasons, which form and anatomize the magnanimity or the ingenuity of this knowledge.

- Truth is perfect because philosophy and mathematics are electronically perfect, both in simulation and stimulation.

- Truth is pure. This is why a work of this nature must function in the rhythm and tedium of philosophical mathematical perfection, protection, and provision, with the use and application of dynamic purity.

- Truth, which is a journey to life and in life, is naturally constant, unbendable, and unpolluted. This is why the principles and policies that drive the functions of philosophical mathematics are intertwined with the rhythm and synergy of truth.

- Truth is above the rhythm of government law, evolution, culture, custom, and tradition. That is why it works perfectly with philosophical mathematics: in order to give illumined knowledge the power to function as rays of light.

- As truth appreciates the functions of one as a letter, philosophy and mathematics equally appreciate the functions of one as an alphabet.

A look at the creations of the past, particularly those that exist at the time of the Oriental wizards, lets one understand that all are patterned at the proximity of truth, with the use of philosophical mathematics as the balanced anatomy of living knowledge. At this point, the task of examining the composition of this knowledge is what this chapter has delved into, and the purpose for this achievement is to enhance human creativity and ensure that creation is purpose-empowered in the use and application of original philosophical knowledge, with its mathematical tenderly, as a way and means by which the incubus of material science with Microsoft technology can be reduced.

It is in the opinion of this work to explain that every act and creation of man that is not in consonance with the rhythm of truth and balanced knowledge will always create a chaotic institution and understanding. This is what the jet age is suffering. The onus task of this museum of knowledge is to ensure that there is a practical and purpose-driven U-turn that will make all and sundry know that any creation or action of man which does not bring the Creator in the forefront will certainly work against man beyond the measure and attitude of that creation. In this respect, knowledge as the activity and power of journey in truth is hereby defined as the function of philosophical mathematics, as the power of science and technology, and as the simultaneous and living signpost of engineering production. This is why all that is compacted in engineering mathematics is practically borrowed from the living functions of philosophical mathematics, which is hereby defined as knowledge in the journey and power of truth. That is why this work is hereby authoritatively denoted as a helping and practical guide to human upliftment.

CHAPTER 11

Is It Logical to Explain This as Being the Way to Immortal Wisdom?

Immortal wisdom as we know is a consummate knowledge without limitation or pollution, devoid of material science and its pollution. It is knowledge that is indeed the power of all things. This knowledge is the immortal wisdom that defines and drives the activities of all things, simply because wisdom is wealth. It is consummate and practically and purposefully ingenious. This is why the likes of Solomon the blessed, Melchizedek, and other known wizards were practically driven with the use and application of immortal wisdom, which is nurtured and natural in absolute ingenuity.

A work to explain the logic and philosophy of this concept must appreciate that wisdom is the way of God, the purpose of life, and the anatomy of all things. That is why philosophical mathematics is hereby defined as a living and consummate wisdom that adheres strictly to the onus task of purity and perfection for which immortal knowledge is always providing the data for its practical functionality.

A lot of people do not know that immortal knowledge is the constituent of human reasoning and the frame and feature of technology. Unfortunately natural scientists and technologists are not inspired in the understanding of what immortal knowledge with its inspired wisdom is all about. At this point we will consider what can be named or what constitutes the province of immortal wisdom, which humanity has not been able to understand regarding its relative impact and importance in the like and likeness of immortal knowledge.

- Man and other creatures.

- The moon.

- The stars.

- The sun.

- The oceans.

- The great rivers.

- The forests.

- The inhabitants of the aquatic province.

- The terrestrial and arboreal elements.

- The natural orchards.

- The unending and diversified deserts.

- All that comes under the umbrella of wild beasts.

- The lilies.

These are logical products of immortal knowledge. It is unfortunate that man does not work to appreciate what can be considered the constituents of immortal wisdom, except for the Twelve World Teachers and the developed adepts who have come to know that logic is a candescent extension of philosophy, and that is why it understands and appreciates the rhythm of mathematics, which is intertwined with the blending glories of philosophy. By blending glories of philosophy, I mean the powers and forces that dynamically direct the functions of philosophy as the mother of science, the father of knowledge, and the extensive institution of Mother Nature, of which man is designated as a philosophical emengard.

In this respect, it is practical to state that the logic of this knowledge, as fulfilled in its factual and pragmatic functions, is borne without compromise, pollution, sentiment, or detriment. Creation as a philosophical mathematical kit is not inhabited by the titanic acts of man, but it is exhibited by the powers and functions of philosophical mathematics, which is logical in rhythm, explanation, and occupation. It is considered to be penetrating the unknown world in order to fashion our actions in accordance with the new knowledge that is being acquired. This knowledge is all about the volumes and veracity of life, about its people, policies, and practises as a tedium of immortal wisdom. It is galvanized with the following philosophic and immortal blending.

- Existing purely as pathway to philosophy.

- Appreciating the logic of knowledge as journey in truth.

- Understanding the rhythm of Mother Nature and being able to convert the rhythm as a motion and as an emotion for practical and functional creativity.

- Utilising the factors and forces of immortal wisdom, which is mathematically driven with the ways of logic in ensuring that the continuity principle (which is a Sagittarius agenda) will be maintained.

- Always being able to interpret, explain, and understand the urgent necessity of the demystification of the oracle of mathematics, which was known and accepted as a sacred and secret institution before the dawn of a new day in human consciousness.

- Always providing science with a lintious signal on how it could galvanize the world as an acceptable and technological cosmogony.

- Always advising technology with engineering production on the urgent necessity of using the laws of balance with standard mechanics in the creative processes with astronomical progress.

- Always giving a dependable, acceptable, and fulfilled warning to the province of Microsoft technology, which is already chaotic, on how to avoid the undue usage of scientific armaments with their technological weapons (which are against the laws of balance in creating and causing constant destruction).

The logic of immortal wisdom is to preserve humanity from the acts and concepts that are not synchronous with the rules and regulations of what made creation the wisdom and expression of Mother Nature. At this point, we can now appreciate that it is logical to explain this knowledge as an immortal wisdom, as the practise of nature's progressive processes, and as the principle by which human actions should be directed. This knowledge is hereby donated and designated as the consummate heartbeat of the totality of all creative forces, of which philosophical mathematics is distinct with practical and purposeful extensions, and of which this ingenious knowledge is symbolic.

A look at the way and manner the science of philosophical mathematics, with its technological prowess, was defined in this work can best be defined and understood as nature's intertwined enabling. That is why this knowledge and its occupation will always ennoble souls to appreciate that creation is the glory of the Creator, whereas philosophical mathematics is the dynamic fountain thrust by which the glories of the creation can best be appreciated with the use and application of truth—which is a rare journey in the philosophical kingdom of immortal wisdom. This is why it is designated as a living knowledge: because it is wisdom without pollution and material waxes.

It is here that wisdom, wish, and will form what is known as the heartbeat of philosophical mathematics, which is practically logical in extension, interpretation, function, understanding, protection, and appreciation. In the following passage, I make a case for the immortality of this wisdom.

> Wish thou what you want to desire, and the will, will pave a smooth way for the alignment of that wisdom. Wish thou what you want to will, and the will with a stronger desire will emancipate you into the province of immortal wisdom, which warehouses the opinion and functions of wish, will, and wisdom that the Microsoft and material technology donated to the world of engineering technology as WWW.

In a nutshell WWW is materially converted from the consummate wisdom of wish, will, and wisdom. Unfortunately, the way and manner humanity is going about this creates a lot of poisons, simply because the material concept is not drawn from the original heartbeat of Mother Nature, who is the lintious incubator of will, wish, and wisdom.

In the practical ontology of this explanation, every journey in philosophical mathematics is designed to take man into a new level of practical purpose. A look at the way and manner the Oriental wizards denote philosophy as journey in truth and pathway to knowledge is capable of making us appreciate that every institution and creation is relatively intertwined with the objective use and application of immortal logic. That logic works to examine, explain, understand, and function as a super-genius intelligence by which man in particular will understand his position in creation as the hallmark of immortal wisdom with the use and application of ingenious inspiration. Inspiration is always there to drive man with the universal and perfect authorities of wish, will, and wisdom, which are hereby denoted in a mathematical, triangular lotus as the power of an appreciative enabling that can be represented as 0 to WWW or A to WWW.

The choice of this understanding is hereby dedicated to man who is naturally created in the super-genic heartbeat of Mother Nature, which is mathematically perfect and philosophically and technologically decent. It is contained in my inspired wisdom to state that immortal wisdom is defined as God in action, which uses man in inspiration to create enduring and purpose-driven acts that are blended with the wisdom of logic, the wish of philosophy, and the will of mathematics as the hallmark of WWW.

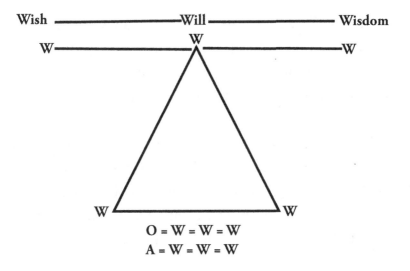

$$O = W = W = W$$
$$A = W = W = W$$

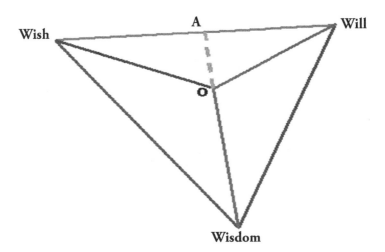

Proof

$$W = A = W = W = W = O$$

$$\therefore \Delta\ WOA = WOA = \underline{W} = \underline{A} = \underline{W} = \underline{O} =$$
$$O \quad O \quad O \quad W$$

$$= \overline{WW} = \overline{WW}$$

∴. **WAW = WAW,** with AOW constant to show its mathematical force with philosophical perfection. This proof can be conversely done using the tedium of ascension order or the descending order of the astronomic system.

It is important to explain that WWW is nothing but the mathematical heartbeat of Mother Nature—which is at work with the motion of the entire waves of the electrical force and current of philosophical creations, which is materially interpreted with a lot of human pollution.

CHAPTER 12

WHY IS LIVING KNOWLEDGE NATURALLY PURE, PERFECT, AND DYNAMIC WITHOUT POLLUTION?

Knowledge that comes from the honest and balanced heartbeat of the greatest monad is living and dynamic because of the level and practicality of its perfection and purity. When knowledge is polluted, it cannot serve its functions; it will never stand the test of time. A polluted knowledge is one which exists at the ephemeral proximity of human elements and thinking with his myopic engineering. It is here that the invention of mortal alchemy is brought into focus. Such knowledge, which has been the way and act of material scientists, usually locks the Creator out of its creations; this is why the knowledge of such highly intellectual humans like Charles Darwin was not galvanized as an educational and academic colloquium, particularly in the annals of human history.

At this point, a living knowledge is one which is borne without the glory of matter, without the myopism of the mental system, without environmental and community pollution. The Creator always hands down such knowledge to man—this is why it is best defined as a rarefied gift. Such a gift is naturally dynamic, ecclesiastically perfect, and lintiously revealing and inspiring. It is such knowledge that the wisest ask for, go for, and preach for, simply because it holds the wisdom of wisdom, the will and wish of the almighty heartbeat of the universal tedium. Such knowledge in the circle of religion is purely recognized and defined as a monumental harbinger of Ebenezer, as the Abel of faith, the Abraham of light, the Moses of the epistles, the called David of all messengers. It is recognized and honoured as the living Christ from the ecclesiastical province, whose functions are to establish the kingdom of his father.

The philosophy of this chapter is to explain that knowledge is not the same, and it can only be assigned through strong desire with perpetual and permanent asking. A living knowledge is naturally dynamic without pollution, and this is the intention of philosophical mathematics and the purpose for the section of its wisdom. Habitual knowledge, which is peculiar and rare, is always working knowingly with the detects and directives of the Creator. That is why the original and clandestine concept is to achieve a desired mindset, in order to remain as a constitutional and universal ornament.

The achievements of the Twelve World Teachers, including the gifted and ingenious humans who originated the foremost world culture and tradition, can best be defined as living knowledge with natural dynamism. The achievements of such philosophical emengards and apostles who established the sacramental foundation of philosophy can best be defined as living knowledge with its dynamic nature of creativity.

When knowledge is rare, it is pure and peculiar. It is ornamented with all millicentia factors, and such knowledge is always at the beacon of the giver. It is here that the achievements of the mystical grail, with its rarefied tranquillity, has remained a challenge and a wakeup call by which humans can appreciate the purpose, policies, principles, and practises of philosophical mathematics, including knowing why the anatomy of a living knowledge, dynamically explained, is recognized as the philosopher's kit, the philosopher's stone, the philosopher's tree. The monumental mathematical ingenuity is hereby naturally configured as the living Ebenezer of this knowledge. This type of knowledge is usually transcendental, trans-mystical, and trans-Oriental. We must appreciate that the wizardry that is configured in this knowledge is usually the way of light, love, and life. It is always being understood by philosophy as the sanctious agenda of the universal conclave.

I have been empowered and propelled for this work, and in my opinion, polluted knowledge is disastrous. Such knowledge is seen in the areas of humans who prepare and manufacture biological and nuclear weapons with their dangerous armaments, which work to destroy that which was created by a living knowledge. It is here that humanity is warned to appreciate the will, the wish, and the wisdom of Mother Nature, who is ornamented and essential in the best production of dynamic knowledge, which works to extol, exhibit, create, understand, and ensure that all in creation is pacified in the ornamented creation of the living monad.

For the record, a living knowledge is one that is pure and perfect in anatomy, dynamic in culture, propagating in action and reaction, and inspiring and infectious in all times. Such knowledge is used in ensuring that humanity maintains a standard growth without melancholy. This is the vision of philosophy and the mission of mathematics, which intertwines its cordial relativity with the vision and mission of Mother Nature. It is here that this work is best denoted as the pen for best minds and the best page for the highly ingenious and inspired. Philosophy is the maker of great thinkers, and it uses mathematics as a scientific and practical enabler. That is why a work to define why knowledge is living, including being naturally dynamic without pollution, must certainly teach and inspire creation in appreciating that the frame of the whole universe is famous in the best utilisation of a living knowledge. The knowledge is naturally dynamic in ensuring the occupation of balanced creativity, which is the honest and apostolic occupation of philosophical mathematics, which ennobles this great work to be designated as a sanctified colloquium of ingenious knowledge.

CHAPTER 13

HOW CAN WE DEFINE THIS KNOWLEDGE AS A PERFECT AND PROVISIONAL SOLUTION TO WORLD CRISIS?

World crisis, which has been rampant since the emergence of the machine age and has continued to our jet age (which is technically non-monad with Microsoft dynamics), requires ingenious and practical knowledge to solve. A living knowledge is philosophically and mathematically driven in presenting facts and figures with the use and application of dynamic exposition and expression. The retrogressive nature of our present civilization, which is affecting humanity in all areas of creation and procreation, is certainly asking for a new agenda in the areas of education, commerce, industry, politics, government, and the social and civil order. The use and application of inspiration is naturally pure and perfect in presenting concepts in the originality of its anatomy. It is evident in the annals of our contemporary civilization that the global universe is becoming saturated and stunted with ideas that cannot move our civilization forward.

The insurgence of material enmity, cannibalism with a high level of brain drain, is a global issue that must be addressed with the use and application of living knowledge. A work to present how this holds the perfect and provisional solution to world crisis must be able to sort out and mention the following points.

- World crisis is chronic and endemic.

- It has become traumatic and infectious.

- It is spreading all over the world, beyond the menace of the equatorially eastern wind.

- It is encouraging the growth of poverty, degeneration, and impoverishment with direct and indirect slavery.

- World crisis is greatly pushing the world to an oblivion, which is not the wish, will, and wisdom of the Creator.

- It is consistently desecrating the image and anatomy of material science.

- It is retarding the growth of wisdom, philosophy, and mathematics at the expense of living knowledge.

- World crisis is critical in ensuring that intelligence is mismanaged and manhandled.

- It is perplexing and dwarfing the growth of mind in relation with the spiritual ideas of time and honest living.

- World crisis has succeeded in creating a stigma that is dangerous in the universal psyche. This is practically evident in the way and manner humans behave without truth, which is against the productivity of philosophical mathematics.

- World crisis has succeeded in destroying the universal social order and the universal economic progress. Honest practise is needed to ensure creation is not detrimentally affected.

- World crisis has given rise to the production of dangerous exterminators like nuclear weapons, biological weapons, and other armaments that work to create hate and slavery among nations.

- It has succeeded in creating and producing strong people in the negative against strong institutions that are the need of the one-world family.

- World crisis, which is being orchestrated by man's inhumanity to man, is a figurative jihad, and most of the effects are certainly seen in the growth and expansion of militancy, corruption, and bribery, including other kinds of cheating against humanity.

The urgent need of using living knowledge in providing perfect solutions to the world's crises requires that the following must be done.

- Humanity at all levels must be managed and driven with the honest teachings of the Bible and other sacred sermons.

- The wisdom of the Twelve World Teachers must be employed as a way and manner of training, re-branding, and institutionalizing the concept of the brotherhood of man's principle.

- The educational system must be detached from high levels of enmity, which have infiltrated it by the Microsoft sector.

- Religions at all levels must be reengineered to seek the face of the Creator against the practises of their ordinances, which most of the time is not in accordance with their universal mysticism.

- Living knowledge must be used at all levels of endeavour and existence, ensuring that humanity is charged in knowing that the world is created for the purpose of mutual and practical co-existence.

- Living knowledge should be used in reducing the menace of material science, which is astronomically in the fast lane and invariably degenerates the growth of truth.

- It should be used to correct the growth of class levels, which is one of the greatest problems that gave rise to the worst world crisis.

- Efforts must be made by all and sundry to protect the growing economies of the world, including ensuring that the growths are protected and empowered in designing an acceptable future for themselves.

- The legal system should be restructured in the best opinion and ability of living knowledge, which will reduce the dangerous use of negative logic in deciding cases that most of the time work against the determination of honest truth.

- The dynamic system of the universe is important to be protected with the use and application of the living knowledge, which is best driven by the consummate standard of philosophical mathematics.

A look at what is happening in our present global universe reveals that humanity has lost track of philosophy, the knowledge of pure and applied mathematics, and the simple and technical ability of ingenious wisdom, which is free for all to use. That is why the honest restructuring of the global universe is a critical and sensitive agenda that is facing world leaders. A look at the production of balanced and living knowledge reveals that this is in fact the only wish of Mother Nature, and one of the peculiar gifts the Creator, in his infinite and illumined grace, gave humanity in order to live in unity in diversity.

Living knowledge as a dynamic examination of philosophical mathematics appreciates that all creation extended from the monumental monad, who is hereby defined as the ingenious strength of all creativity. This is why to live in the context of this knowledge is the only way by which a permanent and provisional solution will be gained, in order to put a peg in the incursion of world crises that have greatly degenerated humanity to the rank and file of the lower animals despite our abilities to sojourn in the areas of Microsoft technology. It must be explained that the growth of famine and starvation must be associated with the system and manner in which the world crises have gradually removed trust among nations. The wrongful use of documentation—both nationally and internationally, and particularly in the areas of international relations, diplomacy, and foreign missions, to name but a few—is a dangerous slap on the face of human engineering. This is enough and must be stopped if humans want to live up to their expectations.

This chapter in its totality is a wakeup call and another version of the sermon on the mountain. It is written for the world of the well informed, which will use it as a Sagittarius gospel that is empathic in living and in explanation with universal examination. It is here that I and all the apostolate are concerned that the best humanity can do in order to actualize the real vision of life is to appreciate with honest adoption that the emergence of strong people create room for world crises and slavery, whereas the presence of strong institutions is always a universal enhancement by which the nobility of humans are guaranteed.

The limitation to manage the universal resources with the use and application of living knowledge, which is the fountain thrust from Mother Nature, is an agenda with simultaneous catalyst that challenges all in creation to appreciate that the introduction to philosophical mathematics, as the tedium of balanced anatomy of living knowledge is hereby dynamically examined with philosophical and mathematical propositions. Those propositions are intended to make the world of creative engineering, dynamic technology, economic analysis (with its engineering mechanism) and governments know that every enabling factor has a structure, and every structure is blended with an anatomy. The choice of how we will live and fare in this ecosystem of love with unity is hereby defined and denoted as the greatest challenge of man. He can make an honest and pragmatic headway with the use and application of philosophical mathematics, which is hereby defined as a luminous crescent that provides the best of light for the universal garden, which man is naturally elected as a permanent and perpetual gardener.

CHAPTER 14

WHY DEFINE BALANCED KNOWLEDGE AS A CONSUMMATE HALL OF IMMORTAL FAME?

It is important to explain that a hall is usually nurtured and structured with the perfect enabling of absolute consummation. We must equally appreciate that since the dawn of consciousness, immortality has remained a consummate hall that has greatly impacted its knowledge in the fame of famous beings. My reasoning to define balanced knowledge as a consummate hall with immortal fame is stemmed on the fact that humanity and the one-world family need to know the following.

- That knowledge is not only power—it is equally love, light, life, and creativity.

- It is the permanent and universal way of the universal One.

- It authored science, which further created engineering and all its segments, and it restructured science, which further incubated technology as the nurtured and gifted heartbeat of its creativity.

- Balanced knowledge authored philosophy, which in turn discovered that mathematics has to be blended and intertwined in order to give wisdom an acceptable assurance. That assurance makes it function in its natural, electrifying mystification with its diversification.

- It is balanced knowledge that gave the first and foremost scholar this conventional and universal name. It is the same knowledge that honoured the immortal works of the great scholars, investors, prophets, seers, sages, and adepts.

- It is equally the first and foremost missionary estate since the dawn of consciousness.

A look at my contributions in defining and diversifying the immortality of balanced and consummate knowledge will let one certainly appreciate that it is a dynamic hall that warehouses the secrets of all ages. Looking at the organograms of the universal anatomy, balanced knowledge is best defined as creation in action, including being the hallmark of the Creator in simultaneous and balanced action. Balanced knowledge at any point in time is pure and perfect; it utilises philosophy and mathematics, which work to enhance the perfection agenda of balanced knowledge both in creation and organization. That is why my definition will be utilised by the balanced wisdom of posterity, which this museum of knowledge is created so that the mistakes of the past can be reduced in order to ensure that a new agenda, which will make man to appreciate himself as the pinnacle of philosophical mathematics, is established.

A look at the present knowledge that humanity is using will certainly appreciate its high level of consummated academics. It is knowledge that is practically driven with Microsoft esteem. That is why it lacks the wish, will, and wisdom of philosophical mathematics, which serve the Orientals both as teachers, chancellors, and the tedium of honest and practical enabling. A lot of people may wonder how and why I am using a whole chapter for this purpose, particularly in this format, and the reason is explained here.

- That man is created a mission with missionary and visionary activities.

- That the universe cannot be managed effectively if the Creator is not brought into the focussed foreground of its agenda.

- That human knowledge is imperfect and improper, while natural knowledge is ingenious, infectious, and illuminating. That is why it is the product of the living agenda of the creative principle, policies, and practises.

It is important for the benefits of the one-world family to appreciate that we cannot do without living knowledge. To know with systematic and balanced adoption that any knowledge that is not the product and directive of the

Creator will only help humanity create more crises, which in turn will relegate them to the background and show them that the act of God is the only act that will ennoble creation to be, pause, and ponder that the glories of the celestial will certainly govern the activities of all in creation.

At this point, we must know that the Creator is a monumental hall, and that is why his arts and crafts are consummate and immortal. We must equally appreciate that his fame is united and natural in impeccable wonders, which reveal that he is famous in everything. A scientific look at the workings of philosophical mathematics reveals that all in creation is contained in the original fame and frame of the excel and exilia of the consummate monad.

Humanity is hereby challenged to know that creation is a hall of fame, an immortal crescent, and a consummate and balanced knowledge. That is why to regenerate in the ability of its esteem is a challenge for all and sundry, including being the honest and practical way by which the immortality of creation can be guaranteed simply because the guarantor, who is one in all, is unique in ensuring that his arts must certainly be in the hall of fame. This work is hereby donated as the jubilee of balanced knowledge.

CHAPTER 15

A Detailed and Dynamic Overview That Blends Part Ten as the Honest and Practical Tedium of Mother Nature

The introduction to philosophical mathematics, designated as the balanced anatomy of living knowledge, has been examined and explained in the opinion, assessment, understanding, and appreciation of the origin. Living knowledge is transcendental, and this is why it is donated as the mystery of science, the power of technology, and the systematic enabling of engineering production.

A look at the chapters, which are splendid and blended with transcendental thoughts, reveals that the anatomy of living knowledge holds the answer to world crisis. They hold the solution to human growth with empathic reasoning, and it can equally be seen as the super-genius foundry of human dynamism. The origin of knowledge, including the past with its functions, reveals that knowledge is the only thing that holds the universe as a comprehensive and composite entity. But unfortunately, humanity—particularly the academic province—has not been able to discover the secrets, including what makes knowledge a sacred sanctum. The academia, including technology and engineering, have not been able to investigate the enabling of technology as a compact and diversified museum and its original estate to exist as an illumined and ingenious archive.

We have the knowledge of what is the philosophy of science, the philosophy of energy, the philosophy of technology. This overview is chosen to exist as a dedicated and documented ornament of the philosophy of knowledge. The chapter that dealt with when and how is a knowledge considered to be living took one to the next segment, which described why philosophical mathematics is designated as the anatomy of a living knowledge. The simultaneous correlation of these chapters can be utilised in assessing the overall esteem of the whole book. But in earnest and with practical honesty, philosophy is perfect with its monumental existence, and with practical revelation. Mathematics is data—and research-driven in ensuring that the course of perfection, including defining all that is considered as the mystery of the anatomy of mathematics, is well galvanized as a standard by which humanity can actually appreciate that life is mathematically driven.

It must be stated that the whole book is driven with what can be understood as a systematic and honest logic. That is why the chapter that deals with whether it is logical to explain living knowledge as being the way and path of immortal wisdom reengineered a thought-provoking answer that gave us the original and fullest meaning of wish, will, and wisdom, which is hereby understood with technical and transcendental science as being another meaning of the World Wide Web.

There is a need to equally attempt to configure in this overview how and why a living knowledge is naturally dynamic without pollution. A look at the injustices of our chaotic civilization will let us accept and believe that we are living in an era of Microsoft pollution that is invariably regenerating a high level of insurgence of armaments, cheating, enmity, lack of focus, vision, and poverty. These issues are visibly seen in the psyche and annals of the universal community. Apart from this, the Microsoft era has committed the global family into what is considered in this context a suicidal torture of technological melancholy. That melancholy makes it difficult for the normal human to reason along the paths of Mother Nature, which in this context is defined as the lead way to living knowledge.

In writing this overview, it is known that the high level of global and economic meltdown is supposed to have been an economic wakeup with astronomical growth, if the province and perfect wisdom of living knowledge is involved in handling human affairs. In this respect, what we are suffering can best be defined as our inherent and negative stubbornness, which we call science and technology, in our attempt to lock the Creator out of his creations. It is here that the wisdom of living knowledge is always knocking at the doors of the universal psyche to involve inspiration as a business necessity and as a scientific and technological guide. This overview, which is hereby donated as a guided colloquium, is hereby given to the one-world family in order to ensure that a serious reversion is made.

It is here that it is authored and am authorized to have the glory of this work and the authority of living knowledge, which is essential for the best objectivity and dynamism of the universal One who is all known, all believed, and all accepted to be the maker of philosophical mathematics.

CHAPTER 16

A GENERAL OVERVIEW ON THE ANATOMY, CONCEPT, AND DYNAMICS OF PHILOSOPHICAL MATHEMATICS, WHICH MAKES IT AN INGENIOUS LIVING KNOWLEDGE

An Oriental weaver once told his admirers that he is the weaver of the Lord's name and not an ordinary weaver for himself. If this legendary maxim is to be understood in the originality of its blending esteem, the dynamic concepts of philosophical mathematics can be redefined as an apostolate and apostolic polyester that is an encyclopaedia, an academic and scholastic Bible. The challenge and the request of the academic sector to give the one-world family the first and foremost book on creation, which is designated and understood as the dynamic concepts of philosophical mathematics, was not borne out of the existence of mental and academic prowess. Based on what is known as existing facts and figures, mathematics is a natural science with its applied philosophy, and philosophy is a natural science with its applied mathematics. That is why the synchronous ornament by which this book exists as a possession pinnacle will help all and sundry in knowing that creation is a dynamic and technical growth process.

A look at the introduction, which exists with the abstract will, practically makes all and sundry redefine the structure of philosophy and mathematics. The manner in which the blending formats, features, and abbreviations exist makes one appreciate that this noble course can be defined as a practical Pascal. Apart from this, the ornamented dynamism that took us to some of the rarefied facts and points of dynamic interest (which are must-know with philosophical understanding and mathematical appreciation) can equally be donated to the one-world family as clandestine and galvanized tutorial.

All the attributes of part 1, with a lot of theorems and proofs, will make the academia reason deeply in accordance with the wisdom of Mother Nature. The practical intrigues and interests that opinionated the harmony of part 1 took us to the original concept of the B-concept of what philosophical mathematics is, and we delved into the C-concept, which made parts 1 and 2 be seen as the beginning of an ingenious abstract. This galvanized into part 3, which is defined as another argument on the metaphysical mathematics of Plato's engineering dynamism.

Practical and honest teachers will certainly believe that this is the systematic scenario that gave rise to the origin of the sine, cosine, and tangent formulas. In this respect, part 3 can best be defined as a philosophical and mathematical rotor that was written to reawaken the industry of mathematics, particularly in the areas of sine, cosine, and tangent appreciation. The way and manner the wisdom of geometry is handled in furthering and featuring assets can best be defined as thanks to the world of geometric and engineering propagation.

Geometry equally let us examine the basics of engineering mathematics, which can best be known as the easiest defined engineering and designated as statistics. The humble explanation of statistics as an engineering databank with its research-driven mechanism took us to part 4, which deals and delves into an introduction to philosophical principles with some mathematical theorems. This part gave us a comprehensive summative that recycled and grilled all that is considered in triangular theorems and their functions, helping us to appreciate the originality of angles, squares, perpendicular lines, and more.

Part 4 cannot be concluded comprehensively if the ornamented prince of mathematics, Pythagoras, was not given a permanent chair in the ingenious hall of pure mathematical reasoning. Its overview started another ascension and ascending drama that delved into part 5, which is an introduction to Pythagoras and philosophical mathematics. A look at the comprehensive existence of this part will let one appreciate that Pythagoras existed during his time as an oracle, a wizard, and a monumental Sagittarius. It is believed that his mathematical and ethical principles and policies are usually donated as the acceptable kingdom of living knowledge, particularly in the areas of algebra and geometry, and including all that can be galvanized and accepted as the tedium of engineering and technological mathematics.

Parts 5 and 6 take us to part 7, which tactfully deals with the demystification of the sacredness of mathematics by the Oriental wizards. In this way, mathematics is scientifically and technically enlisted in the use and application of the following, which starts with mathematics: algebra, geometry statistics, data processing with arithmetic, research methodology, calculus and quantitative analysis—all were treated as segmented, empathic forces. Another area in this segment that co-exists with the existence of mathematics took us to the real knowledge of physics and applied physics, which is translated to a knowledge of transcendental science and astrophysics. Chemistry and applied chemistry, which is the hall and alchemy of chemical technology and chemical and astral engineering, took us to geography and further geophysics.

The area of natural science that made us understand production science, production biomass, and other related components in this segment also took us to appreciate what the chapter defines as the challenges and visions that gave rise to the demystification of mathematics.

The way and manner part 7 took us to parts 8 and 9, with their systematic and renowned overview, can best be defined as seriously challenging and enabling the world of material science, and technology. That is why these two parts succeed in creating sophisticated raw material for the world of science and technology. These parts started with an introduction to philosophical mathematics as the engineering modules and mechanics of balanced constructive engineering, and they ended with a dramatic style, further opening a practical and segmented horizon, which opened a tactful VSAT for part 10, which is hereby donated as a compendium of lessons.

Part 10 lives and exits on its own as the hallmark and anatomy of knowledge. It is natural, purposeful, and practical in providing us, particularly the academic and scientific provinces, how and why the human race must be very careful. All that is contained in this part is best defined as a philosophical mathematical guide that will help the one-world family achieve a simultaneous and constant living.

A look at the overview will certainly challenge all and sundry in knowing that we cannot live in a dynamic world without employing the perfect and purest dynamics of philosophical mathematics. The vision of globalization and the mission of a technological vision can only be established and actualized with the use and application of philosophical mathematics. It is here that the ornamented apostolate—with me as an ingenuous task team leader, and including Rejoice C. Adiele, Queen Nwokocha, and Ubani Ikemba—hereby comprehensively and with all practical agreement raise no arguments to the challenges of Mother Nature that the task to present philosophical mathematics as a dominant act of the universal apostolate must be done with tranquil achievement. This book can best be defined and accepted as the immortal work of pen and pages that will serve the overall universal province as a consummate colloquium of dependable Britannica. Everyone is invited to read and reason in order to acquire the philosophical mathematical understanding contained in the anatomy of living knowledge, as dynamically explained by the original frame and purpose of Mother Nature.

AFTERWORD

The Urgent Necessity of Using the Infinitude Method in Establishing and Solving Mathematical Problems

The infinitude method, which is in synchronous alignment and agreement with the tedium and concept of progressive research methodology, can best be defined as the living anatomy of all balanced knowledge. The science, art, and concept of the infinitude method are nurtured and natural with the use and application of dynamic application mechanism, which is in harmony with a progressive, elongated, creative concept.

In this respect, the infinitude method is urgently needed in solving mathematical problems with the following examples.

- **$0 + 0 + 0 + 0 = 0$** to infinite dimension.

- $1 + 1 + 1 + 1 + 1 + 1 + 1 + 1 + 1 + 1 = 10$, which can be progressively infinite above and beyond the mathematical authorization of a trillion.

- 2 + 2 + 2 + 2 = 8, which can equally lift the dimensions of mathematical progression to the honest mathematical polyester of infinitude dimension.

In other words, 2 × 2 × 2 = 8, which can equally continue in that arithmetic and geometric progression to an infinitude dimension, and which goes to explain that the forces, factors, policies, and principles that fathomed the ingenious factor of philosophical mathematics practically rests on the dynamic concepts of philosophical mathematics.

In this respect, the philosophy and the magnitude of the infinitude method, which is best donated as a mathematical wizard, is best understood in the great impact of transcendental mathematical enabling, which rejects and negates the existence of QED. At this point, how can the infinitude system, with its methods, solve and abolish the numerous mathematical problems? It is important for us to see and welcome the objective answers as a new concept with dynamic understanding that will help the overall universe of human dynamism in knowing that mathematics can only progress and prosper with the use and application of the infinitude concept.

- The infinitude concept, which is original with the philosophy of Mother Nature, works with the mathematical vision of purity and perfection.

- It is a method that understands the objectivity of constancy, assessing and appreciating how and why multiplication, subtraction, addition, and division enhance the performance of the sine, cosine, and tangent formulas.

- The infinitude method is the origin that anatomized all that is conceived and conceptualised in living knowledge. That is why mathematics plays an intertwined role in leading other subjects with philosophy, to the best understanding and appreciation of the infinitude method.

A look at the mathematical coefficient of statistics, data, and research, including all that is configured in the table of values with drawings and graphics, shows that the infinitude method is the original anatomy of mathematics. That is why its structural frame is always empowering the functionality of pure and applied mathematics in relating purposefully to a titan of figures, letters, factors, and factories. The task to demystify QED becomes an acceptable mandate that is functional and futuristic.

Infinitude method as a transcendental science with philosophical mathematics is best known and defined as the heartbeat of mathematics and the mind and brain by which philosophy is seen as the mother of science. The demystification of the oracle of mathematics, as contained in parts 8 and 9, makes us appreciate that mathematics is futuristic in growth and evolution. It is best donated in this concept as the original searchlight for the simultaneous functioning of science, technology, engineering, and all human-related endeavours that are mathematically driven with the use and application of pure and applied philosophy.

For the overall utilisation of the academia, the universe, and the research institutions—with their multifarious factories and industries—the following must be known, used, and appreciated.

- That the infinitude method is the supreme way of the practicality of mathematics at all levels.

- That this mechanism, which is a consummate method, exists with inference and reference, without adulteration and pollution.

- That this method is the pure and perfect origin of science that extends to technology.

- It is equally the consummate origin of pure-alchemy, which is the ancient exponent of chemistry and chemicals—and which St. Germain, in the ability of his enabling, translated into the engineering province of transmutation, which made him to be ornamented as a tutor of alchemical science.

I have donated this concept as a mathematical messiah of pure facts, figures, reasoning and logic, pure ascension, and astronomical growth. That is why the book is always applying the functionality of pure and applied mathematics, which in earnest is the scientific stimulant to scientific engineering mathematics. I donate this work as a hall of fame to empower future mathematicians and scientists in the understanding of natural mathematics, natural science, and natural engineering and technology. I hope that all segments will understand that Mother Nature, which is the honest orbit and enabling of infinitude method, is always at work when humans delve deep in the knowing that all in creation is already invented by the only knower, who is hereby designated as the monumental hall of philosophical mathematics. The keep also gave this book to the famous and glorious posterity, which is coming up with a new agenda that is not driven with the use and application of obscure mathematical methods with their infamous techniques. Those techniques worked to reduce human intellectual prowess in accepting and using the ancient and ambiguous QED, which made mathematics a course in the mockery of knowledge.

For the record, 0 and 1 will always continue in infinitude, whereas the infinitude method will forever continue to infinitum. With this ennobling expression and scientific examination, we will now understand that the infinitude method is the science of balanced and consistent innovation with evolutionary techniques, which makes it function as a comprehensive and consummate Leninger and an absolute colloquium for the use and application of the one-world futuristic family, which is galvanized with the anatomy of the infinitude method.

Quotes

"Every road has a map, and every map is created by the use and application of constructive performance."

—Anthony U. Aliche

"Philosophical mathematics is a databank and engineering kit whose parts and instruments do not fail to actualize the vision of this creation."

—Anthony U. Aliche

"It is necessary that a surgeon should have a temperate and moderate disposition. That he should have well-formed hands, long slender fingers, a strong body, not inclined to tremble and with all his members trained to the capable fulfilment of the wishes of his mind. He should be of deep intelligence and of a simple, humble, brave, but not audacious disposition. He should be well grounded in natural science, and should know not only medicine but every part of philosophy; should know logic well, so as to be able to understand what is written, to talk properly, and to support what he has to say by good reasons."

—Guido Lanfranchi

"Medicine rests upon four pillars—philosophy, astronomy, alchemy, and ethics. The first pillar is the philosophical knowledge of earth and water; the second, astronomy, supplies its full understanding of that which is of fiery and airy nature; the third is an adequate explanation of the properties of all the four elements-that is to say, of the whole cosmos—and an introduction into the art of their transformations; and finally, the fourth shows the physician those virtues which must stay with him up until his death, and it should support and complete the three other pillars."
—Philippus Aureolus Paracelsus

"The sense that the meaning of the universe had evaporated was what seemed to escape those who welcomed Darwin as a benefactor of mankind. Nietzsche considered that evolution presented a correct picture of the world, but that it was a disastrous picture. His philosophy was an attempt to produce a new world-picture which took Darwinism into account but was not nullified by it."
—R. J. Hollingdale

"There is no philosophy which is not founded upon knowledge of the phenomena, but to get any profit from this knowledge it is absolutely necessary to be a mathematician."
—Daniel Bernoulli

APPENDICES

My Philosophical Quotes

- Man is a creative force from the balanced wisdom of electrifying philosophy.

- Philosophy, being the mother of knowledge, is the supreme anatomy of all.

- Every science is electrifyingly compacted into the thinking and balanced wisdom of philosophy. As an elective and electronic science, it is scientifically philosophic.

- Philosophy, as the supreme way of man, is purposefully and pragmatically driven in becoming the way of all.

- The world can neither know peace nor dynamically progress without the use of the electrifying concept of philosophical dynamism.

- Material science is incoherent at all levels of its existence, and hence producing chaotic principles. Philosophy is coherent, compact, edified, and knowledgeable in directing the affairs of all.

+ The universe, which is a philosophical conclave, is driven by the scientific ideology of philosophy.

+ Man as an extension of philosophy is still at the fifth kingdom of balanced and electrifying philosophy.

+ Philosophy is the mother of balanced knowledge and is defined as the structure of all, the secret of light, the sacred wisdom of all ages, and the blessed teacher of immortality.

+ Philosophy, knowing the truth about all things, will remain the only acceptable way of universal civilization, tradition, and religion. It also holds the power in amalgamating all the forces of human dynamism.

+ Both truth and beauty are simplified philosophy, and they do not separate their activities from the true knowledge of philosophical dynamism.

+ Philosophy is the scientific net of the universe and the meta-logic hub of the one-world balance; it is always contesting what is in the knowledge of man and what will be in his future.

+ It is the only science that originated the concept of all websites and all zero powers in the account of the zero; it electrified knowledge with its dynamic balance.

- Man must know that all natural and eternal knowledge originated from the nebula hub of positive philosophy.

- Philosophy is the supreme teacher, which explains what can be known and accepted as the wisdom of a dependable knowledge.

- Philosophy occupies a natural place between theology and science; this is why its extensions are beyond words and thoughts.

- The knowledge of philosophy is the only education that makes man the prince of honest knowledge, including being the pride of dependable wisdom.

To the best of my knowledge and understanding, these quotes can best be defined as an advanced neo-philosophy. At all levels of understanding, they will always remind the apostolic foundation of philosophical engineering that the goal of philosophy is to produce great and balanced thinkers to run the one-world purpose, and to appreciate that the universe is one and must be governed with philosophical love port. That is why at all esteems and extremes, this book has perfectly designated philosophy as the maker of great thinkers.

Poet's Code of Ethics

Inspired by Herbert Spencer and His Aborigines

- To attain the brotherhood of man idea by giving righteous action and goodwill service to every man, instead of taking from him that which he has.
- To discover that all men are extensions of each other, that man is made for man, and that the hurt of one man is the hurt of all men.
- To develop character intelligence and good citizenship by teaching every man from youth how to be a good neighbour and a loyal citizen.
- To discover one's inner self by awakening within him that spark of divine genius, which lies dormant in every man.
- To teach man to think rather than to remember and repeat.
- To realize that work done for the material world should be for man's ennoblement, not for grinding his soul out in the gears of industrial machines.
- To know that man is mind, not body; that he is immortal spirit, not flesh; that he is good, not bad.
- To judge the righteous and religion of any man by what he does to his fellow man—not by his beliefs, doctrines, creeds, or dogmas.
- To give a scientific course of study for comprehension of the Spencer Code of Human Relations as being the inviolate law of God, which man knows he must

obey or else pay the price in personal unhappiness and international wars.

- To teach the scientific meaning of "Seek ye the Kingdom of Heaven within you", "I and my Father are one", "God is Light; God is Love", "What I am, ye also are", and "What I do, ye can also do".

Lao Russell's Code of Ethics for a Living Philosophy

- To bring blessing upon yourself, bless your neighbour.
- To enrich yourself, enrich your neighbour.
- Honour your neighbour, and your neighbour will honour you.
- To sorely hurt yourself, hurt your neighbour.
- He who seeks love will find it by giving it.
- The measure of a man's wealth is the measure of wealth he has given.
- To enrich yourself with many friends, enrich your friends with yourself.
- That which you take away from any man, the world will take away from you.
- When you take the first step to give yourself to that which you want, it will also take the first to give itself to you.
- Peace and happiness do not come to you from your horizon; they spread from you out to infinity beyond your horizon.
- The whole universe is a mirror which reflects back to you that which you reflect into it.
- Love is like unto the ascent of a high mountain peak: it comes ever nearer to you as you go ever nearer to it.

Professor Aliche's Living Code of Ethics for the Modern Generation

- Guard against doing evil, either by omission or commission.
- Stand on a good institution, which will represent you and your generations to come.
- Think good of your neighbour, and the world will think good of you.
- Speak no evil of anybody, see no evil, do no evil, and hear no evil.
- Maintain an ethics of love, with peace to guide you to a good public relation.
- Show love to everybody and ensure that peace governs all your life with truth and love.
- Envy nobody to avoid anybody envying you.
- See the good in others and emulate the best you see in the world.
- Work to perfect your talents and see the beauties in creation as a goal awaiting for you to attain it.
- Help the needy and the helpless, and see to it that your help is without harm to those that need it.
- Develop your faith with sound courage and positive judgement.
- Avoid committing karma so that you and your generations will not suffer from you burden.
- Speak well of your neighbour, yourself, and the world; maintain discipline at all times.
- Shine as a living star in praise of your creator and maintain the spirit of a saint.

- Attune yourself to oneness with the infinite.
- Do your duties without cheating and guide against being cheated.
- From a calmer height of love and wisdom, correct those who err by cheating, killing, and committing other vices of Satan.
- Think positive, be constructive, and use the wisdom of love at all times to be a master builder.
- A living code of ethics for the modern generation is likened unto the ascent of a high mountain peak: it comes nearer to you as you go ever nearer to it.

Specific Treatment of Words and Terms

- Acknowledgement(s)
- all right
- Britannica—a compendium of knowledge
- Cosmogonical—having to do with the cosmic and the cosmos
- Coworker—fellow workers
- the Creator—God
- e-mail—electronic mail
- emengards—gurus, avatars, adepts, or masters
- exilia—anatomy
- fontious
- Godhead—the trinity
- Goras—the real name of Pythagoras (Peter Goras)
- Internet—an international computer network connecting other networks Interconnected, striking
- Iroko—giant forest tree particularly grown in the savannah regions
- Kabbalah—a form of religious ritual in the oriental world
- Lintious—vast, boundless, great
- lumbacite
- rubric—set of instructions written in a book
- metaphysics (but meta-science, meta-religion, etc.)
- millicent,—holy
- monadian—very knowledgeable person
- myopism—narrowness, short-sightedness
- omeganation, omeganization—a united entity that exists for wisdom

- philo-sophia—love for wisdom
- purpose-driven—focused, the aim to achieve
- quantious
- sanctious
- sionsa
- tendamus
- VSAT
- Website—an internet address, a communication line
- willbros

- Double quotes are used.
- Numerals are left as they are because of the mathematical nature of the book.

Other published and yet to be published titles by Anthony Ugochukwu O. Aliche

Published Books

Management/Business

1. *Business Management and How to Achieve Organizational Goals*
2. *The Principles and Policies of Research Methodology*
3. *Contemporary Issues Which Determine the Dynamic Functions of Business Management*
4. *The Role of Ethics in Cooperate Governance*
5. *General Principles of Management and Research Methodology*

Philosophy

6. *Philosophy: The Maker of Great Thinkers*
7. *The Fundamentals of Core Logic*
8. *The Twenty-First-Century Philosophical Caveats*

Religious/Inspirational

9. *Have You Discovered Your Assignment with Destiny?*

10. *A Great Personality Dwells in the Flesh*
11. *Wisdom Is Wealth*
12. *Bring Out the Christ in You*
13. *Whose Vessel Are You?*
14. *What Makes Great Men*
15. *The Best Belongs to All*
16. *God's Gifts Are Irrevocable*
17. *Using Your Gifts to Help Others*
18. *Challenge Is a Catalyst to Success*
19. *With Love All Works Are Easy*
20. *From Problems to Praises*
21. *Success Is My Portion and Who Is Truly Successful*

Science and Technology

22. *The Economic Importance of Information Technology and Information Communication Technology*
23. *The Electrifying Power of Unity: A Dynamic Blueprint for Restoring World Peace and Balance*
24. *The Dynamic Concept of Philosophical Mathematics*

Health

25. *Honey Is Health*
26. *The Healing Powers of Honey*

Metaphysics with Mysticism

27. *What Is beyond Truth Scientifically Explained*
28. *The Mystical Powers of Our Lord's Prayer*
29. *The Mystical Secrets of Our Lord's Prayer*

Literature

30. *30. The Orphan*

Management/Business

31. *The Dynamics of Effective Marketing*
32. *Financial Management and Marketing Strategy*
33. *The Role of Management and Business Administration in a Competitive Environment*
34. *The Dynamic Functions of Business Strategy*
35. *Business Strategy in Our Contemporary Era*
36. *Teach Yourself Research Techniques*
37. *The Evolving Dynamics of Business Strategy*
38. *The Dynamic Challenges of Organizational Growth*
39. *Business Strategy and Entrepreneurship Management*
40. *Understanding the Dynamics of Marketing Management*
41. *The Dynamic Role of Policy and Strategy in Economic Achievements*
42. *Strategy and Business Strategy*
43. *The Foundation of Natural Economics*
44. *Management and Motivation: Its Theory, Principles, and Practises*
45. *The Benefits of Long-Range Planning*
46. *Understanding the Effective Dynamics of Business Strategy*
47. *Effective Planning Key Factor to Business Success/ Management*

48. *Computer Logic for Science and Engineering Students*
49. *The Imperatives of Computer Technology to Our Modern System Analysis*
50. *Creative Management with Computer Technology*

Philosophy

51. *The Importance of Logic and Diplomacy in Conflict Resolution*
52. *Plato's Philosophical Lamentation (Adept Series)*
53. *Akhenaten's Vision for Democracy and Philosophy (Adept Series)*
54. *Philosophy of Order of the Quest*
55. *The Power and Wisdom of the Unknown Philosopher (Columbus Readers/Adept Series)*
56. *Plato's Seven Point Discourses*
57. *Plato's Philosophical Constitution*
58. *Philosophy and Applied Logic*
59. *Introduction to the Anatomy of Transcendental Philosophy*
60. *The Dynamic Functions of Philosophy*
61. *The Scientific Application of Logic for the Benefit of the Internet Age*
62. *The Scientific Application of Logic to Quantum Physics*
63. *Philosophy and Logic for Modern Society*
64. *Philosophy: The Way of Great Thinkers*
65. *The Wisdom of Cosmic Thinking*

66. *Living in the Light of Infinite Truth*

67. *My Love for Knowledge the Aborigines of the Ross Cross*

68. *Great Minds Are Immortals*

69. *The World Is in Need of the Universal One*

70. *Adepts Books of Knowledge*

71. *The Power of Purpose-Driven Prayer*

72. *The Church Is the Way of Sinners*

73. *Melchizedek Mystical Principles*

74. *The Adventure in Wisdom*

75. *The Power of Creative Expression*

76. *Treasures from the Holy Grail (Adept Series)*

77. *The Legislature Is the Trustee of Democracy*

78. *Philosophical Poets, Volumes 1 and 2*

79. *The Encyclopaedia of Philosophy with the Twelve World Teachers*

80. *Living in the Light of the Universal One*

81. *The Benefits of Thinking and Acting Cosmically*

82. *Introduction to the Anatomy of Transcendental Philosophy*

83. *How Does a Philosopher Approach Nature?*

84. *The History and Technology of Logic*

85. *Scientific Logic*

86. *Plato's Thought for a Living Philosophy*

87. *Introduction to Pure Logic*

88. *The Concept of Logical Reasoning*

89. *The Philosophy of Developmental Logic*

90. *Common Logic for Everyday Usage*

91. *The Criticisms of Logic*

92. *The Dynamic Concept of Technical Logic*

93. *Logic for Everybody*

94. *Logical Mathematics for Engineering Students*

95. *The Objective Reasoning of Philosophy*

96. *Philosophical Dialogue in the Internet Age*

97. *Evolution with Philosophical Achievements*

98. *The Philosophy of Pythagoras Mathematics*

99. *Understanding Plato's Wisdom in Philosophy*

100. *The Philosophy of Engineering*

101. *Francis Beacon's Philosophy of Thought*

102. *Evolution and Philosophy*

103. *The Electrifying Wisdom of Philosophy*

104. *The Universe Is an Ethereal Province of Philosophy*

105. *The Philosophy of the Lost Continent*

106. *Mathematical Philosophy*

107. *Introduction to the Philosophy of Engineering Technology*

108. *The Nature and Power of Light*

109. *The Nature of the Errors of Material Science*

110. *The Nature and Powers of the Unseen Forces*

111. *Understanding the Philosophical Basis of Dynamic Mathematics*

112. *The Importance of the Law of Balance*

113. *Plato's Philosophic Thought on Evolutionary Growth/Dynamic*

114. *A Compendium of Immortal Philosophical Quotes of A. U. Aliche*

115. *A Reason in the Light of Philosophy Is a Pure Reason to Life*

116. *Understanding the Perfect Basis of Philosophical Techniques*

117. *Reintroducing Philosophy to the Contemporary Society*

118. *The Meaning of the Science of Philosophy to Man*

119. *Introduction to the Philosophy of Education*

120. *Understanding the Evolutionary Dynamism of Philosophy*

121. *Understanding the Contemporary Dynamism of Philosophy*

122. *Understanding the Bases and Principles of Philosophy as the Supreme Mother and Mentor of All Knowledge (A. U. Aliche's Mystified Dialogue)*

123. *The Power and Wisdom of Philosophical Foundation*

124. *The Cradle of Philosophy*

125. *Journey into Philosophy*

126. *Philosophy the Brain Box of All Knowledge*

127. *The Philosophy of Astrology*

128. *The Philosophical Mystery of Astronomy*

129. *The Man Who Killed Knowledge*

130. *Lone Philosophy*

131. *Time Is the Greatest Philosophy of Life*

132. *Philosophy and Astrology*

133. *The Role of Philosophy and Wisdom in Our Contemporary Era*

134. *Philosophy and the Holy Bible*

135. *Philosophy and the Koran*

136. *The Objective Role of Philosophy and Reasoning*

137. *Philosophy and Evolution*

138. *The Immortality of Leonardo da Vinci*

Religious/Inspirational

139. *Abraham: The Mystical Father of Faith*
140. *The Prodigal Pastor*
141. *The Inspired Life of Jesus (Adept Series)*
142. *Discovering the Fire of Christ in You*
143. *The Dangers of Spiritual Illiteracy*
144. *The Authority of Spiritual Powers*
145. *Divine Healing in a Contemporary Era*
146. *Another Book of the Proverbs*
147. *Spiritual Meditations, Volumes 1 and 2*
148. *Affirmations to Solving Our Daily Problems*
149. *Man Is the Beauty of the Universe, Volumes 1 and 2*
150. *Truth Is Universal, Volumes 1 and 2*
151. *What Is beyond God, Volumes 1 and 2*
152. *The Value of a True Teacher*
153. *Man, Who Do You Worship?*
154. *The Purpose-Driven Workshop*
155. *The World Needs Servant*
156. *The Secrets of Divine Security*
157. *Divine Love Is Everything*
158. *The Secrets of Divine Security*
159. *Favour, Favours the Brave*
160. *When God Calls You a Saint*
161. *The Values and Virtues of Simplicity*
162. *The Values and Virtues of Illumined Parents*
163. *God Has Not Forgotten Man*
164. *Women Are the Aroma of Beauty*
165. *That Two May Be One*

166. *The Miraculous Power of Faith*
167. *The Unequalled Power of Love*
168. *Twelve Positive Guides for Achieving Success*
169. *Man Is a Living Bible*
170. *Man Is My Mission*
171. *Your Success Is in Your Hand*
172. *Divine Virtues That Extol Human Life*
173. *The Values and Virtues of Goodness*
174. *The Price of Hard Work*
175. *Light Does Not Compete with Darkness*
176. *Using Your Meditation Positively to Provoke the Heavens*
177. *What Is Beyond God's Love?*
178. *Your Quality of Life Determines What Happens to You*
179. *Accountability Is the Way of Nature Etc.*
180. *No Man Can Escape Divine Auditing*
181. *Are You in Tune with Your Conscience?*
182. *The Order of Divine Illumination*
183. *Back Your Efforts with Prayers*
184. *How to Achieve Success through the Power of Personal Policy (PPP)*
185. *There Is Joy in Giving*
186. *The Immortality of Man, and Other Lectures*
187. *The Beauty and Power of Spoken Words*
188. *Mission Accomplished*
189. *Arise in Christ*
190. *Who Is the Author of Your Success?*
191. *Man Is Not Man without God*
192. *Who Is a Pastor?*

193. *Evangelism Is a Supreme Task*
194. *The Gospel Message*
195. *The Immortal Role of Destiny*
196. *Who Is My Anchor Hold?*
197. *How to Prolong Your Life with Love*
198. *Truth Is Light Eternal*
199. *Creation Is Creating*
200. *What Is the Worth of Man?*
201. *The Endless Power of the Infinite*
202. *Do You Know Who You Are?*
203. *The Burdens of Karma*
204. *The New World in Born/Creative*
205. *Who Are the Gideons?*
206. *What Is Your Philosophy of Life?*
207. *Has Man Known the Real Meaning of Life?*
208. *The Reality of True Love*
209. *The Challenges of Living with Balanced Love*
210. *You Greatly Owe the Creator Any Day You Live without Service*
211. *What Is the Light of the Stars? (Adept Series)*
212. *The Mystical Mysteries of Birth and Rebirth*
213. *The Power in Creation*
214. *Purify Your Stars with the Use of Nature's Garden (Adept Series)*
215. *Seeing the Powers of Christ in Your Creations/ Work*
216. *Cultists Are Devil Incarnates*
217. *Cultism Is the Problem of One-Man Church*
218. *Purity, Perfection, and Providence Are Divine Attributes*

219. *Kindness Is the Perfect Way of Mother Nature*

220. *In Heaven, Nurtured Souls Are Saints*

221. *How to Blossom with Beauty*

222. *Appreciate Yourself as a True Image of Divine Love*

223. *Learn How to Manage Yourself*

224. *Do Not Lean on the World's Path to Life*

225. *The Influential Currents of Positive Love*

226. *Avoid the Serpent's Love*

227. *Do Not Fear People but Respect Them*

228. *Do Not Trust the Ability of Love*

229. *Trust the Wisdom of Trust*

230. *Love with Caution*

231. *Love with Passion*

232. *Trust the Wisdom of Faith*

233. *Faith Does Not Fear*

234. *How to Make Yourself Relevant*

235. *What Is Your Definition of Real Life?*

236. *Do Not Roam in the Dark*

237. *We Must Know Our Roots*

238. *Do Not Judge Others*

239. *The Avenger of the Spirit*

240. *Self-Mastery Is the Wisdom of the Illumined*

241. *Just Count Your Blessings*

242. *Priests in Prisons*

243. *The Bone of a Hero*

244. *The Dangers of Illiteracy*

245. *Half Education Is More Dangerous Than Complete Illiteracy*

246. *Do Not Call Any Man God*

247. *The Oracle of Greatness*

248. *The Mathematics of Love*

249. *Lots Are Not Loved*

250. *"I Can Try" Is a Giant*

251. *I Am Not an Object of Pity*

252. *The World Around Us*

253. *Do Not Ask for Trouble*

254. *The Holy One*

255. *The Great Truth*

256. *The Drunkard Wife*

257. *How to Protect Your Success*

258. *How to Protect Your Destiny*

259. *Many Errors of Man*

260. *The Dangers of Love*

261. *Women, the First Lucifer*

262. *Living for Christ*

263. *The Reality of Life*

264. *The Call of the Apostles*

265. *Africa Must Be Delivered*

266. *The Secret Destiny of Nigeria*

267. *The Bible and Pastorship*

268. *The Pastor Who Does Not Know the Bible*

269. *The Wisdom of Talking in Parables*

270. *Greatness Cometh from God*

271. *Your Destiny Is Your Website*

272. *Visitors from Above*

273. *Echoes from Above*

274. *Mount Sinai Experience*

275. *The Beauty of Blessings*

276. *My Mission Is Destined Painters*

277. *Heroes of Destiny*

278. *The Power of Destiny*

279. *The Wisdom of Destiny*

280. *The Man Whose Life Inspires Me*

281. *The Will of Life*

282. *Forget the Past*

283. *The Scripture's Wisdom*

284. *Live to Leave Your Footprint*

285. *The Man Who Saw the Angels*

286. *The Message of Love*

287. *God Is a Perfect Match Maker*

288. *The Importance of Keeping Appointment with Prayers*

289. *Prayers Solve All Problems*

290. *To Pray Is to Live*

291. *Life without Prayer Is an Impossible Mission*

292. *The Miraculous Power of Using Sweet Melodies in Healing*

293. *Make Meditation Your Daily Communion*

294. *The Wisdom of the Immortal Words of Jesus Christ Which Earthman Could Not Comprehend*

295. *The Mystical Powers of the Holy Bible*

296. *Satan Is a Loser*

297. *How You Can Input Your Ingenuity for Positive Creativity*

298. *You Cannot Be a Success without Failures*

299. *Your Temporary Failure Is a Stepping Stone to Your Success*

300. *There Is No Royal Road to Success*

301. *How Temporary Failure Is a Good Preparation to Your Success*

302. *You Are Only a Failure when You Decide Not to Make Honest Efforts*

303. *Success If Not Measured in the Amount of Money You Have*

304. *Money Makes You Mad Only When You Allow It to Control You*

305. *Jesus Enriched the Whole World with the Application of His Consummate Wealth*

306. *You Must Strive to Add to the Success of the Universe*

307. *What Will Posterity Remember You For?*

308. *I Am Truly in Love with the Loved Ones*

309. *Only True Love Creates Trust*

310. *Divine Love Creates Balance*

311. *He That Cares for Others Must Be Cared For*

312. *Discipline a Child He Will Understand the Wisdom of Truth*

313. *The Love and Wisdom of Beauty*

314. *Progress Belongs to the Future*

315. *Anointed for Truth*

316. *Man Has Not Given the Creator His Position*

317. *The Emperor of Faith*

318. *From Victim to Victory*

319. *Power from Above*

320. *What Have You Done to Enrich the Universe*

321. *Forgiveness Is Costly*

322. *The World of Effective Creativity*

323. *Thinking for Positive Creativity*

324. *Love Is Expensive*

325. *The Marital Wisdom*
326. *How to Learn from Nature*
327. *What Is Your Degree of Giving*
328. *My Vision Is My Life Teacher*

Worship Series

329. *How to Worship with His Wisdom*
330. *How to Worship with the Holy Spirit (The Great Personality Which Dwells in the Flesh)*
331. *How to Worship with God*
332. *How to Worship with Your Self*
333. *How to Worship with Thy Neighbours*
334. *How to Worship with Yourself and Family*
335. *How to Worship with Your Friend*
336. *How to Worship with Nature*
337. *How to Worship with the Angels*
338. *How to Worship with Light*
339. *The Power and Wisdom of Transcendental Worship*
340. *How to Worship with Your Talents*
341. *How to Worship with Music, Songs, Solos, and Rhythms*
342. *How to Worship in Silence*
343. *How to Worship Alone with Your Wife*
344. *How to Be Thankful to God for Revealing the Secrets of True Worship to You*

Wisdom Series

345. The Life and Wisdom of Pure Souls
346. The World Has Many Eyes
347. Love Is the Supernatural Medicine of Honest Living
348. The Power of Wisdom of the Third Eye
349. The Wisdom of Cosmic Enlightenment
350. The Electric Nature of Wisdom
351. Mysticism and Wisdom in Practise
352. Wisdom the Supreme Knowledge of All Ages
353. How to Develop Your Genius with the Use and Application of Wisdom
354. The Bible and Wisdom
355. Wisdom and the Koran
356. Wisdom and Religion
357. Wisdom and Destiny in Manifestation
358. The Power of Wisdom in Ministration
359. How a Pastor Can Inspire His Knowledge with the Ingenious Wisdom of Inspiration
360. Man Is a Living Foundry of Inspiration
361. Creation Is a Comprehensive Revelation of the Power of Wisdom
362. Wisdom Is Spirit
363. Wisdom Is Consummate
364. The Unlimited Nature of Balanced Wisdom
365. Journey into Wisdom
366. Wisdom Is Perfect Truth in Manifest
367. The Rhythm of Wisdom Is the Creative Pulsation of the Creator

368. *A Mystic Is a Practical User of Ingenious Wisdom*

369. *Wisdom in Service*

370. *Facts about the Wisdom of the Three Wise Men*

371. *The Wish of Wisdom*

372. *The Will of Wisdom*

373. *Living in the Supreme Light of Wisdom*

374. *The Mystical Wonders of the Cosmic Wisdom*

375. *Absolute Wisdom Is Only Acquired through the Balanced Knowledge of True Mysticism*

376. *Wisdom Is a Perfect Midwife*

377. *Wisdom Is a Humble Spirit*

378. *The Wisdom in Creation*

379. *What Man Holds as the Height of Wisdom Is but the Beginning of the Knowledge of True Wisdom*

380. *The Mystical Wisdom of Destiny*

381. *Wisdom Is in the Blood*

382. *The Effective Powers of Wisdom*

383. *How to Marry Wisdom with Knowledge*

384. *Facts about the Wisdom of the Three Wise Men*

385. *How You Can Use Your Wisdom to Influence Others*

386. *Anointed for Wisdom*

387. *Wisdom from the Grail*

388. *Wisdom Is Genius*

389. *The Immortality of Wisdom*

390. *Make Wisdom Your Teacher*

391. *Adopt Wisdom as Your Counsel*

392. *After Trying Virtually Everything, Have You Tried Wisdom?*

393. The Role of Wisdom in One's Life
394. Wisdom Gave Birth to Knowledge
395. Discover What Is the Opinion of Wisdom to Your Destiny
396. How to Use Wisdom to Make Your Destiny Manifest
397. Allow Wisdom to Inspire Your Doctrine
398. Nature Is the Engineer of Living Wisdom
399. Illumined Wisdom Is Man's Heritage
400. Learn How to Make Wisdom an Anchor Point
401. The Immortal Works of Geniuses Are Written in Perfect Wisdom
402. Great Heroes Create with Wisdom
403. Wisdom Was the Unseen Teacher of Christ
404. The Works of the Seraphies Are the Supreme Manifestation of Illumined Wisdom
405. Wisdom Makes One Not to Be a Liability to the Universe
406. Teach Your Friend the Supreme Lessons of Wisdom; in Turn, He Will Teach His Friend the Blessed Lessons of Wisdom
407. Echoes from Wisdom Rules the World
408. No Authority Is Mightier or More Perfect Than Wisdom
409. Wisdom Made Man to Realize That He Is the Image of God
410. Wisdom Makes Me to Realize That God Is Everything in All Things
411. The Power of Leonardo da Vinci Is Derived from Immortal Wisdom

412. *Wisdom Taught Me the Scientific Rhythm of Mother Nature*

413. *All the Powers of Mathematics Are Written with the Immortal Dynamism of Wisdom*

414. *Plato Told the World That Wisdom Is the Wisest Institution*

415. *As Wise and Perfect as It Is, Wisdom Remains the Humble Teacher of All Ages*

416. *Joseph Weeds Told the World That Mysticism Is the Way of Wisdom*

417. *The Pillars of Paradise Are Ornamented with Immortal Wisdom*

418. *The Flame of Wisdom Speak Volumes about the Power of God*

419. *Ask for Wisdom and Do Not Request for Any Other Thing*

420. *Joseph Became an Immortal Hero of Wisdom, and This Is Why Every Shepherd Is a Teacher*

421. *Melchizedek Spoke about the Perfect Nature of Wisdom*

422. *John the Baptist Adopted Wisdom as His Honey*

423. *The Ingredient of Celestial Honey Are Contained in Wisdom*

424. *Teach Little Children How to Love Wisdom*

Metaphysics with Mysticism

425. *Journey in Light*

426. *The Mystical Anatomy of the Universe*

427. *Mysticism and the Bible*

428. *The Doctrine of Mysticism*

429. *He That Is in the Dark Does Not Know the Meaning of Light*

430. *Awake Is a Mystical Invitation for Souls in the Dark Side of Evolution to Transcend into Light (Adept Series)*

431. *The Cross Is a Symbol of Mystical Truth*

432. *The Benefits of Mystical Wisdom*

433. *Psychic Self-Defence*

434. *Understanding the Importance of Your Psychic Powers*

435. *Metaphysics and Astral Physics*

436. *A New Cosmogony of the Astral Plane as Empowered by the Wisdom of Rare Knowledge*

437. *Mysticism the Absolute Way of the Illumined*

438. *Passing the Light to New Era*

439. *Mystics and the Challenges of Mysticism*

440. *Wisdom Beyond the Genius*

441. *Man and His Misconceptions about Meditation, Mystics, and Metaphysicians*

442. *The Encyclopaedia of Metaphysics with the World Teachers*

443. *Meta-Physicist Concept of Astrophysics*

444. *The Twelve Mystic Messengers from Luminous Realms*

445. *The Book of the Great Message*

446. *The Cosmic Basis of the Mystical Doctrine*

447. *The Mystics View of Life*

448. *How to Know the Secrets of Mystical Wisdom*

449. *The Mystical Realities of the Holy Trinity*

450. *The Mystical Secrets of Our Lord's Prayer*

451. *The Mystical Teachings of Our Lord's Prayer*

452. *The Mystical Realities of the Blood of Jesus Christ*

453. *The Cosmic Bible*

454. *In Your Light We See Light*

455. *The Concept of Philosophy of Thought Photography*

456. *The Mysticism of the Gnostics*

457. *The Mystical Powers of the Sermon on the Mountain*

458. *The Ministry of True Mysticism*

459. *How to Live in Mysticism*

460. *The Cosmic Brain*

461. *The Mystical Wisdom of the Gnostics (Adept Series)*

462. *The Aborigines of the Holy Grail (Adept Series)*

463. *The Physical Universe Is Not Yet Wise (Adept Series)*

464. *Discovering the Alchemy of Mother Nature (Adept Series)*

465. *Mystical Adepts Are Our Living Pastors*

466. *Cosmic Databank*

467. *Man Has Not Known the Power of Psychic Wisdom*

468. *The Cosmic Recording Studio*

469. *In Union with Supreme Reality*

470. *The World Has Not Known the Mystical Powers of Truth As the Supreme Basis of All Things*

471. *Why Is Kabala, the Sacred Mysticism of the Jews*

472. *The Brain: The Mysticism of Nature's Compartmentalized Techniques*

473. *Man Is a Mystical Matrix*

474. *Using Your Psychic Powers to Consummate with the Spirit World*

475. *Divine Laws of the Attractive Homogeneity*

476. *Discourse on Mystical Dialogue*

477. *Electricity Is a Cosmic Current*

478. *Using Astral Light to Perform Healing Arts*

479. *The Mysticism of Astral Power/Light*

480. *The Moon Is an Astral Physician*

481. *The Mysticism of Astrology*

482. *Living a Successful Life with Astral Powers*

483. *Wisdom from the Astral Powers*

484. *Purge Yourself from Devilish Acts*

485. *The Cosmic Nature of the Astral Plane*

486. *Understanding the Mysticism of Our Cosmic Existence*

487. *The Universe Is in Search of Its Cosmic Reality*

488. *The Mysticism of Self-Mastery*

489. *Who Am I in the Cosmic Plane?*

490. *Understanding the Mysticism of the Cosmic Clock*

491. *The Mysticism of the Cosmic Studio*

492. *Mystics Are Initiated Mediums*

493. *Mysticism Is the Noble Cause of Cosmic Realization*

494. *Heroes of the Cosmic Plane*

495. *The Call and Initiation of the Gnostics (Adept Series)*
496. *The Life of a Sick Person*
497. *The Philosophy of Supernatural Ethics*
498. *The Occult Bible*
499. *The Metaphysical Bible*
500. *Corrupt Humans Have a Case with the Cosmic Court*
501. *The Divine Judgments of the Cosmic Court*
502. *A Corrupt Person Is an Entangled Soul*
503. *The Power, Wisdom, and Judgment of Retributive Justice*
504. *The Esoteric Book of Wisdom*
505. *Wisdom Is the Power of Self-Transcendence*
506. *Knowledge Comes through Positive Evolution*
507. *Nature Is an Epitome of Harmony*
508. *The Mystery of the Bees*
509. *The Iman Who Does Not Know the Koran*
510. *The Mystery of the Angels*
511. *The Sorrows of a Sick Person*
512. *The Mysticism of the Angels*
513. *Material Technology Is Wicked*
514. *Alchemy and Materialism*
515. *The Orphic Wisdom*
516. *Core Mysticism for Self-Transcendence*
517. *Visitors from the Higher Plane*
518. *Travelling with the Sea of the Spirits*
519. *Baptized but Still Remained a Practical Pagan*
520. *Nature Tells the Story of Birth and Rebirth in a Plain Language*

521. The Functions of Animal Magnetism
522. Lessons from the Grave
523. Communication with the Outside World
524. Occult Science in a Contemporary Age
525. Metaphysical Techniques for Holding Séances
526. Spiritual Discussions with the Dead (Adept Series)
527. Is There Life in the Grave?
528. The Nature of Human Decomposition
529. Experiences of a True Mystic
530. Core Metaphysics (Adept Series)
531. What Are the Problems of World Religion?
532. Echoes from the Spheres
533. Echoes from the Blues
534. The Blue Ridge Brothers
535. Life in the Great Beyond
536. Mystics and Gnostics
537. The Apostles Creed
538. The Sacred Teachings of Mysticism
539. Life in the Dark
540. The Metaphysics of Evolution
541. The Celestial Citation
542. What Is Water Music?
543. Mystical Experience
544. Wisdom Is the Light of Truth
545. Melodies from the Celestial
546. The Power and Nature of Psychic Science
547. The Power of Psychic Science in Christian Doctrines
548. Metaphysics and Psychic Science

549. *The Psychology of Psychic Science*

550. *The Impact of Psychic Science to our Daily Life*

551. *What the Great Master Said about Psychic Powers*

552. *The Man Who Created the First Man*

553. *The Mystic Nature of Melodies*

554. *The Mystical Power of Praises*

555. *How Nature Gave Man Wisdom through Mystical Engineering*

556. *Mysticism and Alchemy*

557. *How Mystics Use Cosmic Powers to Influence the World*

558. *A True Mystic Is a Great Teacher*

559. *The Man Who Saw the First Light (His Trinity Means Light but Not Church)*

560. *He That Knows the Light Knows Everything*

561. *The Positive Use of Supreme Natural Powers*

562. *How to Use Supreme Natural Powers to Enrich the Universe*

563. *How to Be in Mystical Unity with Your Life*

564. *How to Achieve Balanced Success with the Use and Application of Mystical Powers*

565. *How to Heal Others with Cosmic Light*

566. *Cosmic Healing in a Charismatic Age*

567. *Man Has Not Opened the Anatomy of Original and Pure Knowledge*

568. *The Contemporary Issues of Metaphysics*

569. *The Dialogue of Metaphysics with Mother Nature*

570. He That Breaks the Law Shall Be Broken By the Law

571. Law Works with Love

572. Natural Principles

573. The Unknown Wisdom Which Works Wonders

574. Understanding the Power of the Sun

575. Understanding the Wisdom and Creations of the Sun

576. The Mystical Powers of the Planetary Bodies (Adept Series)

577. The Cosmos Congratulates the Cosmic (Orphic Mysticism)

578. Understanding the Electrifying Rhythm of the Waves

579. The Imperatives of the Law of Balance and Balanced Interchange

580. The Mystical Reality of the Blessed Angels

581. The Mystical Powers of the Blessed Rosary

582. Understanding Your Psychic Powers

583. Philosophy and Mysticism

584. The World Is in Need of Balanced and Illumined Thinkers

585. The Wisdom of the Unseen Forces

586. The Immortal Teachings of Jesus Christ

587. The Bible and Astrology

588. The Wisdom of Thinking with Nature

589. The Dynamism of Spirits and System

590. The History of Mysticism

591. The Adoration of the Community Goddess

592. The Volcanic Inhabitants

593. The Mystical Meaning of Cherubim and Seraphim
594. Mysticism and Evolution
595. The Cosmic Thinking of Mysticism
596. Mysticism and the Science of Astrology
597. Astrological Riggs
598. The Oil Wells of Astrology
599. Astrology, Wealth, and Creation
600. Astrological Insight into the Life of All Creations
601. The Illumined Mystic Called Jacob Boehm
602. The Mysteries of Aquatic Mysticism
603. The Mysticism of the Aquatic Nature of Astrology
604. Trends in the Nature, Knowledge, and Wisdom of Astrology
605. Hermetic Astrology
606. Astrology and Yoga Practise
607. The Supreme Wealth and Wisdom of the Unseen Forces
608. The Power of Astrological Wisdom
609. Mother Nature Is the Pride of Astrology
610. The Mystical Powers of the Plants

Science and Technology

611. The First Principle of Science
612. Understanding the Dynamic Creative Process through the Waves

613. The Demystification of Occult Engineering Technology

614. The Anatomy of Marine Technology

615. The Nature, Powers, and Functions of Energy

616. The Dynamic of Environmental Impact on Evolution

617. The Waves Is the Basis for Astro and Astral Physics

618. The Waves Is the Anatomy and Foundry of All Engineering Technology

619. Electricity Is Spirit

620. Science and Life

621. There Is Motion in Everything

622. The Economic Importance of the Internet System

623. Computer Logic for Science and Engineering Students

624. How Research Can Be Used to Expand the Knowledge of Computer Science with Its Mathematical Logic

625. Science Is Giving the World a New Cosmogony

626. Constructive Economic Mathematics with Computer Logic

627. Physics and Electronic Appreciation

628. The Psychology of Human Thoughts

629. The Urgent Necessity of Scientific Balance

630. The Importance of Transcendental Science to Material Science and Technology

631. Man: The Foremost Engine in Creation

632. The Electric Nature of the Brain

633. *The Electronic Nature of Thought with Its Transcendental Mechanism*

634. *Error of the Intellect*

635. *In Tune with Nature's Energy*

636. *Alchemy: The Science and Technology Transmutation*

637. *Understanding the Power of Alchemy*

638. *Alchemy: The Original Structure of All Metals*

639. *Utilising Magnetism to Discover the Concept and Principle of Balance*

640. *Interchange and Exchange Is the Way of Mother Nature*

641. *The Mystery of Electricity*

642. *Geo-Physics for Today*

643. *Experimental Physics*

644. *Rebuilding Our Environment*

645. *The Chemistry of Our Ecosystem*

646. *Assets in Human Existence*

647. *The Dynamics of Engineering Production*

648. *The First Principle of Physics*

649. *The Role of Technology in Economic Development*

650. *Aviation Technology at a Glance*

651. *Modernization and Material Technology*

652. *Astrology the Science of the Eastern Avatar*

653. *The Origin of Matter*

654. *The Problems of Environmental Engineering*

655. *What Causes Climatic Change*

656. *Assessment of Engine Designs*

657. *Introduction to Research and Statistics*

658. *The Concept of Research and Development in Our Contemporary Era*

659. *The Fundamental Concept of Research and Methodology*

660. *The Objective Concept of Environmental Research*

661. *The Contemporary Challenges of Engineering Research*

662. *The Importance of Research and Engineering Technology*

663. *The Economic Importance/Impact of Research and Technology to the Internet Age*

664. *Utilising Research Findings to Enhance the Functions of Engineering Technology*

665. *Utilising Research Technology to Improve the Dynamic Functions of Website Engineering*

666. *The Creative Functions of IT and ICT in a Developing Economy*

667. *The Nature, Logic and Mathematics of IT and ICT*

Health

668. *Research Work on the Pharmacological Importance of Honey (With Subject Matter Honey Health and Health Is Wealth)*

669. *The Medicinal Values of Pineapple*

670. *Philosophy and Pharmacology*

671. *Water: A Natural Healer*

672. *The Healing Powers of Water Hyacinth*
673. *The Mystical Values of Water*
674. *The Herbs and Roots Are Our Natural Physician*
675. *Nature Is Full of Medicine*
676. *The Mystical Wonders of the Plants*
677. *How to Use Natural Medicine to Make Orthodox Medicine More Effective*
678. *The Dynamism of Nature's Chemistry*

Literature

679. *The Price of Love*
680. *Man: The Head of the Family*
681. *The Gentle Breeze*
682. *The Wicked Friend*
683. *The Endless Power of the Pen*
684. *The Oracle of Illiteracy*
685. *The Desert of Buds (Poem)*
686. *The Making of a Great Kingdom*
687. *The Evil Days of the Kings*
688. *Do Not Eat from the Cultist Table (Poetry)*
689. *Politics Is the Game of One with One Hand*
690. *Potters Possess Creative Wisdom (Poetry)*
691. *The Judges Cults Play Your Life with Love*
692. *Ekwubiri the Talkative*
693. *Alice Is a Divine Angel with the Wisdom of Angels' Norms*
694. *Dangers of Deceitful Leadership*

695. *The Agony of a Leader*

696. *The Errors of Wrong Choice*

697. *The Regrets of a Cult Man with His Followers*

698. *How Do You Look at Your Wife?*

699. *The Regrets of Political Adventure*

700. *Tell Me the Stories of Love*

701. *The Queen of Edo Kingdom Who Makes Her Hair in Benin Style*

702. *Kago Must Laugh*

703. *Abel the Bad Boy (The ABC of Literature)*

704. *Travelling with a Boat in the Desert*

705. *My Forests Friends*

706. *My Mum's Soup*

707. *Freedom from My Uncle's Machinations*

708. *My Father's Wasteful Investments on His Brothers*

709. *The Strange Needles of Law*

710. *How Obilor Became the Obi of His Kingdom*

711. *The Quarrelsome Wife*

712. *Disappointed in Marriage*

713. *The Foolishness of Fetish Love*

714. *The Agony of Obi's Princesses*

715. *The Bees and Their Stories*

716. *The Agony of a Police Friend*

717. *Disappointed by the Police Promise*

718. *Still Searching*

719. *The Smokey Love*

720. *Justice at the Jungle*

721. *The Insane Man*

722. *The Agony of a Troubled Wife*

723. *Escape from Police Is Better Than a King Statement with Police*
724. *The Wicked Nurse*
725. *Doctors Are Killers*
726. *Kissing Your Darling Is a Great World of Creative Expression*
727. *My Wife Is a Trouble Maker*
728. *The Unforgiving Husband*
729. *My Husband Is a Drunkard*
730. *The Insatiable Wife*
731. *The Wisdom of Shakespeare*
732. *Escape from Prisons*
733. *The Poor Widow's Dog*
734. *Wanderers from the Desert*
735. *My Mother Made Me to Remember My Wife*
736. *The Man Who Marries for Public*
737. *My Glorious Mother*
738. *The Mother of My Child*
739. *The Woman Who Missed Her Way*
740. *My Mother Is a Millicent*
741. *Women Are Storytellers*
742. *My Wife's Wish*
743. *The Hairy Lady*
744. *The Faithful Servant*
745. *How to Win Your Spouse's Love*
746. *From the World Castle*
747. *My Wife Made Me to Remember My Mother*
748. *The Foolish Maid*
749. *The Pastor Who Employed Prostitute*
750. *The Sweetness of Sin*

751. *Politics Is a Game of Mediocre*
752. *The Odour of a Bad Teacher*
753. *The Character of a Bad Friend*
754. *Love Is Not Force*
755. *Mariam, the Disobedient Daughter*
756. *Squirrels in Their Garden*
757. *The Story of My Pigeon*
758. *My Love for Nature*
759. *The Portrait of Love*
760. *How Eke Ate His Blood (The Story of a Painful Exit)*

Adept Series

761. *The Original Meaning of Kabala*
762. *The Wisdom of Kabala*
763. *The Reasoning for Kabala*
764. *Jesus and the Kabala*
765. *The Mystical Christ with the Key of Kabala*
766. *The Mystical Worship of the Kabala by the Gnostics*
767. *Gnostics and the Kabala*
768. *How Kabala Started the Growth of Mystical Knowledge*
769. *Every Gnostics Come from the Root of Kabala*
770. *Kabala Could Not Be Comprehended By the Uninformed Jews*
771. *True Mysticism Originated from the Practise of Kabala*

772. *Kabala and the Practise of Mystical Rituals*
773. *God the Light and Sound of Original Kabala*
774. *Kabala the Celestial Technology for the Immortal Ones*
775. *Kabala the Ministry of Pure Christian Mysticism*
776. *Kabala the Wisdom of Great Illumination*
777. *Kabala the Origin of Alchemy*
778. *Kabala the Foundry of Mystical Engineering*
779. *Kabala the Faith of Real Mystics*
780. *Occult Kabala*
781. *Metaphysical Kabala*
782. *Grow with the Knowledge and Practise of Kabala and Go to Heaven*
783. *Christianity Does Not Know the Meaning and Language of Kabala*
784. *Kabala Is the Profound Teacher with Consummate Wisdom*
785. *When Will Man Start Thanking Kabala Steadily*
786. *Christ with the Infinite Trinity Started the Kabala*
787. *Life Is the Celebration of the Conquest of Kabala*
788. *Philosophy and Wisdom*
789. *What Is the Himalayas?*
790. *The Agony of the Soul*
791. *The Grail Masters*
792. *The Order of the Gods*
793. *The Immortal Works of Paracelsus*
794. *Echoes from the Angels*

795. *The Mystery of the Serpent*

796. *The Physics of the Luna Eclipse*

797. *Praising the Unseen Powers*

798. *The Mystery of Celestial Light*

799. *The Wisdom of Celestial Life*

800. *The Perfect Soul*

801. *The Supreme Melody*

802. *The Aquarian Life of Our Cosmic Illuminates*

803. *Mysticism and the Hydrogen Age*

804. *Modern Trends in Knowledge and Wisdom*

805. *The Mystical Herbs*

Alchemical Series

806. *Alchemy and Engineering Creativity*

807. *Advancing the Course of Technology with the Use and Application of Core Alchemy*

808. *Nature Is the Foundry of Alchemy*

809. *Alchemy Authored Chemistry*

810. *The Fundamental Issues of Alchemy*

811. *Understanding the Engineering Basis of Alchemy*

812. *Alchemy and Polymer at a Glance*

813. *Metaphysics and Alchemy*

814. *Nigeria Is a Foundry of Alchemical Technology*

815. *Nature Is the Original Anatomy of Alchemy*

816. *Alchemy and Human Anatomy*

817. *The Organic and Inorganic Concept of Alchemy*

818. *Understanding Alchemy with the Organogram of Chemistry*

819. *Alchemy Is the Science and Technology of Mother Nature*

820. *Christianity and Alchemy*

821. *The Anatomy of Occult Science with Alchemy*

822. *Alchemy and Metallurgy*

823. *Alchemy and Transmutation*

824. *Celebrating the Ingenious Creativity of St Germaine: A Case Study of His Science of Alchemy*

825. *Christ, St Germaine, and Alchemy (The Mystical Mysteries of Life)*

826. *Alchemy and Philosophy*

827. *The Foundation of Alchemy with Engineering Technology*

828. *Meta-Mathematical Alchemy*

829. *The Basis of Meta-Material Science with Alchemical Technology*

830. *The Philosophy of Alchemy*

831. *The Contemporary Engineering Issues with the Use and Application of Alchemy*

832. *The Universe Is a Consummate Mechanism of Alchemy*

833. *Alchemy and the Science of Magnetism*

834. *Utilising the Technology and Basis of Alchemy to Appreciate Gravitationalism*

835. *The Definition of Astrophysics and the Relativity to Alchemy*

836. *Aquatic Engineering with Alchemy*

837. *A Chemist: A Humble Servant of Alchemy*

838. *The Mathematical Technicalities of Alchemy*

839. *Understanding Astrology with the Use and Application of Alchemy*
840. *The Physics of Alchemy*
841. *Alchemy the Foundry of Modern Technology*
842. *Alchemy the Creative Engineering of the Oriental Wizards*
843. *Alchemy with Its Holistic Transmutation of Technology*
844. *Alchemy: The Divine Road Map to Investigative Technology*
845. *Who Are the Forerunners of the Science of Alchemy?*
846. *Alchemy and the New Millennium*
847. *Give Me Alchemy, and I Will Give You Creative Invention*
848. *Alchemy and the Jet Age*
849. *How Alchemy Was Used to Advance the Computer Age with Its Technology*
850. *Alchemy the Foundry of Polymer Androbotic Science and Engineering*
851. *Cosmic Alchemy*
852. *Cosmic Union*
853. *Meta-Cosmic Vision*
854. *Engineering and Transmutation*

Grace Series

855. *Are You in the List of Divine Grace?*
856. *Is Your Grace Anointed?*

857. *His Grace Is Sufficient*

858. *His Grace Is Love Eternal*

859. *His Grace Is Driven with Light*

860. *Pray for the Love of the Grace*

861. *All Life Is Fathomed by the Grace*

862. *His Grace Is Our Shelter*

863. *Christ Is the Pinnacle of His Grace*

864. *His Grace Works Wonders*

865. *The Miracle of Supreme Grace*

866. *When Grace Is at Work*

867. *The Power of Working Grace*

868. *The Wisdom of Living Grace*

869. *Creation Is the Power of Grace*

870. *Empower Your Life with Grace*

871. *Wisdom Is the Product of Grace*

872. *Strive to Know What Lied Ahead of Your Grace*

873. *Your Destiny Is the Product of Divine Grace*

874. *Grace Made Abraham to Have Faith in God*

875. *Grace Is God in Action*

876. *Grace Is God in Glory*

877. *Life Is Best Lived with Grace*

878. *Every Prayer Must Be Said with the Wisdom of Grace*

879. *When Grace Speaks the Heavens Are Positively Provoked*

880. *Grace Makes One to Live a Purpose-Driven Life*

881. *The Living Church Is Founded on the Power and Wisdom of Grace*

882. *We Must Ask with Grace*

883. *Every Family Must Dwell on the Absolute Power of Grace*

884. *Grace Empowered Christ to Be Victorious*

885. *The Products of Grace Are Success and Victory*

886. *Grace Cannot Be a Victim*

887. *Strive to Make Grace Your Portion*

888. *No Problem of Life Is Mightier Than the Power of Grace*

889. *Grace Is Life and Light*

890. *Life Lived without Grace Is Vanity upon Vanity*

891. *Man Has Not Known the Power of Grace*

892. *Only Grace Speaks about the Wonders and Wisdom of God*

893. *To Live with Grace Is to Live with the Best*

894. *When We Ask in Grace, We Must Receive in Grace*

895. *Inventions and Science Are the Products and Wonders of Grace*

896. *Divine Love Is Only Guaranteed by Blissful Grace*

897. *Grace Is the Nature of God*

898. *Peace and Unity Is the Product of Grace*

899. *Marriage Is a Divine Symbol of Grace*

900. *Grace Empowered Knowledge to Be the Beacon of True Living*

901. *Grace Is the Greatest Teacher of All Ages*

902. *Grace Is a Divine Current*

903. *Every Prayer Must Be Centred and Driven with the Power of Grace*

904. *Grace Authored the First Man*

905. *Light Is the Product of Grace*
906. *Every Genius Functions with Immeasurable Grace*
907. *The Life of the World Is Life in Manifest*
908. *Nothing in Life Can Quantify the Power of Grace*
909. *Grace Wrote the Bible and Other Immortal Works*
910. *Grace Brought Salvation to the World*
911. *The Worth of Grace Cannot Be Measured*
912. *Grace Is a Consummate Giver*
913. *Grace Is a Peculiar Lover*
914. *Grace Is a Physician*
915. *The Power of Grace Is Universal*
916. *The Spiritual Importance of Grace Is Revealed in the Prayer Which Reads Thus: May the Grace of Our Lord Jesus Christ, the Love of God and the Still Fellowship of Our Lord Jesus Christ, Be with Us Now and Forevermore*

Drama

917. *What Happens when the Pope Hears What?*
918. *My Wife's Pregnancy Is Not from Me*

Women Series

919. *The Lady of Knowledge*
920. *The Wisdom of an Experienced and Virtuous Mother*

921. *The Lady of Love*
922. *The Lady of Light*
923. *The Lady of Wealth*
924. *The Women of Vision*
925. *The Sweet Memories of My Mother*
926. *The Similarities of My Wife and My Mother*
927. *The Man Is the Beauty of the Wife*
928. *The Wife Is the Strength of the Husband*
929. *My Wife Is My Password*
930. *Marriage Is a Divine Testimony with Assorted Experience*
931. *Women and Nature's Medicine*
932. *Women Are Psychically Gifted*
933. *Women Are Natural Angels*
934. *Where Are the African Angels?*

Mathematical Series

935. *The Anatomy of Mathematics*
936. *Evolutionary Mathematics*
937. *The Dynamism of Mathematical Reasoning*
938. *Pythagoras and the Doctrine of Mathematics*
939. *Mathematical Symbolism*
940. *Mathematical Logic*
941. *Mathematical Key Words*
942. *Metaphysics and Mathematics*
943. *Diplomacy and Mathematics*
944. *Modern Mathematics with Economics*
945. *Modern Engineering Mathematics*

946. *Analytical Mathematics*

947. *Mathematics and Data Processing*

948. *Modern Mathematics with the Reasoning of Technology*

949. *Informative Mathematics*

950. *Logic Mathematics*

951. *The Dynamic Functions of General Mathematics*

952. *Creative Mathematics for Technologists*

953. *Wisdom and Mathematics Action*

954. *The History of Mathematics*

955. *Mathematical Mechanics*

956. *Understanding the Metalogic Principles of Mathematics*

957. *Seminar Papers Ranging from Philosophy, Science/Technology, Economy to Governments, Etc.*

Time Series

958. *The Immortality Wisdom of Time*

959. *There Is Time for Everything*

960. *The Gnostical Wisdom of Time (Adept Series)*

961. *The Wisdom and Wonders of Time*

962. *Nature Is the Mother of Time*

963. *There Is No Time in Space*

964. *Time Is a Factor of Creation*

965. *Man Fails Time*

966. *The Glory and Greatness of Time*

967. *Astrology and Time*

968. *What Is the Meaning of Time to a Mystic*

969. *The Bible Is the Holy Book of All Times*

970. *Man Positively Increases with Time*

971. *The Limitation of Time Is Man's Making*

972. *Time and Opportunity Holds the Key for Success*

973. *The Universe Is Governed with the Cosmic Clock (Adept Series)*

974. *The Mystical Symbolism of Time (Adept Series)*

975. *Opportunity Comes with Time*

976. *The Mystery of Time and the Philosophy of Creation*

977. *Try to Show Time the Way*

978. *Always Show Great Kindness with Time*

979. *Try to Appreciate the Spirit That Rules Time*

980. *Make Friends with Time*

981. *Time Will Tell Is a Wise Man's Wisdom*

982. *Few Are Born with the Greatness of Time*

983. *Immortals Do Not Know What Time Is (Adept Series)*

984. *Only Time Makes Us Know That We Come and Go with It*

985. *How to Create a Paradise with the Power and Wisdom of Cosmic Time (Adept Series)*

986. *Who Are the Laurels of Time*

987. *Man Will Soon Experience the Time of the Lord*

988. *The Beauty of Xmas Time*

989. *Nature Has Its Time*

990. *Give Your Time a Fruitful Meaning*

991. *Every Messenger Comes Once in a Time (Adept Series)*

992. *The Jubilee of Time*

993. *How to Appreciate the Value and Seasons of Time*

994. *Give Every Flying Minute Something to Keep in Stock*

995. *Do Not Make Friends with Time*

996. *Man's Concept of Time Is Wrong*

997. *Philosophy Upholds the Wish and Wisdom of Time*

998. *Create Time for Yourself*

999. *Birth and Rebirth Speaks Volume about the Wonders of Time (Adept Series)*

1000. *The Mystical Melodies of Time*

1001. *Great Minds Master the Wisdom and Power of Time*

1002. *The Aquarian Gospel of Time*

1003. *Develop and Increase Your Courage with Time*

1004. *The Aquarian Gospel of Time (Time Series)*

1005. *How Will I Be Received by the Angels at the Time of My Home Call*

1006. *The Mysteries of Eternal Time (Adept Series)*

1007. *He Does Not Fail in His Time*

1008. *The Moonlight Time and Other Events*

Nature Series

1009. *The Creative Power of Nature*

1010. *Nature Is Gentle, Long, Kind, and Peaceful*

1011. *The Dynamic Wisdom of Nature*

1012. *Nature Is the Supreme Fulcrum of the Universe*

1013. *Nature Tells the Story of Birth and Rebirth in a Plain Language*

1014. *The Marvellous Wonders of Nature*

1015. *Nature Is a Perfect Physician*

1016. *Heal Yourself through the Power of Nature*

1017. *Inspiration Is the Monumental Language of Nature*

1018. *Nature Creates with Balance*

1019. *Balanced Interchange Is the Law of Nature*

1020. *Nature Is Homogeneity in Action*

1021. *Try to Think along with Nature*

1022. *A Talk to Nature Is a Talk to All Things*

1023. *I Appreciate Nature for Its Illumined Wisdom*

1024. *Learn to Work Knowingly with Nature (Adept Series)*

1025. *All Great Immortals Listened to Nature*

1026. *Nature's Heartbeat Is Mystical (Adept Series)*

1027. *The Cosmic Wonders of Nature (Adept Series)*

1028. *Nature Is a Symbol of Supreme Intelligence*

1029. *Have You Recognized Your Assignment with Mother Nature*

1030. *Nature Reveals Great Arts to Man*

1031. *Man Has Not Known His Natural Essence*

1032. *Nature Is a Great Teacher*

1033. *Nature Reasons with the Creator*

1034. *The Love of Nature Is a Supreme Gift*

1035. *Man Is a Consummate Anatomy of Nature*

1036. The Nature of the Stars Are Mystical Mystery

1037. All the Planets Love and Respect Nature

1038. The Creative Nature of the Universe Is a Perfect Aestheticism

1039. My Simple Garden Tells How Beautiful and Wonderful Nature Is

1040. Our Flowers Reveal the Greatness of Nature

1041. Nature Is a Great Historian

1042. Nature Is a Living Philosophy

1043. The Science of Nature Is Written on a Simple Marble

1044. Learn How to Respect Nature and in Turn It Will Love You with Perfect Protection

1045. The Man Who Talks with the Flowers Appreciates the Value of Nature

1046. An Angry Man Does Not Reason with Nature

1047. A Meditation Which Nature Is a Direct Communion with the Blessed

1048. Angels Speak about Nature with a Great Homily

1049. Christ Worked with Nature Which Informed Him to Become the Supreme Master of the Universe (Adept Series)

1050. The Universe Is Authored along with Nature's Rhythm

1051. Simply Tells Nature the Truth

1052. Nature Reads Man More than Man Knows Nature

1053. Nature Is a Provisional Encyclopaedia

1054. *There Is a Supreme Partnership between Nature and Wisdom*

1055. *The Fire of Inspiration Is an Eternal Motion (Adept Series)*

1056. *Give Nature a Chance to Perform and It Will Surely Dazzle You with Wonders*

1057. *Man-Woman Electrifying Balance Is Naturally a Monumental Interchange*

1058. *No Human Being Is Capable of Changing Nature, and This is Why Science Cannot Charge or Recharge Nature*

1059. *Nature Is a Consummate Mathematician*

1060. *Nature Started the Act of Creation and This is Why the History of Birth and Rebirth Speak Volumes about Nature*

1061. *A Praise to the Wonders of Nature Is a Glorious Praise to the Consummate Creator of All Ages*

1062. *When You Are Tired Allow Nature to Recharge You*

1063. *Nature Is a Mystical Physician Whose Charges Are Naturally Free*

1064. *How Sweet Is the Reality of Nature and How Wonderful Will Man Be When He Works along with Nature*

1065. *Man Is a Monumental Marble of Mother Nature*

1066. *All Work Which Does Not Invite Nature into It Is as Vain as the Concept*

1067. *A Lion Is Always Asking Nature to Protect Her*

1068. *Nature Is the Brain Box of All Composite Creativity*

1069. *The Supreme Messenger of Light Came from the Abundant Bank of Divine Nature*

1070. *All Great Men Live in Nature's Garden*

1071. *No Man Can Comprehend or Quantify the Wisdom of Nature*

1072. *Nature Sings All the Time with Adoration and Perfect Melodies to the Origin of Its Greatness*

1073. *Mozart the Supreme Master of Water Music Was First Possessed by the Wisdom and Supremacy of Mother Nature*

1074. *Nature Tells Us How Beautiful Our Echoes Are*

1075. *A Look at the Top Mountains Reveals How Powerful, Wonderful, Monumental, and Mysterious Nature Is*

1076. *A Look at the Beach Tells Man That Nature Is Aesthetically Millicential*

1077. *The Man Who Understands Nature Appreciates the Value of Creation*

1078. *The Creator Appeared First to Man at the Garden of Genesis*

1079. *Physical Death at Any Point in Time Tells Us the Value of Nature through the Process of Composition and Decomposition*

1080. *Any Man Who Puts Nature into Action Must Certainly Get to the Mountain Peak without Disturbance*

1081. *Nature Directed the Three Wise Men to Locate Bethlehem*

1082. *Technology Reveals the Greatness of Nature's Aestheticism*

1083. *Engineering at Any Point in Time Is Always Adopting Nature as a Consummate Consultant*

1084. *Logic Told the World That the Power of Nature Is the Supreme Force Which Holds the Concept of the One-World Family*

1085. *Nature Is a Consummate Firmament for All Spiritual Activities (Adept Series)*

1086. *When You Channel Your Thought to Nature, You Must Certainly Conquer the World*

1087. *The Science of Alchemy Cannot Function without Nature*

1088. *Christopher Columbus Is the First Man That Adopted Nature as His Master Which Directed Him to Discover the New World and the Island of Trinidad*

1089. *Agape Love Is a Living Friend of Mother Nature*

1090. *Music and Melody Series*

1091. *Winning Powers of Music*

1092. *The Healing Miracles of Music*

1093. *Using Music to Solve Your Problems*

1094. *The Origin of Music*

1095. *The Miraculous Wonders of Songs and Solos*

1096. *Learn to Live with Music*

1097. *Singers and Angels*

1098. *The Healing Wisdom of Wisdom*

1099. *The Mysticism of Music*

1100. *Authors Are Born Musicians*

1101. *Mozart, Music and Mysticism*

1102. *How You Can Worship with Inspired Melodies*

1103. *Music Was in the Blood of My Grandfather*

1104. *My Mother Was an Endowed Chorister*

1105. *Music Plays a great Role in Attraction Inspiration*

1106. *The Melodies of the Music of the Spheres Is a Living Mystery*

1107. *Wisdom Originated Music and Melodies*

1108. *Ordained Musicians Sing with Inspiration*

1109. *Music Brought Mozart, Beethoven, Bach, and Jimmy Reeves to Fame*

1110. *All the Blessed Angels Are Eternal Musicians*

1111. *The Immortal Words of the Sacrament of Greece Is Written in Mystical Poets (Adept Series)*

1112. *No One Can Be Happy without Music*

1113. *The Man Who Played His Music on the Mountaintop*

1114. *Most Geniuses Are Musicians*

1115. *The Story of Kant Village Is Narrated in Music*

1116. *Solomon Wrote the Songs of Songs and the Songs of Sorrows with Music*

1117. *The Mystical Powers of Music Made Paul and Silas to Become Friends to the Lions*

1118. *The Man Who Authored the First Music Showed Men the Way to Heaven*

1119. *Heaven Is a Celebration of Celestial Melodies*

1120. *The Power of Praises Brought Me Closer to My Creator*

1121. *Use the Power of Music to Will Your Way Through*

1122. Music Makes You to Rubbish Challenges

1123. Great Amnesty Goes with Songs and Melodies

1124. The Man Who Wrote the Sacred Songs and Solos Is a Mystic

Faith Series

1125. Faith Is Real

1126. Faith Is Life

1127. Faith Is Purposeful and Powerful

1128. Faith is Courageous

1129. Faith and Life

1130. Faith and Belief

1131. Faith and Church

1132. Faith and Doubt

1133. Faith and Vision

1134. Faith and Prophecy

1135. Faith and Perception

1136. Faith Is Wisdom in Divine Manifestation

1137. Living Faith Brings Man Closer to His Creator

1138. Abraham Is Faithful to God

1139. Faith Must Function with Work

1140. All Great Souls Lived for Faith

1141. Jesus Is the Supreme Victory of Faith

1142. You Must Put Your Faith into Work

1143. Victory Is the Product of Faith

1144. You Must Have a Living Garden of Faith

1145. All Faithful Men Are Spiritually Tireless

1146. Jesus Built the Church on Faith

1147. *The Mission of Calvary Was Destroyed with the Power of Faith*

1148. *Great Works Are Written on the Victory and Vision of Faith*

1149. *Man, Where Is Thy Faith?*

1150. *I Treasure the Ordination of Faith*

1151. *A Blessed Man Is a Victor of Faith*

1152. *Faith through Abraham*

1153. *Living Faith Is a Great Manner from Above*

1154. *Life Is a Messenger of Living Faith*

1155. *The Reality of the Blood Is Only Guaranteed*

1156. *Lord, Teach Me to Know Thee with the Wisdom of Faith*

1157. *Faith Builds the Greatest Empire*

1158. *The Mystical Symbol of Paradise Is Faith (Adept Series)*

1159. *Faith Is the Archive and Architect of Immortal Creations (Adept Series)*

1160. *Faith in God Gave the World Name and Power*

Marriage Series

1161. *The Psychology of Marriage*

1162. *Marriage: Nature's Greatest Business Necessity*

1163. *Understanding the Manner and Nature of Marriage*

1164. *Marriage Manifests God in Reality*

1165. *What Couples Must Know about the Beauty of Partnership*

1166. *The Wisdom and Beauty of a Dynamic Marriage*

1167. *Women Are Special Gifts to Marriage*

1168. *The Man Who Loves His Wife Loves Himself*

1169. *Abraham Told the World That Sarah Is a Gifted Millicent*

1170. *No Man Can Be a Success without the Power of a Balanced Marriage*

1171. *Marriage Makes Us to Remember Our Eternal Roots*

1172. *No Ceremony Is as Powerful as Marriage*

1173. *Every Woman Is Proud of Her Marriage No Matter the Challenges*

1174. *Noble Women Appreciate Their Husbands in the Beauty of the Angels*

1175. *My Mother Told Me That My Loving Dad Is Her Greatest Companion*

1176. *The Simplicity of the Man Makes the Woman Large and Proud*

1177. *Women Naturally Tell the Story of Marriage with Humble Smiles*

1178. *The Woman That Understands the Challenges of Marriage Is a Practical and Purpose-Driven Manager*

1179. *The Law of Homogeneity Started with Marriage*

1180. *The Power of a Praying Wife Is a Great Protection to the Man*

1181. *Keep Your Wife Going She Will Keep You Moving*

1182. *A Man's Success Makes the Wife Happy*

1183. What Youth Should Know about Marriage

1184. Every Couple Must Strive for Eternal Marriage

1185. God Established the Constitution of Marriage for the Glory of His Name

1186. He That Does Not Respect Marriage Does Not Know God

1187. The Joy of Family Circle Is the Beauty of Life

1188. My Wife Is Possessed with Permanent Charisma and Beauty

1189. My Wife Is My Bank

1190. My Wife Is My Dependable Asset

1191. Women Are Crafted with Permanent Aestheticism

1192. Our Modern Marriage Must Emulate Abraham and Sarah

1193. My Wife Makes Me Appreciate Nature Greatly

1194. Polygamy Is the Greatest Canker of Marriage

1195. Every Woman Needs Caring

1196. Mystics Cherish Their Wives and Divine Love (Adept Series)

1197. Celibacy Contravenes the Law of Marriage and Divine Love

1198. How Would the World Be If Everybody Practises Celibacy?

1199. Children Are the Beautiful Fruits from the Garden of Marriage

1200. Marriage Is an Eternal Garden

1201. Marriage Is a Divine Temple

1202. To Be Enriched with Marital Blessings Is to Be Enriched with Everything

1203. Every Woman Aspires for Marriage

1204. Only Marriage Explains the Power and Wisdom of the Continuity Principle

1205. We Must Pray for Couples without Issues

1206. To Adopt a Child in Marriage Is Equally a Blessing

1207. He That Hurts a Woman Hurts the Immaculate Womb of Mother Nature

1208. When a Woman Cries, Nature Is Usually Angry

1209. Men Must Respect the Position of Their Wives

1210. My Wife Has a Meridian Power

1211. Nature Makes Us Appreciate How Caring She Is through Woman

1212. Mothers Are Millicents to Marriage (MMM)

Agricultural Series

1213. Agricultural Engineering in Our Contemporary Era

1214. The Economic Importance of Agriculture

1215. The Importance of Fishery in Economic Development

1216. Problems Facing Agriculture in Our Contemporary Era

1217. The Importance of Soil Conservation

1218. Pest Control and Agricultural Development in Third-World Countries

1219. The Importance of Research and Development in the Agricultural Sector

1220. *Understanding the Dynamics of Forest Preservation*

1221. *The Economic Importance of Agricultural Products*

1222. *Utilising the Goldmine of Agricultural Technology to Enhance the Goldmines of Our Economy*

1223. *Understanding the Economic Importance of Crop Research*

1224. *The Importance of Climatic Change to Agricultural Development*

1225. *Improving the Standard of Our National Economy with the Use and Application of Agricultural Technology*

1226. *Providing Employment Opportunities with the Use and Application of Agricultural Development*

1227. *General Agricultural Science for Higher Institutions*

1228. *The Importance of Plant Technology in Agricultural Development*

1229. *General Agricultural Science for Higher Institutions*

1230. *The Economic Importance of Bee Farming*

1231. *Problems Facing Agricultural Development in Our Contemporary Era*

Thought Series

1232. *Only in Positive Thinking Lies the Secrets of Creation*

1233. *Where Are the Thinking Men of Africa?*

1234. *God Is the Only Consummate Thinker*

1235. *Learn How to Think Inwardly*

1236. *Geniuses Think Inwardly*

1237. *Positive Thinking Makes Me Feel Great*

1238. *Great Arts Are the Living Concepts of Balanced Thinking*

1239. *Learn How to Think Positively*

1240. *What Is the Power of Positive Thinking?*

1241. *Positive Thinking Is the Keyword for Success*

Others

1242. *The Contemporary Challenges of Democracy and Diplomacy*

1243. *The Evolutionary Imperatives of Democracy and Diplomacy*

1244. *The Universe Is a Comprehensive Anatomy of Divine Grace*

1245. *How to Work with the Absolute Power of the Mystical Christ*

1246. *The Wisdom of the Mystical Christ*

1247. *How to Realize the Nature of Your Mystical Self*

1248. *Anatomy of Agricultural Economics*

1249. *Occult Economics Importance*

1250. *The Importance of Aquatic Technology*

1251. *The Importance of Aquatic Technology for Developing Technology for Developing Nations*

1252. *The Technology of Nature's Engineering Modulations*

1253. *Logical Mathematics for Engineering Experts/ Students*

1254. *Understanding the Physics of Nature's Motion*

1255. *The Practical Philosophy of Physics*

1256. *System Development and Management*

1257. *The Principle of Industrial Technology*

1258. *The Theory and Practise of Industrial Technology*

1259. *The Principles of Industrial Physics*

1260. *Understanding the Techniques of Industrial Chemistry*

1261. *Understanding Chemistry in Our Contemporary Era*

1262. *Practical Industrial Physics*

1263. *Our Elemental Friends*

1264. *The Dynamics of Environmental Management*

1265. *The Agonies of the Pope*

1266. *A. U. Aliche's Celebration and Exaltation of Wisdom (His Poetic Verses)*

1267. *The Dynamics of Technical Mathematics in Economic Development*

1268. *Contemporary Technical Mathematics*

1269. *General Mathematics for Technical Education*

1270. *The Economic Importance of Technical Education*

1271. *Evolving Technical Mathematics*

1272. *The Material Problems of Physics*

1273. *The Philosophy of Physics with The Dynamics of Electronics*

1274. *The Dynamic Philosophy of The Brain with Its Networking with The Spirit Mind*

1275. *The Simultaneous Rhythm of The Physics of The Brain*

1276. *The Original Errors/Misconceptions of Physics*

1277. *The Anatomy of Cosmic Physics*

1278. *The Anatomy of The Physics of Matter*

1279. *The Physics of Natural Medicine*

1280. *Astral Physics with Nature*

1281. *Hard Facts about Philosophy and Physics*

1282. *Philosophy and Physics Is for Developmental Technology*

1283. *Philosophy, Technology, and Physics*

1284. *Pure Physics*

1285. *Alchemy and Physics*

1286. *Alchemy of Physics*

1287. *Alchemy, Physics, and Technology*

1288. *The Crafting Technology of Pure Physics*

1289. *How Philosophy Utilised Physics, Chemistry, and Biology to Develop Medicine*

1290. *The Anatomy of the Physics of Medicine*

1291. *Contemporary Physics*

1292. *Alchemy: The Mother of Physics and Chemistry*

1293. *Developmental Alchemy*

1294. *Meta-Mathematical Physics*

1295. *Meta-Mathematical Alchemy*

1296. *The Dynamics of Cosmic Foundry*

1297. *Aeronautic Physics*

1298. Aquatic Physics

1299. The Language of Flowers

1300. The Reason to Depend on Natural Medicine

1301. The Immortal Credentials of Mother Nature

1302. Have You Heard the Music/Melodies of Waters

1303. Waterfalls are a Symbol of the Waves

1304. How to Water the Spent to Be Creative/Fruitful

1305. The Age of Innocent/Perfect (Adept)

1306. Man, Wisdom, and Philosophy

1307. You and Philosophy

1308. Traditional Wisdom

1309. Traditional Philosophy

1310. The Hole in Your Home

1311. The Dynamism of Business Strategy

1312. The Objective Goal of Management Strategy

1313. Management Strategy

1314. What Is Business Risk?

1315. The Analysis And Benefits of Management Risk

1316. Garden from the Grail

1317. Astral Garden

1318. The Greatness of Grace

1319. Astrological Messengers

1320. The Wisdom of The Grail

1321. Astrological Wisdom

1322. Wishes from Beyond

1323. The Organogram of the Cosmic Clock (Adepts)

1324. The Mystical Talent

1325. The Destiny of Mystics

1326. Avatars Are Destined Mystics/Messengers

1327. Messengers from Destiny

1328. *The Mysticism of the Tarots*

1329. *Psychic Mysticism*

1330. *The Invisible Wisdom*

1331. *Peter and Paul: The Pillars of Christianity*

1332. *The Sinful Paul before Convert*

1333. *The Invitation of Priesthood*

1334. *The Mystery of Pentecost*

1335. *The Insane Priest*

1336. *Carnage in The Church*

1337. *The Unknown Planets (Adept Series)*

1338. *Is Calvary Real in The Context of Religions*

1339. *The Voice of Melodies*

1340. *Pentecostal Hypnotism*

1341. *The Shameless Face of Pentecostalism*

1342. *Dangers of Using Talisman*

1343. *The Secret Doctrine of Cosmic Wisdom (Adept Series)*

1344. *The Sacred Doctrine of True Mysticism (Adept Series)*

1345. *The Sacred Destiny of Nigeria/Africa*

1346. *How the Mountains Communicate with the Astral Region*

1347. *The Inner Power of Great Mountains (Adept Series)*

1348. *Heroes of Wisdom*

1349. *Emperors of Destiny*

1350. *Destined for Greatness*

1351. *Knowing and Working with Your Star*

1352. *The Melodies of Destiny*

1353. *The Greatness of Wisdom*

1354. Who Is Truly Wise

1355. The Man Who Is Plagued with Wisdom

1356. The Immortal Wisdom of Destiny

1357. Why I Am Not Alone

1358. The Lonely Journey

1359. The Reward of Greediness

1360. Only Grace Made Him What He Is

1361. I Cannot Withstand You

1362. Echoes from the Stars

1363. The Road to Astrology

1364. The Wisdom of Psychic Creativity

1365. What Is the Destiny of Wisdom

1366. The Hood of Anger

1367. Embraced with Wisdom

1368. Remembered for Truth

1369. The Clement of the Rosary

1370. Liars from the Altar

1371. Rays of Beauty

1372. The Meaning of Living

1373. Lessons from Inspiration

1374. The Meridian Beauty

1375. Foundry of Love

1376. Nature Revealed Christ to Me

1377. Secret Wisdom from Nature

1378. The Wisdom of the Twelve World Teachers

1379. The Secret Destiny of Wisdom

1380. The Destiny of Love

1381. The Destiny and the Power of Truth

1382. The First Kingdom Man

1383. The First Human Evolution

1384. The First Human History

1385. The First Human Citizen

1386. The First Human Civilization

1387. Wisdom Is the Light of Truth

1388. Wisdom in the Light of Destiny

1389. The Importance of Information Management in a Globalized System

1390. The Dynamics of Industrial Engineering

1391. The Reality of Information Technology with Systems Analysis

1392. Data Appreciation with Management Technology

1393. Data Application with Information Case Studies in Data Processing

1394. Murders in the Forest

1395. The Forest Captains

1396. The Forest Goddess

1397. The Forest and the Sea Dog

1398. The Bees

1399. The Forest Beast

1400. The Forest Visitor

1401. Mohammed's Wish (Adept Series)

1402. Minds at Sunset

1403. The Sailors Tales

1404. Contemporary Metaphysics

1405. Developmental Metaphysics

1406. Dynamic Meta-Science

1407. Investigative Logic

1408. Objective Metascience

1409. Non Contemporary Metaphysics

1410. Abstract Logic

1411. Abstract Philosophy

1412. Abstract Wisdom

1413. Wisdom and Metaphysics

1414. Metaphysical Forces

1415. Tales from the Sailors

1416. The Mermaid Wisdom

1417. The Will of the Ancestors

1418. Jungle in the Sea

1419. The Whale and Other Animals

1420. The Crocodile Monitor

1421. Perfect Wish

1422. Perfect Wisdom

1423. Perfect Will

1424. Prolonged Promise

1425. Tales from the Seas

1426. The Mountain Melodies

1427. The Joy of the Inner Being

1428. Athan and the Friend

1429. A Friend's Deceit

1430. Pigeons on the Roof

1431. The Eagle's Beauty

1432. The Eagle's Wisdom

1433. The Meaning of a Friend's Letter

1434. The Lair's Tales

1435. Moon in the Cold Weather

1436. The Tortoise and the Shell

1437. The Tortoise and His Wisdom

1438. The Animal Forum

1439. Years of Silence

1440. The Wisdom of Silence

1441. Journey in A Mission

1442. Udo the Fine Boy

1443. The First Teacher of Christ

1444. Knowledge from the Grail

1445. Wisdom from the Grail

1446. When Mercy Is Received

1447. Buddha's Wish (Adept Series)

1448. The True Meaning of Earth Ma

1449. Mundane Forces at Work (Adept Series)

1450. Life in the Beyond (Adept Series)

1451. Secret Revelation

1452. Sacred Wisdom

1453. Ordained Thought

1454. The Wish of His Wish (Adept Series)

1455. Cosmic Melodies

1456. Cosmic Birds

1457. Abraham: The True Father of Mysticism

1458. Abraham's Faith in Mystics Christ

1459. Christianity and Mysticism

1460. The Mystical Light

1461. The Garden of Grace

1462. Orpheus Message (Adept Series)·

1463. Lord Bacon's Illumined Wish

1464. Weeds of Destruction

1465. Rancour in Society

1466. The Chief Priest

1467. The Oracle of the Mystical Wisdom

1468. Rays of Grace

1469. The Kindergarten Saint

1470. The Priest Who Impregnated the Reverend Sister's Mother

1471. Quest for Mondian Wisdom (Adept Series)

1472. Wisdom in Thinking

1473. Madness of Governance

1474. Heroes of Knowledge

1475. The Illiterate President

1476. The History of Birth

1477. Brains from Above

1478. The Dog Who Made Love to the Master's Wife

1479. The House Wife and Her Maid

1480. Impediments Which Work against Our Assignment with Destiny

1481. The Judge of the Jungle

1482. The Justice of Wisdom

1483. The Judgment of Truth

1484. Fracas from the Mind

1485. Creating for Creation

1486. Gates to Nobility

1487. Music from the Guardian Angel

1488. Power from the Soul

1489. The Blessings of Truth

1490. A Mischief Maker Who Murders Himself

1491. Arrows from Heaven

1492. Remember You Must Be Judged

1493. The Hungry Goddess

1494. Many Sins of the Gods

1495. Ranti the House Boy (Story)

1496. The Shepherd's Story

1497. The Pains of Disobedience

1498. Fighting for Your Right

1499. Late Love Makers

1500. My Mother's Goat

1501. The Goat and the Gate (Story)

1502. My Husband Refused to Give Me Love

1503. The Cause of Quarrel

1504. Magic from the Oracle

1505. The Deceit of the Native Doctor

1506. The Pastor Who Sold His Church for a Bigger Business

1507. Crisis from Criticism

1508. The Agony of a Critic

1509. The Rat in My Room

1510. Have You Discovered Your Talent?

1511. The Beast of The Year

1512. Sinful People Have Taken Over the World

1513. Beings with Righteous Action Has Taken Over the Universe

1514. Avatars in Our Midst

1515. The Blessings of True Mysticism

1516. My Divine Heritage

1517. Understanding with Might Thinking

1518. Understanding Your Horoscope

1519. Pistol in My Pocket

1520. The Wish of My Grandmother

1521. Who Gave Me the Name Onyembarachiobi?

1522. The Prayer of My Family

1523. Is Jonah a Mystic?

1524. Biblical Mysticism

1525. Live Stories from the Florence

1526. *Tales from the Wise Bird*

1527. *Questing and Requesting to Know (Journey in Wisdom)*

1528. *The Quest of The Maid*

1529. *Understanding and Appreciating the Power of the Spirit*

1530. *The Suffering of the Hero*

1531. *Great Is the Eternal Rock of Ages*

1532. *The Kings Must Be Judged*

1533. *How Kind Is Nature*

1534. *The Miracle of Nature's Creation*

1535. *Importance of the Spirit*

1536. *The Importance of the Human Thought*

1537. *How to Work with Your Spirit*

1538. *The Corporative Sprit*

1539. *The Stubborn Man Who Falls in Love with A Hard and Difficult Lady*

1540. *Working with Light*

1541. *The Ennobled Mind*

1542. *The Mundane Occult Metaphysics*

1543. *Is Creation by Accident*

1544. *The Cosmic Master Plan*

1545. *The Cosmic Architect*

1546. *The Cosmic Designer*

1547. *My Mother's Kitchen*

1548. *Poison in the Mind*

1549. *Developmental Education*

1550. *Experimental Metaphysics with Developmental Meta-Science*

1551. *Noble Institute for Planetary Philosophers (NIPP)*

1552. *Noble Institute for Planetary Teachers(NIPT)*

1553. *Royal Institute for Planetary Scientists RIPS*

1554. *Developmental Technology*

1555. *Developmental Astrology*

1556. *Developmental Technology*

1557. *Developmental Engineering*

1558. *Developmental Mathematics*

1559. *Developmental Physics*

1560. *Developmental Chemistry with Reference to Dynamite Forces*

1561. *Politics and Religion Are Jointly Promoting the Sinful Activities of Material Kingdom (Jesus Truly Wept)*

1562. *Contemporary Political Thought/Movement*

1563. *The Immortality of Natural Law*

1564. *The Future of Alchemy And Technology*

1565. *Alchemy: The Mother of Technology*

1566. *What Is the Technology of Alchemical Transmutation*

1567. *What Wisdom Taught Alchemy*

1568. *The Alchemy of Wisdom*

1569. *The Alchemy of the Human Spirit*

1570. *The World Is Longing for Truth*

1571. *Have You Discovered Your Talent?*

1572. *The Mystic's Way of Reasoning*

1573. *Akhenaton: The First African Unknown Mystic*

1574. *Akhenaton's Request to the Almighty God*

1575. *The Rosicrucian, Akhenaton*

1576. Akhenaton's Mandate to Africa

1577. Why He Is in the Immortal Record of the Twelve World Teachers

1578. Bible Is a Living Wisdom

1579. Bible of Wisdom

1580. A. U. Aliche's Balanced Book of Wisdom

1581. A Consummate Colloquium on the Immortal Blessings and Powers of Ingenious Wisdom

1582. My Memo in Honour and Memory of The Supremacy of Wisdom

1583. Akhenaton: The Hero of Global Mysticism

1584. The Challenges of Modern Mystics

1585. The First Celestial Mother

1586. The First Ever Known Mystics

1587. The First Ross Cross

1588. The First Sanctum Worshippers

1589. Sanctum Worshippers in the Light of Cosmic Doctrine

1590. The Mystical Reality of the Miracle of the Three Loaves of Bread

1591. The Mystical Doctrine of the Sea Sanctum

1592. Cosmic Experience

1593. The Mystical Flowers

1594. Gardeners in the Threshold

1595. The Cosmic Revelation of the New Atlantic

1596. Messengers from Light

1597. Who Are the Called Ones?

1598. The Blessing of Cosmic Experience

1599. How Wonderful Is the Mystical Path

1600. Serpents in the Land

1601. Mystically Enriched for Cosmic Assignments

1602. Mystical Baptism in the Light of Cosmic Wisdom

1603. The Fisherman and His Fishes

1604. How Beautiful Is the Experience of Marriage to a Mystic

1605. Can Women Make Heaven?

1606. Women and the Psychic Reality

1607. Messengers from the Cross

1608. The Cosmic Current

1609. How to Remain in Worship with the Cosmic Sanctum

1610. How the Gnostics Received Their Wisdom

1611. The Meaning and the Definition of Cosmic Waves

1612. What Is a Blessed Day to Man

1613. The Mystical Noah Who Has Not Been Explained

1614. How Beautiful Is Cosmic Transition

1615. Lessons in the Light of Destiny

1616. Lessons in the Wisdom of Destiny

1617. The Power of Active Thoughts

1618. Learn to Listen to Your Thoughts

1619. Elders Must Be Crucified for Refusing to Uplift the Younger Ones

1620. The Wickedness of Elders

1621. Some Elders Are Heathens

1622. Tribalism Is a Curse to Real Human Development

1623. Leave Me Alone

1624. Alone She Left

1625. The Regrets of a divorced Wife

1626. My Sickly Wife

1627. My Naughty Girlfriend

1628. Enmitied with Caring

1629. Enriched with Love

1630. My Husband Is Unpredictable

1631. Lao Tse: The Called One

1632. Marvels from Heaven

1633. The Beautification of Eden

1634. The First Ever Written Bible

1635. In Tune with Nature

1636. The Unholy Wish of Nicodemus

1637. Ho to Live Your Life in the Light of Truth

1638. My Personal and Humble Wish (Mystical Will Adept Series)

1639. Have You Realized How Much God Has Invested in You?

1640. You Must Return the Glory of His Investment to Only Him

1641. The Life of the Most Wicked Christian

1642. The Wicked Man

1643. The Mandate of Light

1644. I Respect My Creator with Reverence

1645. Immortal Biology

1646. Occult Biology

1647. How to Make Life work for You

1648. How to Make Nature Bless You

1649. The Beauties of Wisdom

1650. Imprisoned Life

1651. Small Minds Celebrate Nothing

1652. Larger Minds Celebrate Great Achievement

1653. Sending Your Thoughts across the Continent

1654. *The Biggest Dream of My Life*

1655. *The Greatest Vision of My Life*

1656. *What Will Posterity Say About Me*

1657. *The Will of My Soul*

1658. *The Rhythm of Celestial Melodies*

1659. *How the Cosmic Determines and Directs All Causes*

1660. *Reasoning Along the Light*

1661. *What Informed the Creativity for Making Me the Way I Am*

1662. *The Cosmic Desire of Every Soul*

1663. *The Marital Desire of Every Man*

1664. *The Grave Cannot Hold the Saint*

1665. *The Grave Is Necessary for the Sinners*

1666. *Christ Proved That the Grave Has No Power over the Saint*

1667. *Man Is Nature's Medium*

1668. *Rays from the Sanctum*

1669. *Cosmic Rays*

1670. *The Dynamism of Mystical Rays*

1671. *Echoes from the Cosmic World*

1672. *Born to Be a True Reveller*

1673. *My Master's Principle*

1674. *The Cosmic Nature of the Thought*

1675. *You Must Be a Mystic*

1676. *The Wisdom of Cosmic Revelation*

1677. *Finally the Revealer Is Born*

1678. *The Wisdom of the Global Teacher*

1679. *Understanding the Powers and Wisdom of Unity*

1680. *My Days Are Not Numbered*

1681. *The Day I Met My Destiny*

1682. The Day I Met My Talent

1683. Teachings from Destiny

1684. How to Find Your Way Back Home-

1685. Natural Things Supersedes Artificial Things

1686. The Man Who Discovered the Secrets of Mother Nature

1687. The Cost of the Future Fulfilment

1688. How We Can Make the Future Meaningful

1689. Technology Is Not Meant for the Uninformed

1690. Academic Knowledge Does Not Produce Technological Wizardry

1691. God Is the Supreme Maker of True Miracles

1692. Research in a Developing Era (13 August, 2012)

1693. The Dynamics of Research in An Internet Age

1694. Research and Globalization

1695. The Problems of Research in the Cooperate Era

1696. Evolution and Research

1697. The Problems of Technical Research

1698. The Importance of Mathematical Research

1699. Obstacles Towards the Development of Scientific Research

1700. The Meaning and Purpose of Research

1701. The Problems of Engineering Research

1702. What Is Your Wisdom about Life?

1703. What Is Your Definition of Life?

1704. Have You Developed Your Vision?

1705. What Do You Know about Your Life?

1706. Embraced by Wisdom

1707. Loved by Wisdom

1708. Blessed for Wisdom

1709. Enriched by Wisdom

1710. True Greatness Comes from Wisdom

1711. What Is Your Definition of True Wisdom

1712. What the Angels Say about Wisdom

1713. Light from Wisdom

1714. Power from Wisdom

1715. Healed by Wisdom

1716. Rays of Wisdom

1717. Alone with Wisdom

1718. He Has a Tremendous Wisdom

1719. Man, Wisdom, and God (MWG)

1720. Only Created to Manifest Wisdom

1721. Secret Knowledge from Wisdom

1722. Every Man Should Know the Wisdom of His Power

1723. Man Learn How to Pattern with the Creative Power of the Creator

1724. The Power of Nature's Decision

1725. The Wonders of Nature's Kingdom

1726. The Beauty of Natures' Talent

1727. What Does Man Call Nature

1728. Great Melodies from Nature

1729. Thinking along with Nature

1730. The Power of Nature's Decision

1731. The Wonders of Nature's Wisdom

1732. What Does Man Call Nature?

1733. The Beauty of Nature's Talent

1734. Great Melodies from Nature

1735. Thinking along with Nature

1736. Thinking along the Path of Cosmic Light

1737. What Does Man Know about Cosmic Rays?

1738. Has Science Known That the Power of the Solar System Is Controlled by the Cosmic Foundry?

1739. Why Is All Light Originated from the Cosmic Light?

1740. Have We Known How Powerful the Cosmic Rays Are?

1741. Who Made the Cosmic Light?

1742. Why Is the Cosmic Wisdom the Power of Cosmic Messengers?

1743. How And Why Do All Mystics Seek for the Anointing of the Cosmic Power?

1744. Is Every Mystic a Cosmic Messenger?

1745. Why Is Christ Known as a Consummate Mystic?

1746. What Is the Cosmic Sanctum?

1747. Who Is a Cosmic Illuminate?

1748. How Many Comes to Earthman in an Era?

1749. How Many Has the World Had Since the Dawn of Consciousness?

1750. How Can Man Free Himself from Mortal Slavery?

1751. What Is Cosmic Initiation?

1752. Why Is John Known as a Mystic Messenger?

1753. John Prepared the Way for a Cosmic Messenger. Why?

1754. The Power, Face, Force, and Wisdom of Divine Justice

1755. Why Divine Justice Does Not Accept Manmade Judgments

1756. Why Intellectual Courts Do Not Understand the Perfection of the Cosmic Laws

1757. How Incorrect Are the Judgments of Earthly Courts?
1758. Man Is an Ornamented Factory
1759. What Can Posterity Benefit from You?
1760. Are You in Tune with Your Mind?
1761. Those That Work in Spirit
1762. The Benefits of Mind Power
1763. The World Is Loaded with Anarchy
1764. The World Is Expecting a Living Daniel
1765. Man Needs Eternal Rescue
1766. Servants Are True Messengers
1767. The Gospel of God Is Truly

Value Series

1768. How to Values Your Values
1769. How to Value Your Wife
1770. How to Value Your Friend
1771. How to Value Your Work
1772. How to Value Your Family
1773. How to Value Your Nature
1774. How to Value The Blessed Christ
1775. How to Value Your Destiny
1776. How to Value Your Prayer Period
1777. How to Value Your Wisdom
1778. How to Value Your Creativity
1779. How to Value Your Knowledge
1780. How to Value Your Environment
1781. How to Value Your Principles
1782. How to Value Your Time

1783. How to Value Grave

1784. How to Value Your Desire

1785. How to Value Your Gifts

1786. How to Value Your Miracles

1787. Placing Value on Your Philosophy

1788. How to Value Your Kindness

1789. How to Value Your Charity

1790. How to Value the Cross

1791. How to Value Your Creator

1792. How to Value Your Ambition

1793. How to Value Your Teacher

1794. How to Value Your Relations

1795. How to Value Your Relationship

1796. How to Value Your Mind

1797. How to Value Spirit

1798. How to Value Soul

1799. How to Value Body

1800. How to Value Thinking

1801. How to Value Your Thoughts

1802. How to Value Your Creative Genius

1803. How to Value Your Imaginations

1804. Your Destiny Is Your Asset

1805. What Did the Orientals Tell the Modern

1806. Has the Modern World Utilised the Wisdom of the Orientals?

1807. Why Is Abstract Knowledge the Power of All Wisdom?

1808. Who Revealed Truth as an Immortal Teacher to All Ages

1809. What Is the Past Era Awakening?

1810. Why Is Civilization Not Civilized?

1811. Has Man Known the Truth About Creation?

1812. Why Is Man Named Man?

1813. The World Has Few Honest Teachers

1814. Religion Has Suffered Creation

1815. The Universe Is a Living Candle

1816. The Value of Marriage in Christ

1817. The Dynamic Power of Beauty

1818. The Magnifying Wisdom of Beauty

1819. Why Does Beauty Require Truth?

1820. The Excellent Nature of Beauty

1821. The Symbolic Rhythm of Beauty

1822. The Natural Reflection of Beauty

1823. Why Is Beauty the Symbol of Dynamic Immortality?

1824. Why Is Creation Beauty Driven?

1825. Has Man Known What Balanced Beauty Is?

1826. The Nature of Celestial Beauty

1827. Who Is the First Beautiful Creation (Light)?

1828. Why Is Beauty Light Driven?

1829. The Nature, Power, and Wisdom of Agape Love

1830. Man Does Not Know the Meaning and Usage of Agape Love

1831. Christ Is the Symbol of the Mystical Agape Love

1832. The Agape Teachers of Christ

1833. The Mystical Powers of Agape Love

1834. Man Is an Agape Love Light

1835. The Economic Importance of Astral Economics

1836. Nature Is a Perfect Miller, and This Is Why He Mills without Chaff

1837. The Dangers of Using Material Mental System

1838. Does God Think?

1839. The World Is Becoming More Dynamic Than the Speed of a Jet

1840. The World Is Wicked to Wisdom

1841. The Reality of Karma

1842. Karma and the Bible

1843. Karma and Mysticism

1844. The Holiness of the Grail

1845. Spiritual Sickness

1846. The Forgotten Soul

1847. How Does the Spiritual Come into the World?

1848. How Is the Creator from the Spirit?

1849. Mundane Spirit

1850. The Joy of Working with a Pure and Dynamic Spirit

1851. The Cruel Sinner

1852. The Purpose of Earth Life

1853. The Benefit of Earth Life

1854. What Is the Purpose of Spiritual Evolution?

1855. The Mystical Reality of Grace

1856. The Spiritual Power of Grace

1857. What Is the Mystical Meaning of the Bible?

1858. What Is Creation in the Light of Truth?

1859. Who Created Creation?

1860. The Mystical Importance of Ancient Wisdom

1861. The Mystical Temple

1862. Temple Activities in the Mystical Tradition

1863. Who Is Jacob Boehm? he Illumined Mystical Shoe Maker

1864. Is There Any Relativity between Mysticism and Magic?

1865. What Is Magical Christianity? (Great Work from the Adept World)

1866. Why Is Pentecostalism the Greatest Lucifer in Christianity?

1867. Why Is Christ Not a Pentecostal?

1868. Pentecostalism Is Wicked: The Spiritual Progress and Growth of the Soul—How, Why, and for What Purpose?

1869. The Mystical Power of Truth

1870. The Journey in Light

1871. How to Reason and Create with Your Mystical Ingenuity

1872. They That Refuse Positive Growth

1873. Money Cannot Afford Peace, Progress, and Power

1874. The Power of Living with Cosmic Constitution (Adepts)

1875. The Wish and Wisdom of Cosmic Constitution

1876. What Is Material Science in the Mystical Tradition?

1877. The Positive Will

1878. The Power and Greatness of Positive Will

1879. The Mysticism of Will Power

1880. Arrest Your Ego

1881. The Positive Dream of My Life

1882. What Is a Man's Secrets?

1883. Poverty Is Material Bondage

1884. Those That Ask for the Gift of Wisdom

1885. The Willing Spirit

1886. Many Woes of Wicked People

1887. The Power and Beauty of Practical Simplicity

1888. Even the Gentiles Asked for His Grace

1889. I Bless the Day I Met the Angel of Wisdom

1890. Loving and Living in Wisdom

1891. For Life to Be Meaningful,—Only Ask for Wisdom

1892. The Power of Mystical Wisdom

1893. Always Know How to Chatter a Positive Cause

1894. I Dreamt for Life, God Gave Me Grace

1895. The Wish of the Positive Man/Mind

1896. Water Series

1897. The Man Who Talks with Waters

1898. How to Live a Loving and Peaceful Life

1899. The Mysteries of the Waters

1900. The Mysticism of the Waters

1901. The Blessings of the Water

1902. The Mystical Wisdom of the Water

1903. The Healing Power of the Water

1904. The Wisdom of the Astral Water

1905. The Cosmic Spirit of the Waters

1906. The Creative Nature of Water

1907. Water and Environment

1908. The Climatic Nature of the Waterfalls

1909. Water in Our Gardens

1910. How Everything Is Water Formed

1911. The Electronic Nature of Water (Adepts)

1912. Water: The Unknown Power in Different Planets

1913. The Value of Water

1914. The Economics of Water

1915. How Water Discovered and Formed the Best Technological Foundry

1916. The Engineering Dynamics of Water

1917. The Technological Power and Riches of Water

1918. Bermuda Triangle: An Astral, Water-Logged Planet/Zone

1919. Water: A Cosmic Blessing

1920. Water: A Mystical Agent

1921. My Spiritual Love for Water

1922. How to Live Dynamically with Water

1923. How to Protect Our Environmental Water

1924. What Michael Faraday Did with the Celestial Water before the Emergence of Electricity

1925. Every Direction of Water Leads to Glory

1926. The Mystical Wisdom of Christ Came the Aborigines of the Living Waters

1927. How to Make Your Grace Flow from the Abundance of Living Waters

1928. The Wisdom of the Water Spirits

1929. Why Great Things Love the Waters

1930. Why Is Water a Great Home

1931. How Water Formed the Different Sizes and Levels

1932. The Mirage of Waters Baptism

1933. Man Has Not Known His Natural Relativity with the Waters

1934. How to Live in the Mystical Presence of the Living Waters

1935. A Great Lesson from the Waters

1936. The Sea Captain in Praise for His Love of Waters

1937. How Tony Ugo Met the Electrifying Wisdom of the Waters

1938. The Mystical Aborigines of the Waters

1939. Who Are the Water Inhabitants?

1940. The Cleansing Power of the Waters

1941. The Sacred Power of the Waters

1942. Every Creation Has the Wisdom of the Water

1943. Water: A Highly Rated Asset

1944. Water Wells and Other Forms/Segments

1945. Water in the Hydrogen Age: Its Mystical Truth

ABOUT THE AUTHOR

It may not be necessary to project the author from the horizon of academic excellence, but it is important to explain and express the author as a non-obnoxious incubator. Studies in natural law, natural science, and living philosophy, with a high specialization in meta-science before proceeding to study concept therapy—all with the blessed giftedness of Lao Russell's scholarship—quantified and qualified him as a genius in that rare sector of a doctor of science.

He has other academic recognitions from Columbus International University, Virgin Island, and the College of Metascience, England. He has equally done comprehensive and unquantifiable research, particularly in the areas of science, technology, and philosophy. In the areas of public and civil experiences, this gifted monad has served as director in many blue-chip companies, including the banking and ceramic industries. He is currently with the World Bank—sponsored project in Nigeria in Abia State.

His universal enabling, particularly in the area of research, prompted him to be appointed by the world headquarters of the College of Metaphysics as the Nigerian chairman with an instruction to supervise all the African subchapters of the institution.

He is married with a gifted wife, and he has talented children. This author, whom admirers and universal readers respect and define as a modern-day Leonardo da Vinci, is in a nutshell churning out books and works with the use and application of inspiration that he has recognized and utilised as a business necessity. He lives in Aba, Nigeria.

This book, *The Dynamic Concept of Philosophical Mathematics*, is one of those works which he honestly and decisively dedicated as a perfect and practical lenninger to help all and sundry in appreciating that the world is progressive with its growth processes. This is why it is ever in need and in search of balanced and electrifying knowledge, of which the dynamic concepts of philosophical mathematics is hereby given as a compact and consummate incubator for the harmonious realization of this knowledge.

ABOUT THE BOOK

The honest structure of this book can best be understood by humans who appreciate that there is an intertwined and inter-electronic relativity that exists between philosophy and mathematics, with a bias towards assessing and accepting the positive dynamics of other relative sciences. This book, which is already accepted as among from the honest hall of an ingenuous masterpiece, is systematic, practical, educative, and research driven with a lot of scientific and philosophic technicalities.

That is why the book is presented and represented to all the dynamic functions of science, technology, engineering, and creative and applied arts. It also demystifies the oracle of mathematics.

The view and honest concept of demystifying the concepts of mathematics—including nullifying the acceptance of QED, which was prevalent and dominant during the Oriental era—did not negate or reject the monumental contributions of these mathematical wizards. It exists with a compendium of ideas in order to create in the mind of scholars the idea that the infinitude method can best be utilised as a segmented lubric of the understanding of the natural issues: that man deserves to work with inspiration as a monumental business necessity.

In this respect, the tedium of this work is practically and dynamically achieved as a wakeup call to new ideas, as an inventor of new horizons, and as a research-driven, philosophical mathematical apostolate. It is here that it is finally and famously donated to the world of ingenuous knowledge as an ornamented scholar whose views and concepts are naturally brilliant with mathematical and philosophical vision in action. This book is best donated as the monumental and acceptable university of man.

AUTHOR'S PICTURE

Printed in the United States
By Bookmasters